Woody Plant
Biotechnology

NATO ASI Series

Advanced Science Institutes Series

A series presenting the results of activities sponsored by the NATO Science Committee, which aims at the dissemination of advanced scientific and technological knowledge, with a view to strengthening links between scientific communities.

The series is published by an international board of publishers in conjunction with the NATO Scientific Affairs Division

A	Life Sciences	Plenum Publishing Corporation
B	Physics	New York and London
C	Mathematical and Physical Sciences	Kluwer Academic Publishers
D	Behavioral and Social Sciences	Dordrecht, Boston, and London
E	Applied Sciences	
F	Computer and Systems Sciences	Springer-Verlag
G	Ecological Sciences	Berlin, Heidelberg, New York, London,
H	Cell Biology	Paris, Tokyo, Hong Kong, and Barcelona
I	Global Environmental Change	

Recent Volumes in this Series

Volume 205—Developmental Patterning of the Vertebrate Limb
edited by J. Richard Hinchliffe, Juan M. Hurle, and Dennis Summerbell

Volume 206—Alcoholism: A Molecular Perspective
edited by T. Norman Palmer

Volume 207—Bioorganic Chemistry in Healthcare and Technology
edited by Upendra K. Pandit and Frank C. Alderweireldt

Volume 208—Vascular Endothelium: Physiological Basis of Clinical Problems
edited by John D. Catravas, Allan D. Callow, C. Norman Gillis, and Una S. Ryan

Volume 209—Molecular Basis of Human Cancer
edited by Claudio Nicolini

Volume 210—Woody Plant Biotechnology
edited by M. R. Ahuja

Volume 211—Biophysics of Photoreceptors and Photomovements in Microorganisms
edited by F. Lenci, F. Ghetti, G. Colombetti, D.-P. Häder, and Pill-Soon Song

Series A: Life Sciences

Woody Plant Biotechnology

Edited by

M. R. Ahuja

Federal Research Centre for Forestry and Forest Products
Institute of Forest Genetics and Forest Tree Breeding
Grosshandsdorf, Germany

Plenum Press
New York and London
Published in cooperation with NATO Scientific Affairs Division

Proceedings of a NATO Advanced Research Workshop
on Woody Plant Biotechnology,
held October 15-19, 1989,
in Placerville, California

Library of Congress Cataloging-in-Publication Data

NATO Advanced Research Workshop on Woody Plant Biotechnology (1989 :
  Placerville, Calif.)
    Woody plant biotechnology / edited by M.R. Ahuja.
        p.   cm. -- (NATO ASI series. Series A, Life sciences ; v.
  210)
    "Proceedings of a NATO Advanced Research Workshop on Woody Plant
  Biotechnology, held October 15-19, 1989, in Placerville, Calif."-
  -T.p. verso.
    Includes bibliographical references and index.

    1. Forestry biotechnology--Congresses. 2. Trees--Biotechnology-
  -Congresses. 3. Woody plants--Biotechnology--Congresses.
  I. Ahuja, M. R., 1933-   . II. North Atlantic Treaty Organization.
  Scientific Affairs Division. III. Title. IV. Series.
  SD387.B55N37 1989
  634.9--dc20   91-24161
                                                           CIP

ISBN-13: 978-1-4684-7934-8   e-ISBN-13: 978-1-4684-7932-4
DOI: 10.1007/978-1-4684-7932-4

© 1991 Plenum Press, New York
Softcover reprint of the hardcover 1st edition 1991

A Division of Plenum Publishing Corporation
233 Spring Street, New York, N.Y. 10013

PREFACE

        This volume is based on a workshop on Woody Plant
Biotechnology held at the Institute of Forest Genetics, USDA
Forest Service, Placerville, California, USA, 15-19 October,
1989. This workshop was organized by the IUFRO (International
Union of Forestry Research Organizations) Working Party
S2.04-07 - Somatic Cell Genetics -, and supported by the NATO
Scientific Affairs Division, Advanced Research Workshop
(ARW 692/89) Programme.

        This was the second workshop of the IUFRO Working Party
on Somatic Cell Genetics. The first meeting of this Working
Party was held at the Institute of Forest Genetics and Forest
Tree Breeding, Federal Research Centre for Forestry and Forest
Products, Grosshansdorf, Federal Republic of Germany.

        The purpose of the present workshop was to bring together
scientists from different countries of the world for discussions
in the area of woody plant biotechnology.

        Tissues from woody plants, in particular forest trees,
are in general difficult to grow and differentiate in vitro.
However, recent advances in tissue culture technology have
paved the way for successful culture of organs, tissues, cells,
and protoplasts of woody plants. By employing juvenile tissues,
plant regeneration has been accomplished in a number of woody
plant species. On the other hand, clonal propagation of mature
trees, in particular conifers, is still very difficult by
tissue culture.

        Clonal fidelity of in vitro regenerated plants is an
important research area under active investigation. True-to-
type propagules are not always produced by employing tissue
culture technology. Although clonal fidelity would be important in clonal forestry programmes, variation originating in
tissue culture, termed somaclonal variation, would be useful
for the recovery of desirable genotypes, for example, disease,
salt or frost resistance. Once new useful variants have been
isolated and established, their clonal fidelity would be
necessary for producing true-to-type propagules.

        Foreign genes have been transferred in woody plants by
employing Agrobacterium vector systems. In addition, a number
of new approaches employing biolistic microprojectile delivery
systems have been explored, in recent years, for direct transfer
of foreign genes in woody plants. Expression of reporter genes

                                                              v

and herbicide tolerant genes have been monitored in woody plants. The fate of foreign genes introduced into long-lived tree species, on a long-term basis, needs to be fully investigated.

Biotechnological approaches have been employed for preservation of woody plant germplasm. These include slow-growth at low temperatures (4-15°C) of tissues, microshoots, and plantlets in vitro, or cryopreservation (storage in liquid nitrogen at -196°C) of tissues, somatic embryos, meristems, dormant buds, or seeds. Following cryopreservation, the material is retrieved and tested for viability and growth by in vitro or in vivo techniques.

A number of problems remain to be resolved by future research. These include growth and differentiation, and maturation and rejuvenation of tissues in woody plants. Gene transfer is still in its infancy in the tree species. We need to learn more about the genome structure and molecular biology of tree species for a better understanding of cellular differentiation and expression of genes in woody plants.

More than 60 participants from 17 countries of the world participated in this workshop. In order to take advantage of sunny California weather, coffee breaks and lunches were arranged outdoors on the beautiful grounds of the Institute of Forest Genetics, Placerville. On the afternoon of 17 October, 1989 an excursion was organized to view the conifers on the high Sierra ranges and to visit Lake Tahoe. On that fateful afternoon, a major earthquake shook California. San Francisco and nearby areas in Berkeley and Orinda were not too far from the epicenter. Several buildings and one segment of the San Francisco-Oakland Bay Bridge were badly damaged. We were more than 400 km from the earthquake zone, but our thoughts and wishes were with those affected by this calamity.

It would be appropriate to express my gratitude to Bohun Kinloch, Project Leader, Institute of Forest Genetics, USDA Forest Service, Berkeley and Placerville, for accepting our proposal to hold the workshop in Placerville. I would like to thank Dave Neale (local liaison), Clair Kinlaw, Kim Marshall, Chris Nelson, and Mike Campbell for excellent local organization. Helpful suggestions from Janie Curtis, Editor, Plenum Publishing Corporation, are also appreciated. And finally, I am indebted to my wife Ila for help in editorial chores and understanding and patience during the work on this volume.

M. R. Ahuja

Grosshansdorf, Germany

CONTENTS

JUVENILITY, MATURATION AND REJUVENATION

SOMATIC EMBRYOGENESIS

POSTER PRESENTATIONS

WOODY PLANT BIOTECHNOLOGY: PERSPECTIVES AND LIMITATIONS

M. R. Ahuja

Federal Research Centre for Forestry and Forest Products
Institute of Forest Genetics and Forest Tree Breeding
Sieker Landstrasse 2, 2070 Grosshansdorf Germany

ABSTRACT

Woody plant biotechnology has come of age. Basic techniques in the biotechnology of other plants also work in woody plants. Regeneration of plants, clonal fidelity, somaclonal variation, protoplast regeneration and somatic hybridization, gene transfer and molecular genetics, and preservation of germplasm are briefly reviewed. A number of questions, issues and problems in the application of biotechnology to woody plants are discussed.

INTRODUCTION

Biotechnology means different things to different people. To some it means the beginning of a new era full of promises. To others it means traversing through an uncertain and risky future domain. Well, I do not wish to go into the semantic of these viewpoints. Instead I shall attempt to review what we might consider biotechnology in the context of the present workshop. Of course, some consider that there are two types of biotechnology:
- Classical Biotechnology, which includes regeneration by tissue culture,
- New Biotechnology, which includes molecular biology and genetic engineering of plants.

I believe that this division is artificial, and classical biotechnology and new biotechnology are different aspects of the same technology. There is sufficient overlap between the two and at a cellular level one blends into the other. At any rate, knowledge about the biology and regeneration of plants are prerequisites for plant biotechnology.

The next question we may address is: What are the broad areas of biotechnology that are under active investigation in the woody plants? I shall briefly examine these areas and reflect on their current status.

REGENERATION, MICROPROPAGATION AND CLONAL FIDELITY

Broadly speaking, micropropagation may be accomplished by in vitro culture of tissues leading to:

a. Differentiation of microshoots followed by rooting of microshoots to regenerate plantlets.

b. Differentiation of somatic embryos and somatic seedlings.

Although plants have been regenerated by in vitro technology (7, 13),
there are still a number of questions that need to be addressed, particularly in the areas of genetic fidelity, maturation states of donor tissue
and economics of regenerants. In other words, are the regenerants true-
to-type? What is the efficiency of regeneration? Can we micropropagate
from both juvenile and mature tissues of woody plants? Can we produce
plants in large numbers by in vitro technology? What kind of cost est-
mates do we have for in vitro regenerants as compared to stecklings or
seedlings? Have we adequately field-tested in vitro produced regenerants?
Are there early maturation problems with micropropagated plants? Will
clonal forestry reduce genetic variation and diversity, which is so essential for the survival of the species? We probably have answers to some
of these questions. Others must await further investigations and evaluation. Some of the presentations in this workshop will address these and
related questions on micropropagation. Clonal propagation via somatic
embryogenesis offers prospects for mass cloning at low costs. A prerequisite to commercial scale clonal propagation would be high-frequency
somatic embryogenesis, genetic fidelity, and normal growth and development of somatic seedlings and somatic plants.

Here I am introducing a couple of new terms to distinguish between
plants produced by a zygote or somatic cells . The lineage from zygotic
or somatic cells is shown in Figure 1. Since we will be dealing more and
more with plants produced by asexual means, it might be useful to make
some distinctions between sexually or asexually produced plants. Under
this terminology, a zygote gives rise to an embryo, which in turn develops into a seedling and plant. Comparable stages of development in a
somatic cell undergoing somatic embryogenesis would be somatic embryo,
somatic seedling and a somatic plant. On the other hand, a somatic cell
undergoing organogenesis, gives rise to either microshoots or roots,
followed by the development of plantlet and eventually the so-called
plantling.

In order to distinguish the progeny of in vitro produced regenerants
from that of a sexual progeny, the former have been designated R1, R2 as
contrasted to F1, F2 in the sexually reproduced plants (12). I proposed
that the progeny of the regenerants produced following somatic embryogenesis be termed SR1, SR2 and so on, where SR stands for somatic embryogenesis regenerants.

At this stage a few remarks are in order regarding the regenerants
produced by cotyledonary approach or by somatic embryogenesis. Cotyledon
is a terminally differentiated organ in plants. It's primary function is
storage of food material to be made available to germinating seedlings.
The cells in the cotyledon are normally not programmed to undergo differentiation. However, under in vitro conditions involving short or long
exposure to higher levels of phytohormones, the cotyledonary cells can
be induced to differentiate into merestemoidal tissues. These differentiate into shoots, that can be rooted to yield plantlets. By employing
cotyledonary approach, large numbers of plants have been produced in
radiata pine (Pinus radiata) in FRI, Rotorua, New Zealand (9, 29), loblolly pine (Pinus taeda) in N.C. State University, Raleigh (10) and
Douglas fir (Pseudotsuga menziesii) in Weyerhaeuser, Tacoma (26). Although details of field trials on the plants derived from cotyledonary
cultures are not available in all these species, it seems that early
maturation states (21) and other developmental problems need to be fully

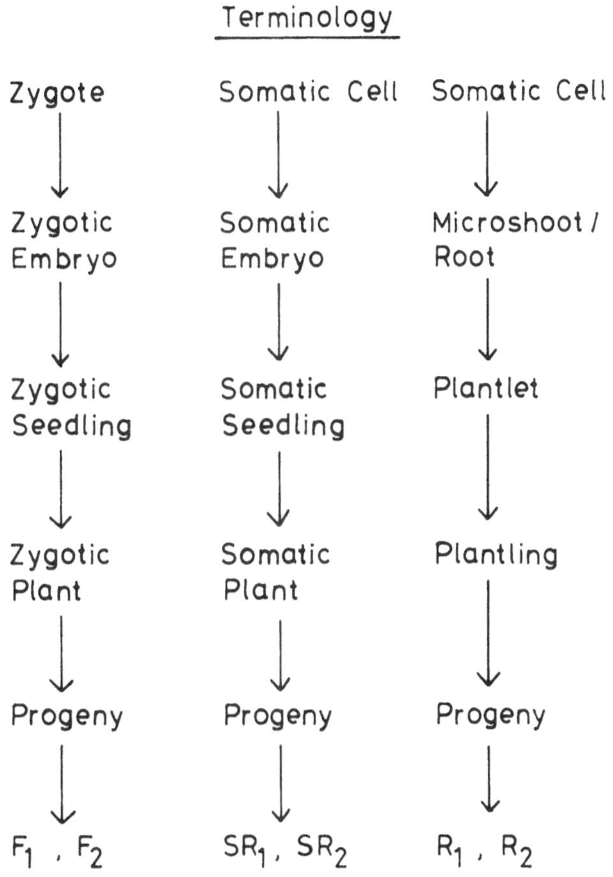

Zygote	Somatic Cell	Somatic Cell
↓	↓	↓
Zygotic Embryo	Somatic Embryo	Microshoot / Root
↓	↓	↓
Zygotic Seedling	Somatic Seedling	Plantlet
↓	↓	↓
Zygotic Plant	Somatic Plant	Plantling
↓	↓	↓
Progeny	Progeny	Progeny
↓	↓	↓
F_1 , F_2	SR_1, SR_2	R_1 , R_2

Figure 1. Terminology proposed for plants derived from a zygotic cell or a somatic cell. The plant obtained from a zygote is termed zygotic plant, while those derived from somatic cells may follow either of the two pathways: one derived via somatic embryogenesis is termed somatic plant, while the other obtained via organogenesis is termed plantling. The progeny derived from crosses between zygotic plants and, between plantlings have been termed F1. F2, and R1, R2 respectively. It is proposed that the progeny obtained following crosses between somatic plants be termed SR1, SR2, in order to distinguish between the modes of origin of the regenerants.

addressed. These may include Phenodeviants or somaclonal variation involving chromosomal abnormalities and gene mutations. Because of long generation time, progeny tests to determine the nature of gene mutations are generally not feasible in the forest tree species. For this reason, we require a dependable in vitro technology that yields clonal material that is morphologically and genetically stable.

Another problem with the cotyledonary micropropagation is that the plants regenerated by this approach are derived from unselected bulked seed, representing different genotypes. At this stage it is difficult to predict which of these seeds will eventually give rise to an elite tree. Therefore, clonal propagation using cotyledonary culture or other juvenile tissues has a drawback because the clones are derived from unselected and untested genotypes. For this reason micropropagation from organ culture (bud meristem) of mature selected trees offers the advantage of capturing maximum genetic gain.

As regards somatic embryogenesis, plants do not normally produce them under in vitro conditions. When plants are wounded or severely pruned, they differentiate new meristems and shoots, and these can be rooted to produce new plants. Under normal circumstances, plants do not produce somatic embryos to deal with man-made and environmental challenges. Nevertheless, the plant genome does contain genetic information for the differentiation of somatic embryos. It is possible to activate the somatic embryognesis development program only under stress conditions of in vitro (involving certain phytohormones). Although the genetic information for somatic embryos may be present in all cells, only certain types of cells, for example, zygotic embryonic cells seem to be more competent to develop somatic embryos, usually involving a callus intermediate stage. The embryogenic callus is usually transparent white or brown, and mucilaginous, indicating that the callus is highly stressed. Although, structurally the zygotic and somatic embryos seem similar, there might be some subtle differences in the expression of their genetic programs. For example, it has been reported that plants derived by somatic embryogenesis in Pinus stay in a quasi-dormant state after a few years of growth (30).

Genetic fidelity in somatic embryos is another important question. How stable are the somatic embryogenesis derived plants? Now that several systems have been developed in the woody plant somatic embryogenesis, we should be having answers to some of these questions in the near future.

SOMACLONAL VARIATION

Clonal fidelity is one of the main concerns in the micropropagated plants. True-to-type propagales, whether from microshoot regeneration or somatic embryogenesis generated plants are the expected product of clonal forestry. But do they always occur? The answer is that tissue-culture generated variation termed somaclonal variation has been frequently observed in long-term cultures (3, 16). While it may be undesirable in the clonal forestry programs, somaclonal variation offers prospects for recovery of desirable regenerants in woody plants.

Let me use coastal redwood, Sequoia sempervirens, for illustrating some of the issues and problems in regeneration of plants. Redwood has been introduced as an exotic species in Europe, including Germany. Redwood is sensitive to frost and sub-zero temperatures. However, because of its polyploid nature (2n=66, a hexaploid with a basic number 11), and adequate genetic variability in this species, frost-hardy or cold-tolerant trees have been observed. We had a very cold winter in 1984-85 with temperatures reaching 20 degrees below freezing point (-20 C). In a redwood plantation near Cologne, Germany, most of the trees were badly damaged by the cold winter weather. However, some trees survived the unfavorable weather. The owner of this plantation Mr. Nieheus, sent us twigs from 3 trees in 1985-1986, that had survived the cold spell. We cultured bud explants from these 3 trees (2 trees were approximately 23 years of age and one was 8 years old) on a modified woody plant medium (19), supplemented with low levels of 6-benzylaminopurine, BAP, (0.5-1.0 mg/l) and naphthaleneaxetic acid, NAA, (0.02 mg/l). We have established shoot cultures from the three cold-hardy trees of coastal redwood. Rooting was a problem in the initial cultures. However, after some 20 passages on a low BAP medium, microshoots could be induced to root on an auxin (indole-3-butyric acid and NAA) supplemented medium, albeit at a low frequency of 20 percent. Studies are in progress in our lab to improve the rooting frequency in the three Sequoia clones.

We are currently testing the performances of micropropagated redwoods under field conditions. A total of 760 in vitro propagated plants from the three clones were moved outside the nursery last year. There were differences in the plant architecture between clones and within a clone. For example, more than 90 percent of the plants in clone 3 were green (the rest were bronze-coloured) and exhibited profuse budding all over the plant. On the other hand, less than 10 percent of the plants from clones 1 and 2 were green: the remaining 90 percent showed bronze phenotype, and sparse budding on the plants. The main stem in the green plants was strong and erect; whereas, in the bronze plants it was weak and needed support. At this stage it is not clear whether the observed variation within a Sequoia clone is somaclonal (heritable) or non-genetic. We also do not know which one of these traits is associated with cold-hardiness. In vitro and in vivo studies are in progress in our lab to test the cold-hardiness in these 3 clones of Sequoia sempervirens.

These results illustrate that we still do not fully comprehend the problems of growth and differentiation in trees. The understanding and optimization of morphogenesis would be important for application of in vitro technology to mass cloning of forest tree species.

The somaclonal selection approach has been employed at the USDA Forest Service, Rhinelander, St. Paul, the Iowa State University, and perhaps other centres, for isolating useful variants in poplar against diseases and for herbicide tolerance. The in vitro selection against diseases would constitute an approach to biological control of diseases without polluting the environment. Somaclonal variants have been recovered in poplar for herbicide tolerance (22). The commercialization of herbicide tolerant plants in poplar would probably lead to extensive application of herbicides to kill the broad-leaved weeds that compete with poplars in the first few years. The introduction of herbicide tolerant plants in the forest lands, which seems acceptible in the U.S., is a highly emotional issue in Germany and other European countries. The environmentalists are concerned that extensive application of herbicides would pollute water and soil in the forests and surrounding areas and thus disturb the biological and ecological equilibrium.

Nevertheless, somaclonal selection of variants remains an attractive and a viable approach for recovery of desirable (non-controversial) genotypes, without resorting to very expensive gene transfer technology.

PROTOPLAST REGENERATION AND SOMATIC HYBRIDIZATION

Protoplasts have been isolated and cultured in a number of woody plant species (1, 2, 20). However, the success rates in regeneration of plants from protoplasts of woody plants have been sporadic (2, 27). Sustained cell divisions and colony formation have been reported in protoplast cultures of several tree species (15, Lang and Kohlenbach, This Proceeding). Successful plant regeneration from protoplasts have been reported in poplar, Populus spp (28), loblolly pine, Pinus taeda (14), and white spruce, Picea glauca (11).

Somatic hybridization involving fusion of protoplasts offers a new approach for obtaining hybrids between sexually incompatible plants. Initially, it was oversold by protoplast zealots as a promising avenue for gene transfer in plants. As it turned out, this was a short-sighted proposition. Although, interspecific and intergeneric somatic hybrids have been produced in several plant species (1, 2, 27), genetic stability of these hybrids have been questioned. Besides, regeneration from protoplasts and selection of desirable genotypes from protoplasts cultures (protoclonal variation) or following protoplast fusion still remains a problem area in a number of economically important plant species, in particular tree species.

Protoplasts seem to have value for detection of transient expression of alien genes and their subsequent integration process. However, we have to invest a lot of research in the isolation of viable and regerable protoplasts from conifers and hardwoods. Protoplast regeneration from juvenile cells is valuable, but the ultimate aim of this research would be isolation and regeneration of protoplasts from mature trees. Maturation problems are common to protoplasts and cells, and need to be addressed at a molecular level.

GENE TRANSFER AND MOLECULAR GENETICS

In the initial stages of gene transfer, it was thought that protoplasts are essential for introduction of foreign genes in plants. This may still be true in certain cases of direct gene transfer in plants. However, in the Agrobacterium mediated gene transfer systems, plant cells/tissues can be employed as Ti plasmids carrying foreign gene(s) can go through the plant cell wall. Some of the recent technologies employing microprojectile bombardment with DNA coated tungsten pellets allow transfer of DNA in protoplasts as well as cells. I do not wish to dwell on the genetic transformation in woody plants as there will be a number of presentations on this topic in this workshop. This area of biotechnology has generated most excitement. But, at the same time, it should be pursued with caution.

It is known that genes from microorganisms express their products when introduced into plant cells. However, sometimes there are problems in expression of foreign genes in plants, including woody plants. Forest trees have long generation cycles with an extended vegetative phase ranging from one to several decades. Foreign genes may be expressed immediately, or remain silent for a long time, or they may be lost during the long vegetative phase of the tree species. Genetic variability associated with foreign gene transfer has been reported and may arise by several different mechanisms (4, 5, 6). Whether genetic transformation releases more genetic variability than conventional hybridization (interspecific or intergeneric) has not been fully examined. By designing appropriate promoters, optimal to enhanced expression of foreign genes in woody plants is currently under active investigation. We still do not fully understand how genes express in their native cells or foreign cells. Genome organization, regulation and expression of genes is, therefore, one of the main thrusts of plant biotechnology.

6

Genes involved in commercially important products or traits from microorganisms, higher plants or even animals, when available, will be employed in future for recovery of novel genotypes by employing genetic engineering techniques. But first we have to overcome several problems. Isolation and characterization of commercially important genes of woody plants remains a technically challenging problem. Information on genome structure and genetic maps is still lacking in woody plants. Biotechnology may be helpful inunravelling the genetic architecture of tree species.

PRESERVATION OF GERMPLASM

Because of forest decline world-wide, preservation of forest tree germplasm has attracted much attention in recent times. In this respect, conservation of genetic diversity, on the one hand, and preservation of specific genotypes, on the other hand, would be necassary. Genetic diversity is essential for survival of forest tree species under changing environmental and stress situations. Therefore, gene conservation and preservation of biodiversity would become increasingly important in woody plants. Approaches to preservation of forest tree germplasm have been discussed in recent publications (8, 17, 18, Ahuja, this volume). Briefly, these include in situ and ex situ preservation measures. In situ measures involve preservation on site, that is, preservation of stands by multiplication via natural regeneration from seed and/or vegetative propagation. The ex situ mesures involve preservation away from the original site of the stand, either under field or controlled conditions. Under field conditions, the germplasm may be preserved in seed or clonal orchards. Ex situ measures under controlled conditions involve storage of seed, vegetative parts in gene banks at specific temperature regimes. For example, seeds of forest tree species have been stored at temperatures ranging from 4 C to -20 C. Plant tissues have been stored in vitro at temperatures ranging from 15 C to 4 C for various lengths of time. Seeds and dormant buds (Ahuja, this volume) and plant tissues have been stored in liquid nitrogen (cryopreservation). Recent approaches for preservation of forest tree germplasm at low temperatures, inluding -196 C (liquid nitrogen) will be presented during the workshop.

PERSPECTIVES AND LIMITATIONS

Let me summarize. Woody plant biotechnology has come of age. Basic techniques in biotechnology of other organisms will generally work in the woody plants. However, there are a number of limitations. Firstly, growth, differentiation and regeneration of woody plants are not well understood, and therefore need further physiological, biochemical and molecular studies. Because of long generation cycles, traditional genetic and breeding techniques are handicapped for gene mapping. No major genes have been isolated or mapped in the woody plants. The current focus is on genes that control resistance against diseases. By employing restriction fragment length polymorphisms (RFLPs), gene mapping has been initiated in loblolly pine (Pinus taeda) at the Institute of Forest Genetics, USDA Forest SErvice, Berkeley and Placerville, California (23, 24). One can learn a lot from the human genome mapping programme currently under active investigation in USA (31), where RFLP technology is also pivitol for gene mapping. In this regard, rapid sharing of information between different national and international groups would be desirable. Therefore, it is imperative to initiate a scientific network to examine the infrastructure and the problem areas of woody plant biotechnology at an international level. Periodic meetings between these groups are vital to the progress in this area.

I should mention here that recently United States Department of Agriculture has published guidelines for the release of biotechnologically engineered material in the field (25). The European Economic Community (EEC) is also working on these guidelines and plans to announce them for genetically engineered material in 1992. Therefore, by 1992 both USA and Europe will have responsible guidelines for the release of genetically engineered materials. The regulatory problems for the microorganisms are not yet resolved. As regards deployment, more than 50 fiels tests of biotechnologically produced plants are currently in progress in the United States. The fiels testing does not seem to be as controversial issue as it was a few years ago. Following field testing, USDA or related Government agencies in the United States will be in a position to release this information to the rest of the world (Stan Krugman, pers. comm.). That would be one step forward, but certainly not the complete story on the impact of genetically engineered plants/microorganisms on the environment in a long run.

And finally, what are the products of woody plant biotechnology? What do have to offer It is not an inditement. We do not work with agricultural crops. Ours is a long range proposition. However, we may have to revise some of our expectations in due course of time. These are hard decisions to make. Where do we go next? What can be accomplished and what is not feasible, only future research in this exciting, but at times uncertain domain, will determine the future course of woody plant biotechnology.

ACKNOWLEDGEMENTS

I would like to thank Stan Krugman for discusions and exchange of ideas in woody plant biotechnology.

REFERENCES

1. Ahuja, M.R., 1982, Isolation, culture and fusion of plant protoplasts. Silvae Genet. 31:66-77.
2. Ahuja, M.R., 1984, Protoplast research in woody plants. Silväe Genet. 35:32-37.
3. Ahuja, M.R., 1987, Somaclonal variation. In: Cell and Tissue Culture in Forestry, vol. 1., J.M. Bonga and D.J. Durzan (Eds.), Martinus Nijhoff Publishers, Dordrecht, pp. 272-285.
4. Ahuja, M.R., 1988, Gene transfer in forest trees. In: Genetic Manipulation of Woody Plants, J.W. Hanover and D.E. Keathley (Eds.), Plenum Press, New York, pp. 25-41.
5. Ahuja, M.R., 1988, Gene transfer in woody plants: perspectives and limitations. In: Somatic Cell Genetics of Woody Plants, M.R. Ahuja (Ed.), Kluwer Academic Publishers, Dordrecht, pp. 83-101.
6. Ahuja, M.R., 1988, Molecular genetics of transgenic plants. In: Frans Kempe Symposium 'Molecular Genetics of Forest Trees', J.E. Hällgren (Ed.), Swedish University of Agricultural Sciences, Umea, pp. 127-145.
7. Ahuja, M.R. and Muhs, H.J., 1985, In vitro techniques in clonal propagation of forest tree species. In: In Vitro Techniques - Propagation and Long-Term Storage, A. Schäfer-Menuhr (Ed.), Martinus Nijhoff Publishers, Dordrecht, pp. 41-49.
8. Ahuja, M.R. and Muhs, H.J., 1989, Biotechnologische Verfahren zur besseren Evaluierung, Erhaltung und Nutzung von forstlichen Genressourcen. Berichte über Landwirtschaft, 201. Sonderheft. Verlag Paul Parey, Hamburg, pp. 414-422.
9. Aitkin-Christie, J., Singh, A.P. and Davies, H., 1988, Multiplication of meristematic tissue: a new tissue culture for radiata pine. In:

Genetic Manipulation of Woody Plants, J.W. Hanover and D.E. Keathley (Eds.), Plenum Press, New York, pp. 413-432.

10. Amerson, H.V., Frampton, L.J., Mott, R.L. and Spaine, P.C., 1988, Tissue culture of conifers using loblolly pine as a model. In: Genetic Manipulation of Woody Plants, J.W. Hanover and D.E. Keathley (Eds.), Plenum Press, New York, pp. 117-137.

11. Attree, S.M., Dunstan, D.I. and Fowke, L.C., 1988, Plantlet regeneration from embryogenic protoplasts of white spruce (Picea glauca). Bio/Tech. 7:1060-1062.

12. Chaleff, R.S., 1981, Genetics of Higher Plants Cambridge University Press, Cambridge.

13. Dunstan, D.I. and Thope, R.A., 1986, Regeneration in forest trees. In: Cell Culture and Somatic Cell Genetics of Plants, vol. 3, I.K. Vasil (Ed.), Academic Press, New York, pp. 223-241.

14. Gupta, P.K. and Durzan, D.J., 1987, Somatic embryos from protoplasts of loblolly pine proembryonal cells. Bio/Tech. 5:710.712.

15. Kirby, E.G. and David, A., 1988, Use of protoplasts and cell cultures for physiological and genetic studies of conifers. In: Genetic Manipulation of Woody Plants, J.W. Hanover and D.E. Keathley (Eds.), Plenum Press, New York, pp. 185-197.

16. Larkin, P.J. and Scowcroft, W.R., 1981, Somaclonal variation- a novel source of variability from cell cultures for plant improvement. Theor. Appl. Genet. 60:197-214.

17. Ledig, F.T., 1988, The conservation of diversity in forest trees. Bioscience 38:471-479.

18. Melchior, G.H., Muhs, H.J. and Stephan, B.R., 1986, Tactics for the conservation of forest gene resources in the Federal Republic of Germany. For. Eco. Manag. 17:73-81.

19. Lloyd, G. and McCown, B.H., 1981, Commercially feasible micropropagation of mountain laurel (Kalmia latiflora) by use of shoot tip cultures. Proc. Int. Plant Prop. Soc. 30:421-427.

20. McCown, B.H. and Russel, J.A., 1987, Protoplast cultures of hardwoods. In: Cell and Tissue Culture in Forestry, vol. 2, J.M. Bonga and D.J. Durzan (Eds.), Martinus Nijhoff Publishers, Dordrecht, pp. 16-30.

21. McKeand, S.E., 1985, Expression of mature characteristics by tissue culture plantlets derived from embryos of loblolly pine. J. Am. Soc. Hort. Sci. 110:619-623.

22. Michler, C.H. and Haissig, B.E., 1988, Increased herbicide tolerance of in vitro selected hybrid poplar. In: Somatic Cell Genetics of Woody Plants, M.R. Ahuja (Ed.), Kluwer Academic Publishers, Dordrecht, pp. 183-189.

23. Neale, D.B., Tauer, C.G., Gorzo, D.M. and Jermstad, K.D., 1989, Restriction fragment length polymorphism mapping of loblolly pine: methods, applications, and limitations. In: Proc. 20th Forest Tree Imp. Conf. Charleston, South Carolina, pp. 363-373.

24. Neale, D.B., Ahuja, M.R., Alosi, M.C., Devey, M. E., Jermstad, K.D. and Marshall, K. A., 1991, Applications of high density restriction fragment length polymorphism maps in forestry. New Forests (in Press).

25. Purchase, H.G. and Mackinzie, D.R. (Eds.), 1990, Agriculture Biotechnology. Introduction to Field Testing. Office of Biotechnology, U.S. Department of Agriculture, Washington, D.C.

26. Timmis, R. and Ritchie, G.A., 1984, Progress in Douglas fir tissue culture. Proc. Int. Symp. Recent Advances in Forest Biotechnology, Travers City, Michigan, pp. 37-46.

27. Russel, J.A., 1991, Advances in the protoplast cultures of woody plants. In: Micropropagation of Woody Plants, M.R. Ahuja (Ed.), Kluwer Academic Press, Dordrecht (in Press).

28. Russel, J.A. and McCown, B.H., 1988, Recovery of plants from leaf protoplasts of hybrid-poplar and aspen clones. Plant Cell Rep. 7:59-62.

29. Smith, D.R., 1986, Radiata Pine (Pinus radiata D. Don). In: Biotech-
    nology in Agriculture and Forestry, vol. 1, Trees I., Y.P.S. Bajaj
    (Ed.), Springer Verlag, Berlin, pp. 274-291.
30. Thorpe, T.A. and Hasnain, S., 1987, Micropropagation of confers:
    methods, opportunities and costs. In: Proc. Tree Improvement-
    Progressing Together, E.K. Morgenstern and T.J.B. Boyle (Eds.),
    Canadian Tree Improvement Association, Ottawa, pp. 68-82.
31. Watson, J.D., 1990, The human genome project, past, present and future.
    Science 248:44-49.

CONTROL OF REGENERATION AND CLONAL FIDELITY

IN VITRO PROPAGATION OF CONIFERS: FIDELITY OF THE CLONAL OFFSPRING

J.M. Bonga

Forestry Canada Maritimes Region
PO Box 4000, Fredericton, NB, E3B 5P7, Canada

ABSTRACT

Conifer propagation in vitro is still largely restricted to propagation of juvenile, in most cases embryonic, material. So far, the main method of regeneration has been rooting of adventitious or axillary shoots. However, this method has found little commercial application because, 1) the propagation rates are too low, 2) the cost is too high, and 3) propagation is often not true-to-type. Fortunately, methods have been developed recently to induce somatic embryogenesis in cell cultures derived from embryo explants. Somatic embryogenesis is better suited for mass propagation than rooting of adventitious or axillary shoots. Therefore, it is anticipated that commercial propagation of juvenile conifer material will soon be possible. Mature trees are far more difficult to clone by in vitro techniques than juvenile specimens. Our work with about 30-year-old Larix decidua trees has resulted in large-scale adventitious shoot formation, but little rooting. Some of the cultures have produced yellow, slimy cell masses which were similar in appearance to embryogenic masses obtained from embryo or megagametophyte explants. They contained bundles of long cells associated with small cells, but none of these developed into recognizable embryos. The potential applications of cloning in vitro in tree improvement schemes are discussed.

PROPAGATION IN VITRO OF JUVENILE MATERIAL

Cloning of a conifer by in vitro techniques was first reported in 1975 (18). Rooting of adventitious or of axillary shoots produced in vitro (micropropagation) has since then been achieved for many conifer species. However, micropropagation has, so far, found little commercial application.

The first reason for this is that the rate of formation of adventitious shoots is, except in a few cases, e.g., Pinus radiata, too low to be effective. In fact, for many species higher propagation rates can be achieved by successive cycles of rooting of cuttings taken from seedlings than by micropropagation. For example, with Larix sp. from 200 to 300 rooted cuttings can be produced from one seedling in one year (Park, Forestry Canada, Fredericton, pers. commun.). This is a number of propagules per ortet, that is, for most conifer species, difficult to match by micropropagation. In some species, especially Pinus sp., seedlings can be sprayed with cytokinin to induce development of fascicular shoots. This increases the number of shoots available for rooting and thus makes rooting of cuttings even more competitive with micropropagation (11, 13).

Woody Plant Biotechnology, Edited by M.R. Ahuja
Plenum Press, New York, 1991

A second problem with micropropagation is that there is a shoot initiation, a shoot elongation, a rooting, and an acclimation phase, each phase requiring a different nutrient medium and extensive handling. This makes the process labor intensive, and thus expensive. For example, micropropagated Pinus radiata plantlets are twice as expensive as rooted cuttings (7).

Finally, rooted cuttings, at least if obtained from young seedlings, often develop into plants that show a better resemblance to the ortet than plants that have developed from micropropagules. A common problem with conifer micropropagules is slow initial growth and the early appearance of mature characteristics (5, 15, 19). For example, after 7 years of field testing, propagules obtained from embryos of Pinus radiata showed accelerated aging, strong apical dominance, and reduced diameter growth (17). Some abnormalities are probably caused by the number of adventitious roots formed on the adventitious or axillary shoots, and by the distribution pattern of the roots on the shoots. The root number presumably affects vigor of the plantlet, while an uneven root distribution can cause plagiotropism (19).

Although abnormalities are common, field tests with micropropagules are not always negative. For example, micropropagules of Pseudotsuga menziesii have, after some delay in growth in the first season, performed as well as seedlings (15). In this field test the micropropagules performed better than rooted cuttings. However, it should be pointed out that the cuttings used in this test were obtained from 8-year-old ortets; rooted cuttings from younger ortets probably would have performed better.

Over the last few years two new developments have occurred that may circumvent some or most of the above problems. These new techniques promise to clone conifers at a lower cost, and at a faster rate than traditional micropropagation or rooting of cuttings. The first of these new techniques was initially developed for Pinus radiata. It is based on the regeneration of shoots from rapidly multiplying, meristematic nodules in suspension cultures. Because these nodules are developing in suspension cultures, their production is amenable to automation (4). An even more promising achievement is the development of protocols for somatic embryogenesis; such protocols are already available for many conifer species (see elsewhere in this volume).

Embryogenesis has a number of advantages over micropropagation. In contrast to micropropagation, which produces multiple shoots which have to be separated and then individually rooted, embryogenesis produces propagules with a shoot and primary root in one step. Therefore, embryogenesis is less labor intensive and thus is expected to be cheaper than micropropagation once the process is scaled-up to an industrial production level. Furthermore, while adventitious and axillary shoots are generally produced on solid media, somatic embryogenesis can be initiated in suspension cultures. Suspension cultures have the advantage that they can easily be automated and thus are more suitable for commercial, mass-culture than cultures on solid media.

For most species for which somatic embryogenesis has been reported, results of field tests have not yet been published. Therefore, we do not yet know if propagation will generally be true-to-type, or if problems similar to those experienced with propagules obtained by micropropagation will occur. However, the limited, preliminary field test information that is available at the moment, indicates that the propagules are mostly normal. For example, initial results of field tests of spruce plants obtained by somatic embryogenesis, show that these plants have good vigor and no obvious abnormalities (Webster and Roberts in this volume).

Because somatic embryogenesis is still restricted to juvenile material, the initial application will mainly be rapid clonal propagation of embryos from seeds obtained in breeding experiments. It could provide an alternative to the use of large seed orchards, which currently are the means to provide improved planting stock in large quantities (7, 17).

PROPAGATION IN VITRO OF MATURE MATERIAL

Proper evaluation of trees for clonal selection is generally not posssible before the trees have reached at least about half of the rotation age. Therefore, the term "mature" will only be used for trees that have reached half rotation age or more, and are sexually mature.

Micropropagation of mature conifers has only been reported for a few species, e.g., Pinus radiata (10) and Sequoia sempervirens (8). However, propagation in these cases was on a small scale. Furthermore, field test results have not been published, and, therefore, it is not yet known if micropropagules of mature conifers are capable of normal or near-normal growth.

Cloning of mature conifers has a function in tree improvement schemes that is different from that of cloning of juvenile material. Cloning of juvenile ortets can effectively capture the improved traits obtained by breeding. These improved traits are mostly traits with a high heritability, i.e., they are obtained primarily by selection among additive variance, breeding being less effective for non-additive variance. Cloning of mature trees, on the other hand, captures all variance, including variance with low heritability. Furthermore, in short term breeding only one or a few traits are improved simultaneously, and subsequently introduced into a large population by cloning. By cloning superior mature ortets, all desired traits are captured at once and are introduced into a large clonal population, no matter how many of these traits there are and how low their heritability.

Once micropropagation of mature conifers is possible, it remains to be established whether the propagules are true-to-type, not only initially, but over an extended period of time. For example, the initial growth rates could be high, as expected, but could slow down considerably later due to an unexpected, heavy, premature cone production. Furthermore, extensive tests will initially be required to determine to what extend the superior characteristics of the ortet are based on its genetics or on its environment.

In conclusion, in future tree improvement schemes there is room for cloning of juvenile as well as of mature material. The preference of one over the other will depend on the degree of natural variation in the species, the effectiveness of breeding of the species, and the degree of breeding that has already occurred. In the short run, cloning of juvenile material can be expected to have a greater impact than cloning of mature material, particularly because of the advances that have been made recently in somatic embryogenesis, in genetic engineering and in cryopreservation. Nevertheless, cloning of mature conifers, once posible, could make a valuable contribution to tree improvement programs.

IN VITRO REGENERATION OF MATURE LARIX

For a number of years we have attempted micropropagation of approximately 30-year-old trees of Larix decidua. Shoot explants of these

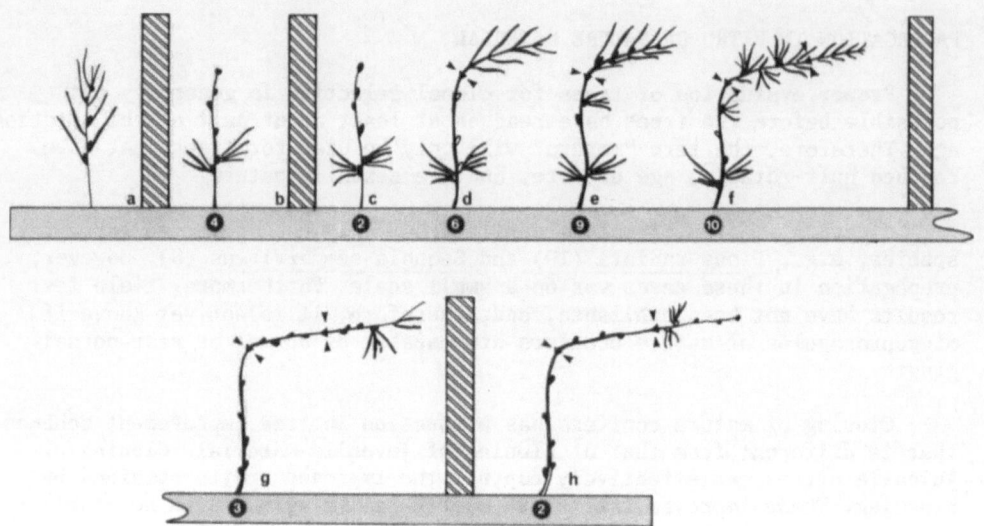

Fig. 1. Diagram showing the growth pattern of a rooted adventitious shoot obtained in a culture of a bud explant of a mature <u>Larix</u> <u>decidua</u>. The vertical bars indicate a 6 - 8 week cold treatment; the numbers indicate weeks after the last cold treatment. Further details see text.

Fig. 2. Photograph of the plant shown in Fig. 1a.

trees produce adventitious shoots either on the explant, or on a green, shiny, subcultured callus (6). However, in contrast to adventitious shoots derived from embryos, which elongate and root readily and produce normal looking plants after transfer to soil (Bonga, unpublished), shoots obtained from explants of mature trees show little stem elongation and root poorly.

Of the several thousands of adventitious shoots that we have produced over the years from buds of the mature trees, only one has formed a plantlet capable of surviving transfer to soil (6). This potted plantlet was kept in a growth cabinet at 20C and 16/24 hr photoperiod. Initially the plantlet grew like a normal seedling (Fig. 1a, and Fig. 2), but then became dormant. This dormancy, and later ones, was broken by keeping the plantlet at constant low temperature (4C) for 6-8 weeks and low intensity light (50 $\mu E\ m^{-2}\ s^{-1}$) for 8/24 hr. Four weeks after the plantlet was returned from the coldroom to the growth cabinet, it started to grow again, but in an abnormal fashion. Only the two bottom lateral buds flushed, each forming a short shoot (Fig. 1b). Two weeks after the second cold storage period, these same two bottom shoots flushed again (Fig. 1c). After 6 weeks, the terminal bud, which had remained dormant during the previous growth period, flushed and formed a long shoot, which, however, was plagiotropic (Fig. 1d). After 9 weeks the top lateral bud on the old growth flushed and formed a short shoot (Fig 1e, and Fig. 3). One week later the terminal bud of the new long shoot, and several of its lateral buds flushed, with the terminal bud forming a long shoot, and the lateral buds short shoots (Fig. 1f). Three weeks after the third cold period only the terminal and one lateral bud on the latest long shoot flushed, each forming a short shoot (Fig. 1g). Two weeks after the next cold period only two lateral buds on the latest long shoot flushed (Fig. 1h). However, the needles of these flushing buds were short and albino, and died after a few weeks. A few weeks after that, the plantlet was dead.

Fig. 3. Photograph of the plant shown in Fig. 1e.

These observations showed that plagiotropism was not the only abnormality. The synchrony of bud flushing was severely affected, which suggests that the hormonal control mechanisms in the propagule were severely disturbed. It is unlikely that the environmental regime, in which the plantlet was alternately kept in the growth cabinet and in cold storage, was responsible for the abnormal growth habit of the plantlet. Grafts of scions of the ortet, which were exposed to the same growth cabinet-cold-room cycle, flushed properly and grew very well under this regime.

Recently we have obtained axillary shoots (Fig. 4) instead of the adventitious ones that have appeared exclusively in earlier cultures (6). The experiment that produced these axillary shoots differed from the earlier experiments in that elongating long-shoot stem sections collected the last week in June were used as explants instead of buds collected between the late summer and early spring. The axillary shoots showed better stem elongation than the adventitious shoots obtained earlier. This is not surprising, because axillary shoots of other species, e.g., _Picea abies_ (20), have shown better stem elongation than adventitious shoots as well. Rooted axillary shoots of _Pinus brutia_ behaved better in field tests than rooted adventitious shoots. After one year of field testing, many plantlets of adventious bud origin were bushy or plagiotropic while those of axillary origin looked like normal seedlings (1). Axillary shoots are, therefore, clearly preferred.

Fig. 4. An axillary shoot on a stem explant of a mature _Larix decidua_.

To date, there are no reports of somatic embryogenesis in explants from mature conifers. However, recently we have obtained cell masses in our bud cultures that were similar in appearance to the embryogenic cell masses obtained from embryo and megagametophyte explants of _Larix decidua_ (2, 3). They were light yellow in color, slimy, and soft and were attached to larger masses of caulogenic (not embryogenic) green callus. The slimy, yellow cell mass grew slower than the embyogenic masses of megagametophyte or embryo origin. Furthermore, they only temporarily survived separation from the green callus and subculture. A few of the yellow cell masses were fixed, partially cleared and stained, and lightly pressed under a coverglass for microscopic observation. They were largely composed of long cells with large nuclei, and small clumps of small, densely cytoplasmic cells. In some of the cell masses the long cells were oriented parallel in bundles, with each bundle being attached to a clump of small cells (Fig. 5). These structures were similar to ones in haploid embryogenic cultures (see Figs. 15 and 16 in (2)), many of which eventually turned into proper embryos. However, the structures in Fig. 5 did not develop further.

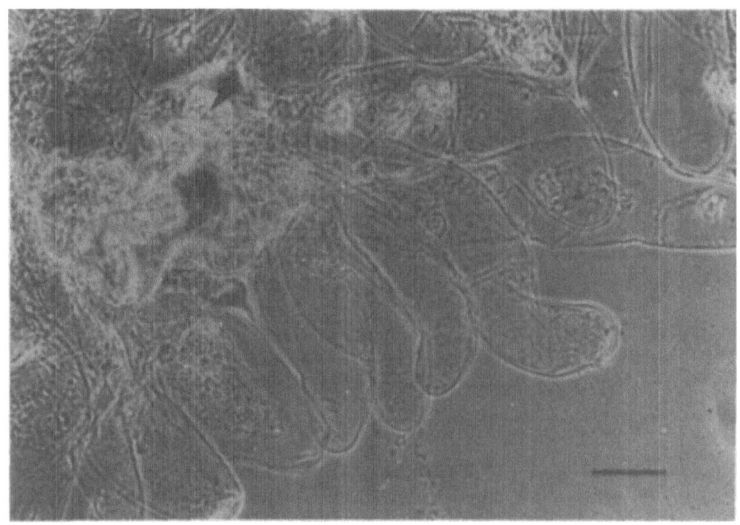

Fig. 5. A partially cleared squash preparation of a yellow, slimy cell mass
obtained from a culture initiated from a bud of a mature Larix
decidua tree. Note parallel arrangement of the long cells. Arrow =
group of small cells. Bar = 30µm.

In the first experiment where the yellow, slimy cell masses appeared,
the effect of pretreating the explant with DMSO was investigated. This
experiment was carried out because it has been reported that DMSO stimulates
polyembryony in ferns (16), embryogenesis in spring wheat (14), and cell
division in protoplasts of Pinus lambertiana (9). In this experiment the bud
explants were soaked in 5% DMSO for 20, 40, or 80 minutes prior to transfer
to 1/2 strength Litvay medium (6) minus growth regulators. The first time
this experiment was carried out (17 March 1987), several of the DMSO treated
explants formed the yellow, slimy callus (Table 1).

Table 1. The appearance of green callus capable of adventitious shoot
formation, and of yellow, slimy cell masses showing some of the
characteristics of early embryogenic cultures, after a soak in 5%
DMSO. The experiment was carried out with buds collected 17 March
1987 and subsequently repeated with buds collected 22 November
1988. Forty bud explants of Larix decidua were used for each soak
duration for each collection date. Each number indicates the number
of explants out of 40 that responded. Culture duration 3 months.

Duration of DMSO soak	March 1987		November 1988	
	Green callus	Yellow, slimy mass	Green callus	Yellow, slimy mass
20 min.	15(7)[*]	3	3(2)	0
40 min.	11	3	9(2)	0
80 min.	10	1	–	–

[*] ( ) = number of green calli forming adventitious shoots

19

In a repeat experiment, started on 22 November 1988, the yellow cell masses did not appear (Table 1).

In a third experiment, the culture conditions of which differed from those of the experiments shown in Table 1, the yellow cell masses appeared again. This experiment was carried out with buds collected on March 14, 1988. These were cultured on 1/2 Litvay medium; after three months, 20 cultures were transferred to GMD medium (12) without growth regulators. Three of these cultures produced the yellow, slimy cell masses. In a fourth experiment, with material collected 7 April 1988, one such cell mass developed, together with 7 green, shoot-forming calli (40 explants), on the 1/2 Litvay medium.

The 1/2 Litvay medium is the medium that we have used routinely for many years for adventitious shoot induction. Normally the yellow cell masses do not show up on this medium. Neither have they appeared on the media that we normally use to initiate embryogenesis in megagametophytes or embryos (3).

So far we have not been able to obtain the yellow callus reproducibly. However, it is an interesting coincidence, and possibly a significant one, that all three experiments in which the yellow cell masses appeared were started with buds collected in late March to early April. In future experiments we will have to establish if collection date is indeed one of the determining factors in the formation of the yellow cell masses. In conclusion, we cannot claim to have obtained embryogenic cultures from explants of mature trees; far from it. However, the appearance of bundled long cells attached to small clumps of small cells in the yellow cell masses, indicates that the early embryogenic stages may have been induced. The fact that this has occurred in three separate experiments is encouraging.

REFERENCES

1. Abdullah, A.A., Grace, J., and Yeoman, M.M., 1989, Rooting and establishment of Calabrian pine plantlets propagated in vitro: Influence of growth substances, rooting medium and origin of explant. New Phytol. 113: 193–202.
2. Aderkas, P. von, and Bonga, J.M., 1988, Formation of haploid embryoids of Larix decidua: early embryogenesis. Amer. J. Bot. 75: 690–700.
3. Aderkas, P. von, Klimaszewska, K., and Bonga, J.M., 1990, Diploid and haploid embryogenesis in Larix leptolepis, L. decidua, and their reciprocal hybrids. Can. J. For. Res. 20: 9–14.
4. Aitken-Christie, J., Singh, A.P., and Davies, H., 1988, Multiplication of meristematic tissue: A new tissue culture system for radiata pine, in: Genetic Manipulation of Woody Plants, J.W. Hanover, and D.E. Keathley, eds., Plenum Press, New York, pp. 413–432.
5. Amerson, H.V., Frampton, L.J., Mott, R.L., and Spaine, P.C., 1988, Tissue culture of conifers using loblolly pine as a model, in: Genetic Manipulation of Woody Plants, J.W. Hanover, and D.E. Keathley, eds., Plenum Press, New York, pp. 117–137.
6. Bonga, J.M., and Aderkas P. von, 1988, Attempts to micropropagate mature Larix decidua Mill., in: Somatic Cell Genetics of Woody Plants, M.R. Ahuja, ed., Kluwer Academic Publishers, Dordrecht, pp. 155–168.
7. Carson, M.J., 1986, Advantages of clonal forestry for Pinus radiata: Real or imagined?. N.Z. J. For. Sci. 16: 403–415.
8. Franclet, A., Boulay, M., Bekkaoui, F., Fouret, Y., Verschoore-Martouzet, B., and Walker, N., 1987, Rejuvenation, in: Cell and Tissue Culture in Forestry, Vol. 1, General Principles and Biotechnology, J.M. Bonga, and D.J. Durzan, eds., Martinus Nijhoff Publishers, Dordrecht, pp. 232–248.
9. Gupta, P.K., and Durzan, D.J., 1986, Isolation and cell regeneration of

protoplasts from sugar pine (<u>Pinus</u> <u>banksiana</u>). Plant Cell Rep. 5: 346–348.

10. Horgan, K., and Holland, L., 1989, Rooting micropropagated shoots from mature radiata pine. Can. J. For. Res. 19: 1309–1315.

11. Inglis, J.E., 1984, The effects of some growth substances on the promotion and rooting of interfascicular shoots in <u>Pinus</u> <u>caribaea</u> Morelet. Commonw. For. Rev. 63: 115–120.

12. Mohammed, G.H., Dunstan, D.I., and Thorpe, T.A., 1986, Influence of nutrient medium upon shoot initiation on vegetative explants excised from 15- to 18-year-old <u>Picea</u> <u>glauca</u>. N.Z. J. For. Sci. 16: 297–305.

13. Norton, M.E., and Norton, C.R., 1986, An alternative to <u>in</u> <u>vitro</u> propagation – axillary shoot enhancement on whole plants. J. Hortic. Sci. 61: 423–428.

14. Qureshi, J.A., Kartha, K.K., Abrams, S.R., and Steinhauer, L., 1989, Modulation of somatic embryogenesis in early and late-stage embryos of wheat (<u>Triticum</u> <u>aestivum</u> L.) under the influence of (+)-abscisic acid and its analogs. Plant Cell Tissue Organ Cult. 18: 55–69.

15. Ritchie, G.A., and Long, A.J., 1986, Field performance of micropropagated Douglas fir. N.Z. J. For. Sci. 16: 343–356.

16. Sheffield, E., 1984, Effects of dimethyl sulfoxide on the gametophytic and sporophytic phases of <u>Pteridium</u> <u>aquilinum</u> (L.) Kuhn. Ann. Bot. 54: 531–536.

17. Shelbourne, C.J.A., Carson, M.J., and Wilcox, M.D., 1989, New techniques in the genetic improvement of radiata pine. Commonw. For. Rev. 68: 191–201.

18. Sommer, H.E., Brown, C.L., and Kormanik, P.P., 1975, Differentiation of plantlets in longleaf pine (<u>Pinus</u> <u>palustris</u> Mill.) tissue cultured <u>in</u> <u>vitro</u>. Bot. Gaz. 136: 196–200.

19. Timmis, R., and Ritchie, G.A., 1988, Epigenetic effects in clonal propagation, <u>in</u>: Proceedings of the Tenth North American Forest Biology Workshop, Vancouver, July 20–22, pp. 12–31.

20. Tsogas, M., and Bouriquet, R., 1982, Propagation de l'épicéa par culture <u>in</u> <u>vitro</u> d'embryons et de plantules. Ann. Rech. Sylvicoles, AFOCEL, pp. 345–367.

THE TRANSITION BETWEEN SHOOT REGENERATION COMPETENCE AND

CALLUS DETERMINATION IN INTERNODAL STEM EXPLANTS OF *Populus deltoides*

Stephen G. Ernst and Gary D. Coleman

Department of Forestry, Fisheries and Wildlife
University of Nebraska, Lincoln, NE 68583-0814 USA

ABSTRACT

Experiments were conducted to monitor the competence status of internodal stem explants of 15 *Populus deltoides* genotypes in *in vitro* culture. The focus of this study was to investigate the transition from shoot regeneration competent to callus determined growth when in the presence of the inducer zeatin. Shoot regeneration competence and callus determination were measured by transferring explant tissue from callus inducing medium (CIM: WNA medium supplemented with 0.5 mgl^{-1} 2,4-D) to shoot inducing medium (SIM: WNA medium supplemented with 0.5 mgl^{-1} zeatin). Transfers from CIM to SIM were made at 1, 2, 4, 6, 8, and 10 day intervals. The number of regenerated shoots per explant and the percent of explants regenerating at least one shoot were determined after 60 days.

Three general explant competence responses were observed among the 15 *Populus deltoides* genotypes: (1) two genotypes were initially competent, with little increase in shoot regeneration by culture on CIM before transfer to SIM; (2) seven genotypes were not initially competent for shoot regeneration, but competence was acquired by initially culturing the explants on CIM before transfer to SIM, and this resulted in marked increases in shoot regeneration; and (3) six genotypes were not initially competent, and showed only slight competence enhancement after initial culture on CIM and produced relatively few adventitious shoots. The competence state transition from high levels of shoot regeneration to callus determination was very marked for the initially competent and competence acquired genotypes. Explants cultured on CIM for 6 days before transfer to SIM produced a relatively large number of shoots. However, explants subjected to CIM for 8 days before transfer to SIM produced few if any shoots, and the explants became determined for callus growth regardless of how long they remained on SIM. Genotypic responses to the different treatments will be discussed, in addition to preliminary results of analysis of protein differences associated with the competence state changes.

INTRODUCTION

The classic studies of Skoog and Miller (16) led them to suggest that the interactions of auxin and cytokinin provide a mechanism for the regulation of morphogenic events in plants. Morphogenic patterns in several plant species have since been manipulated *in vitro* by altering the auxin and cytokinin ratios, but the basic underlying mechanism controlling *in vitro* organogenesis remains obscure. The endogenous and exogenous factors required to alter cellular commitment are still largely unknown in relation to plant development (10), and *in vitro* model systems are probably the best approach to studying such events (11). Thorpe (17) has postulated that the process of *in vitro* morphogenesis requires an initial cellular dedifferentiation of the explant tissue before the cells are capable of responding to the specific morphogenic signals.

The process of *in vitro* morphogenesis has been divided into the discrete developmental events of competence, induction and determination (2). Competence is the ability of cells to respond to an inductive stimulus. Induction is the process by which a stimulus results in a unique developmental fate. Cells, tissues or organs that are determined exhibit a preordained developmental fate which cannot be altered even when exposed to a contrasting inductive signal. The study of competence, induction and determination requires the experimental manipulation of a stable system in which the developmental fate of cells can be manipulated (14; 13).

By experimental manipulation of the competence state of the explant tissue, it has been possible to temporally define the developmental stages of competence and determination. Examples of the effect of experimentally manipulating competence state include the *in vitro* organogenic response of *Medicago sativa* L. (19), *Convolvulus arvensis* L. (2, 3), *Pinus strobus* L. (8, 9), *Arabidopsis thaliana* (L.) Heynh. (7; 18), and *Populus deltoides* Bartr. ex Marsh. (6). It has been suggested that recalcitrance to manipulation *in vitro* results from a lack of cellular competence (4). In addition, the competence state was shown to vary in both stability and duration among genotypes of *Convolvulus arvensis* subjected to the same experimental treatments (3).

Based on the results of a previous study (6), the objectives of this study were to (1) classify different genotypes of *Populus deltoides* according to competence response, (2) among responsive genotypes, determine the uniformity of the transition from shoot regeneration competence to callus determined growth, and (3) identify developmental state-specific protein markers for either shoot regeneration competent or callus determined tissue.

MATERIALS AND METHODS

Competence Experiments: The specifics regarding the competence treatments and *Populus deltoides* genotypes are described elsewhere (6). Briefly, internodal stem explants from 15 genotypes of *Populus deltoides* were initially cultured on WNA medium supplemented with either 0.5 mgL^{-1} zeatin (zeatin control) or 0.5 mgL^{-1} 2,4-D (callus inducing medium, CIM). Additionally, a treatment without added plant growth regulators served as the WNA basal control. After 1, 2, 4, 6, 8, or 10 days culture on CIM, the internodal stem explants were then transferred to WNA medium supplemented with 0.5 mgL^{-1} zeatin (shoot inducing medium, SIM). After 60 days in culture the mean number of regenerated shoots per internodal stem explant was determined for each treatment replication for each genotype within an experiment. Additionally, the percent of explants regenerating at least one adventitious shoot was determined for each treatment within each genotype and experiment.

Protein extraction: Protein changes during the transition from shoot regeneration competent to callus determined growth was studied by sampling proteins from five internodal stem explants per treatment from genotype 301 (competence acquired) for the following treatments: 1, 2, 4, 6, 8, and 10 days CIM/20 days SIM. To further investigate the protein changes associated with cytokinin transfers, proteins from internodal stem explants of genotype 301 were sampled from the 2, 4, and 10 day CIM treatments after transfer for 10, 15, and 20 days to SIM. Also, proteins from internodal stem explants of genotypes 175 (immediately competent) and 297 (recalcitrant) were sampled from the SIM control, 2 days CIM/20 days SIM, and 10 days CIM/20 days SIM treatments. All internodal stem explant samples were frozen on dry ice and stored at -20°C until protein extraction.

Proteins were extracted from the internodal stem sections using the phenol partition method (15). The pellet was air dried and resuspended in Laemmli's (12) lysis buffer (0.0625 M Tris-HCl pH=6.8; 2% [w/v] SDS; 10% [v/v] glycerol; 5% [v/v] 2-mercaptoethanol; [Sigma, St. Louis, MO]) and placed in boiling water for 5 minutes to denature the proteins. Protein concentrations were determined and the proteins were either used immediately for electrophoresis or stored frozen (-20°C) in 50 ul aliquots.

Protein quantification: The concentration of proteins was determined using a modified bicinchoninic acid procedure (1). In this procedure a 100 ul protein sample (a 1:10 dilution of the stock protein sample) was mixed with 900 ul of double distilled water and 100 ul of sodium deoxycholate (0.15%, w/v) (Sigma, St. Louis MO). After standing at room temperature for 10 minutes, 100 ul of trichloroacetic acid (72%, w/v) (Sigma, St. Louis MO) was added. The mixture was vortexed briefly and then centrifuged at 3000x g for 20 minutes. The supernatant was removed and 2 mls of bicinchoninic acid reagent (Sigma, St. Louis MO) were added and incubated for 30 minutes at 37°C. Absorbance was measured at 562 nm using a Bausch and Lomb Spectronic 710 spectrophotometer. Protein concentrations were determined by comparison to a bovine serum albumin (BSA) standard curve.

Protein electrophoresis: Proteins were separated in 7.5% to 20% linear gradient polyacrylamide gels (16cm x 16cm x 1mm) essentially as described by Laemmli (12) except for the addition of 0.1M sodium acetate to the anodal buffer (5). The gels were loaded with 40 ug of total protein and electrophoresis was at 15 mA per gel for 1100 volt-hours. Upon completion of electrophoresis, the gels were immediately fixed in 40% methanol, 10% acetic acid. Protein molecular weights were determined using molecular weight standards (Bio-Rad, Richmond, CA).

Silver amine staining: Fixed polyacrylamide gels were silver stained using a commercially available kit (Bio-Rad, Richmond CA) according to the manufacturer's directions.

RESULTS AND DISCUSSION

The 15 *Populus deltoides* genotypes could be categorized into three different competence types based on response to the competence treatments (Table 1): (1) two genotypes (56 and 175) were initially competent, with little increase in shoot regeneration by initial culture on CIM before transfer to SIM; (2) seven genotypes (10, 53, 171, 174, 179, 298, and 301) were not initially competent for shoot regeneration, but competence was acquired by initial culture on CIM before transfer to SIM, resulting in marked increases in shoot regeneration; and (3) six genotypes (54, 177, 178, 296, 297, and 300) were not initially competent, and showed only slight competence enhancement after initial culture on CIM with relatively few adventitious shoots produced. Genotypes that regenerated the greatest number of adventitious shoots per explant consistently resulted in the greatest percentage of explants that regenerated shoots (Table 1). In addition, for the nine genotypes that produced adventitious shoots *in vitro*, the transition from shoot regeneration competence to callus determined growth was fairly consistent and distinct (Table 1). Generally, after 8 or 10 days on CIM before transfer to SIM, the explants became determined for callus growth, and no shoots were produced regardless of how

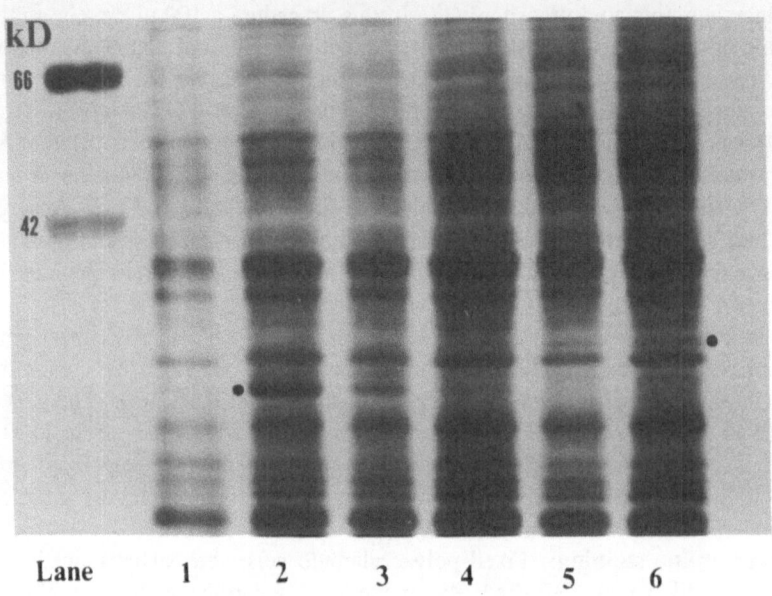

Figure 1. Silver-stained 1D SDS/PAGE gel of proteins extracted from *Populus deltoides* internodal stem explants after culture using the following competence treatments: 0 days CIM/20 days SIM (lane 1), 2 days CIM/20 days SIM (lane 2), 4 days CIM/20 days SIM (lane 3), 6 days CIM/20 days SIM (lane 4), 8 days CIM/20 days SIM (lane 5), 10 days CIM/20 days SIM (lane 6). The closed circles indicate the position of the 32 kDa and 35 kDa proteins referred to in the text.

Table 1. Mean number of shoots regenerated, standard error of the mean (in parentheses), and percent of explants regenerating adventitious shoots, among internodal explants of the nine responsive *Populus deltoides* genotypes after being subjected to each of the eight competence treatments.

Genotype	Basal	Zeatin	Time (days) on CIM before transfer to SIM					
			1	2	4	6	8	10
10	0.50 (0.19) 21.7%	4.45 (0.28) 91.7%	8.97 (0.49) 100%	6.62 (0.79) 98.3%	7.07 (0.16) 100%	5.77 (0.67) 90.0%	5.38 (1.15) 98.3%	3.02 (0.92) 88.3%
53	0.57 (0.224) 33.3%	1.18 (0.35) 58.3%	2.12 (1.27) 63.0%	0.63 (0.14) 38.5%	2.28 (0.48) 71.7%	3.32 (0.91) 81.7%	0.22 (0.11) 6.7%	0.00 (0.00) 0.0%
56	0.17 (0.06) 15.0%	2.22 (0.27) 85.0%	1.37 (0.90) 52.5%	0.07 (0.07) 5.0%	0.00 (0.00) 0.0%	0.00 (0.00) 0.0%	0.00 (0.00) 0.0%	0.00 (0.00)b 0.0%
171	0.17 (0.07) 13.6%	1.90 (0.29) 100%	3.28 (0.44) 100%	4.55 (0.46) 100%	4.61 (0.60) 100%	4.25 (0.58) 100%	2.14 (0.76) 75.9%	0.22 (0.17) 13.3%
174	0.03 (0.02) 3.3%	1.27 (0.27) 63.3%	1.90 (0.05) 96.7%	1.68 (0.31) 83.3%	1.28 (0.40) 61.7%	0.45 (0.03) 25.0%	0.23 (0.09) 11.7%	0.08 (0.08) 5.0%
175	0.21 (0.06) 12.5%	3.57 (0.27) 95.9%	1.15 (0.54) 46.7%	1.60 (0.28) 56.0%	1.33 (0.16) 41.3%	0.73 (0.21) 25.3%	0.00 (0.00) 0.0%	0.00 (0.00) 0.0%
179	0.28 (0.12) 26.7%	2.32 (0.15) 100%	2.60 (0.69) 90.0%	1.51 (0.14) 67.8%	1.23 (0.22) 66.7%	1.98 (1.37) 53.6%	1.42 (0.12) 65.0%	0.48 (0.34) 26.%
298	0.17 (0.12) 11.7%	0.46 (0.12) 30.5%	0.11 (0.00) 7.9%	1.17 (0.59) 51.8%	1.19 (0.12) 64.9%	0.19 (0.12) 15.8%	0.60 (0.17) 44.6%	0.72 (0.29) 46.7%
301	0.00 (0.00) 0.0%	0.30 (0.16) 18.3%	1.15 (0.64) 53.3%	1.25 (0.51) 60.0%	3.82 (0.35) 95.0%	2.35 (0.43) 71.7%	0.92 (0.45) 33.3%	0.00 (0.00) 0.0%

long the explants remained on SIM. Therefore, with an additional two days on CIM before transfer to SIM (i.e., 6 versus 8 days on CIM), the explant lost competence for shoot regeneration and became callus determined.

The protein changes associated with the transition from shoot regeneration competence to callus determined growth included a 32 kDa protein that increased in relative concentration only in explants subjected to those treatments that resulted in shoot regeneration. For genotype 301, this included only the 2, 4, and 6 day CIM/20 day SIM treatments (Figure 1). For immediately competent genotype 175, the 32 kDa protein was present for the SIM control treatment, but not for the 10 day CIM/20 day SIM treatment, and for recalcitrant genotype 297, the 32 kDa protein was not observed (data not shown). The 32 kDa protein was detected as early as 15 days after transfer to SIM (data not shown), which corresponds to when we first observe promeristemoids forming in fixed thin sections. Additionally, a 35 kDa protein was detected in the 8 and 10 day CIM treatments which were cultured for an additional 20 days on SIM, but not in the 2 and 4 day CIM/20 day SIM treatments (Figure 1).

Using serial transfers to differing culture media, the developmental stages of competence and determination were defined for internodal stem explants of nine responsive genotypes of *Populus deltoides*. This system has demonstrated that events during the first ten days of culture have significant effects upon the organogenic fate of internodal stem explants. Such a system will aid in the elucidation of the early events of morphogenesis. Correlation of the developmental events defined in this system with physiological, biochemical and molecular changes should prove valuable in improving our understanding of *in vitro* and *in vivo* responses of plant cells. Further analysis of the developmental state-specific protein markers should also help elucidate the molecular events of organogenesis, and provide markers for the early screening of the regeneration potential of cultured plant tissue.

NOTE: A manuscript describing in greater detail one- and two-dimensional PAGE analysis of the proteins associated with shoot and callus determined tissue has been submitted to Plant Science.

REFERENCES

1.   Brown, R.E., Jarvis, K.L. and K.J. Hyland. 1989. Protein measurement using bicinchoninic acid: elimination of interfering substances. Anal. Biochem. 180:136-139.
2.   Christianson, M.L. and D.A. Warnick. 1983. Competence and determination in the process of *in vitro* shoot organogenesis. Develop. Biol. 95:288-293.
3.   Christianson, M.L. and D.A. Warnick. 1985. Temporal requirements for phytohormone balance in the control of organogenesis *in vitro*. Develop. Biol. 112:494-497.
4.   Christianson, M.L. and D.A. Warnick. 1988. Organogenesis *in vitro* as a developmental process. Hortscience 23(3):515-519.
5.   Christy, K.G. Jr., LaTart D,B. and H.W. Osterhoudt. 1989. Modifications for SDS-PAGE of proteins. Biotechniques 7(7):692-693.
6.   Coleman, G.D., and S.G. Ernst. 1990. Shoot induction competence and callus determination in *Populus deltoides*. Plant Science (accepted).
7.   Feldmann, K.A. and M.D. Marks. 1986. Rapid and efficient regeneration of plants from explants of *Arabidopsis thaliana*. Plant Sci. 47:63-69.
8.   Flinn, B.S., Webb, D.T. and W. Newcomb. 1988. The role of cell clusters and promeristemoids in determination and competence for caulogenesis by *Pinus strobus* cotyledons *in vitro*. Can. J. Bot. 66:1556-1565.

9.   Flinn, B.S., Webb, D.T. and W. Newcomb. 1989. Morphometric analysis of reserve substances and ultrastructural changes during determination and loss of competence of Eastern white pine (*Pinus strobus*) cotyledons *in vitro*. Can. J. Bot. 67:779-789.

10.   Goldberg ,R.B. 1988. Plants: Novel developmental processes. Science 240:1460-1467.

11.   Hicks, G.S. 1980. Patterns of organ development in plant tissue culture and the problem of organ determination. Bot. Rev. 46(1):1-23.

12.   Laemmli, U.K. 1970. Cleavage of structural proteins during the assembly of the head of bacteriophage T4. Nature 227:680-685.

13.   McDaniel, N.C. 1984. Competence, determination, and induction in plant development. In: Pattern Formation a Primer in Developmental Biology. Macmillian, New York. pp.

14.   Meins, Jr., F. and A.N. Binns. 1979. Cell determination in plant development. Bioscience 29(4):221-225.

15.   Schuster, A.M. and E. Davies. 1983. Ribonucleic acid and protein metabolism in pea epicotyls I. The aging process. Plant Physiol. 73:809-816.

16.   Skoog, F. and C.O. Miller. 1967. Chemical regulation of growth and organ formation in plant tissues cultured *in vitro*. Symp. Soc. Exp. Biol. 11:118-140.

17.   Thorpe, T.A. 1980. Organogenesis *in vitro*: Structural, physiological, and biochemical aspects. Int. Rev. Cytol. Suppl. 11A:71-112.

18.   Valvekens, D., M. Van Montagu, and M. Van Lijsebettens. 1988. *Agrobacterium tumefaciens*-mediated transformation of *Arabidopsis thaliana* root explants by using kanamycin selection. Proc. Natl. Acad. Sci. USA 85:5536-5540.

19.   Walker, K.A., Wendeln, M.L. and E.G. Jaworski. 1979. Organogenesis in callus tissue of *Medicago sativa*. The temporal separation of induction processes from differentiation processes. Plant Sci. Lttrs. 16:23-30.

MICROPROPAGATION OF SILVER BIRCH (Betula pendula Roth.) AND CLONAL FIDELITY

OF MASS PROPAGATED BIRCH PLANTS

K. Jokinen and T. Törmälä

Kemira Oy, Espoo Research Centre
P.O.Box 44
SF-02271 Espoo Finland

ABSTRACT

A commercial scale micropropagation technology was developed in Finland for the silver birch (Betula pendula), which is an increasingly important raw material source for the forest industry.

Genotype affected more the success of initiation and multiplication
rate than e.g. age of the mother tree or season. The clonal fidelity of the
in vitro propagation system based on axillary proliferation has been high.
When producing plantlets using adventitious tissue culture system one distinct variant was discovered. The field performance of the micropropagated
plantlets during the first season in the nursery was slightly better than
that of conventional seedlings.

INTRODUCTION

The silver birch is one of the three important forest tree species in
Finland. Its wood is valuable especially for liquid packing boards, for
high quality printing papers and for the plywood industry. The growth of
silver birch is relatively fast and it flowers at a young age. It is
practically the only northern forest tree, which can be bred effectively.
Through conventional breeding techniques remarkable increase in the growth
rates has been achieved.

Generally birch cuttings are difficult to root and consequently in
vitro propagation methods have been investigated (3). Micropropagation
allows the rapid multiplication of valuable trees for breeding or even for
direct reforestation purposes. Micropropagation methods fall into three
categories: adventitious or axillary shoot proliferation and somatic
embryogenesis (2). Plantlets of birch have been produced using all the
three methods as reviewed recently by Meier-Dinkel (4). Considering the
difficulties involved in somatic embryogenesis (Jokinen unpublished) adventitious and axillary methods seem to be the most promising in vitro methods
for the mass propagation of birch. The adventitious way may, however, be
sensitive to somaclonal variation because of the prolonged callus stage
during the multiplication phase (2). The benefit of the adventitious propagation technique in comparison with the axillary one is the much greater
multiplication rate (1).

The aim of this study was to develop a micropropagation method, which could be applied to several mature silver birch genotypes. An important target was also the evaluation of the clonal fidelity of micropropagated birch plants in field experiments. In addition to the axillary system also adventitious way of micropropagation was applied.

MATERIALS AND METHODS

The silver birch individuals (genotypes) were selected for micropropagation experiments on the basis of growth rate and wood quality. The genotypes originated from southern Norway, eastern Finland and northern Finland. The age of the trees varied from 20 to 100 years.

The basic methods for the initiation of culture, regeneration of buds, shoots and roots used in this study are mostly similar to those described by Särkilahti (6). Instead of MS mineral salts WPM medium (5) supplemented with 4uM benzyladenine and 2 % sucrose was used for initiation and multiplication of the cultures. For rooting WPM medium with IBA 1 uM and 2 % sucrose was used.

In the regeneration experiments 30 buds of each genotype were taken and cultured in vitro in order to evaluate the probability of regeneration (% of buds forming shoots). Also the influence of the season was tested by sampling buds throughout the year at the intervals of ca. four weeks.

The multiplication rate experiment was based on axillary production of new shoots. The multiplication rate is defined as the number of nodes produced from one node after four weeks of culture (one cycle). The multiplication rate given for a genotype is the mean of 25 nodes. The experiment continued for three cycles.

In an experiment aimed at comparison of the methods of multiplication i.e. adventitious or axillary only one genotype (JR1/1) was used. After the establishment of the culture in the multiplication phase this genotype tends to form continuously also adventitious buds and shoots at basal parts of the growing microshoots in addition to axillary shoots. The experiment with the adventitious system was run for seven cycles prior to the elongation of the shoots. In the axillary system only the nodes of the microshoots used as the multiplication unit. Approximately 3000 and 7000 plantlets were produced for field experiments with the adventitious and axillary ways of propagation, respectively.

In a nursery experiment the performance of micropropagated plantlets was evaluated using conventional seedlings as controls. The seed originated from the same areas as the mother trees of the micropropagated plantlets. The height of 50 plants of each population was measured during the growing season.

RESULTS AND DISCUSSION

Over 60% of the birch genotypes selected for their growth rate and wood quality could be regenerated on the standard culture medium (Fig. 1). The age of the stock plant (genotype) did not affect the ability of birch to grow in vitro suggesting that other factors are more significant in the mature birch. There are, however, possibilities to increase the number of genotypes, which can be regenerated by modifying the composition of the media, especially by applying different kinds of cytokinins for genotypes difficult to regenerate (Jokinen, unpublished).

The shoot apex of the birch was isolated most successfully during the autumn and winter months, but the initiation could also be done in other times of the year (Fig 2). The most laborious time is summer, when the meristems are very succulent and fragile. When the genotype is in general hard to regenerate the time of the year is rather unimportant as in the case of K-2673 shown in Fig. 2. Although it was possible to regenerate the genotype K-2673 once it grew very poorly in vitro and was hard to multiply.

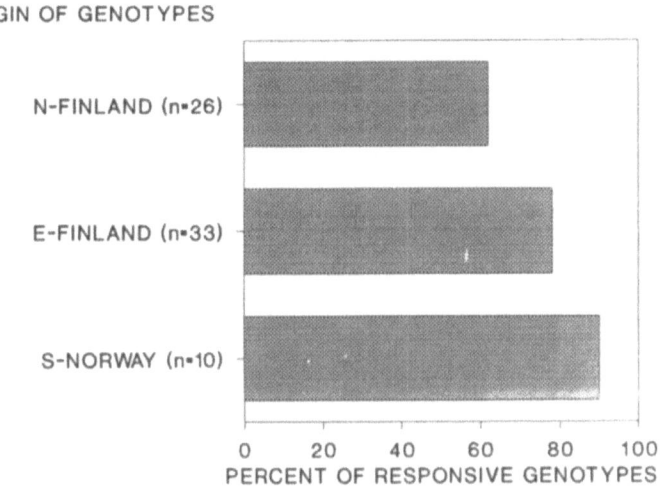

Fig 1. The regeneration probability of birch genotypes of different origin.

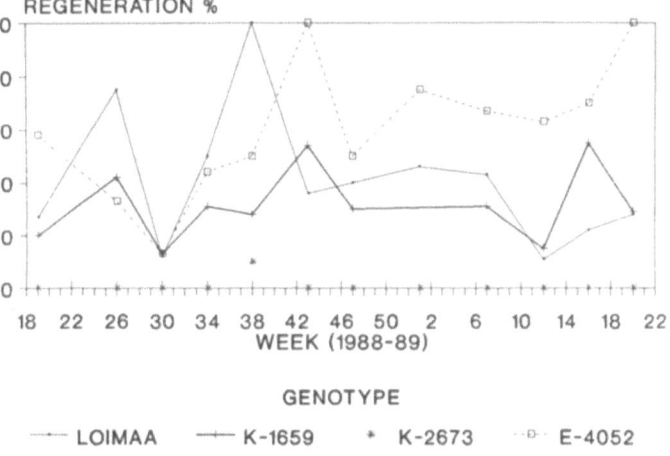

Fig 2. The regeneration probability of birch in relation to the genotype and the time of the year

The multiplication rate of the birch in stabilized cultures depended greatly on the genotype (ca. 100 genotypes tested) varying from two to twenty.

The age of the stock tree was less important. The multiplication rate of a given genotype was very stable from cycle to cycle. The time of the year of the establishment of the culture did not affect the multiplication rate of a given genotype. These results indicate that the multiplication rate is a very stable character and it can be used as a general indicator of the growth of birch in vitro.

The results suggest further that the growth of a genotype in vitro is determined by the basic physiological processes governed by the genome and influence of the season on the internal state of the plant is less important. The dormancy of buds can be overcome by in vitro treatments, if the genotype is in general competent for growing in vitro.

In the experiment, which compared the adventitious and axillary methods of propagation, there occurred one birch variant population of 250 plantlets among 3000 adventitiously propagated plantlets. The variant ("snake birch") differed drastically from the rest of the adventitiously propagated plantlets, which looked phenotypically identical to the plantlets produced by axillary method. The snake birch had much smaller leaves and shorter stem than the normal plants. The growth habit was horizontal or creeping. After one year in the nursery new in vitro cultures were established in order to evaluate the stability of the variant. It was compared with normal plants of the same genotype both in vivo and in vitro. The results confirmed the preliminary observations on the in vivo growth of the snake birch (Table 1).

Table 1. Morphological characters of snake birch variant compared with normal plants of the same genotype (JR1/1) grown for six weeks in the greenhouse. Both clones were propagated in vitro. (n=20, x $\pm$ S.E.)

	Snake birch	Normal
Height of the plants	15.2 $\pm$ 0.5	39.0 $\pm$ 1.0
Fully expanded leaves		
length	5.5 $\pm$ 0.3	12.1 $\pm$ 0.6
width	4.0 $\pm$ 0.3	9.9 $\pm$ 0.4

The variant had higher multiplication rate, was longer and heavier and produced more shoots than normal plants and was thus probably more competitive than the normal plants during the in vitro multiplication. The relatively high proportion of the variants can be explained by the occurrence of the putative mutation in the first cycles of multiplication and the high multiplication rate of the variant. Although it is presently impossible to state the frequency of variation or the number of cycles needed for increased probability of variation, this case indicates, however, there exist risks of variation, even if a relatively low number of plants is produced adventitiously during a rather short time in vitro culture. So far there is no indication of variants among the about 400 000 plantlets (100 genotypes) produced through the axillary system.

Micropropagated plantlets developed somewhat faster in the nursery than the seedlings during the first growing season (Fig. 4). The seeds originated from the same area as the mother trees of the micropropagated plantlets suggesting that increased average growth can be achieved by selecting the most productive mature trees. However, the first year growth does not yet reliably demonstrate the performance of micropropagated plantlets in the forest and by 1989 300 000 plantlets have been produced for the establishment of large field experiments.

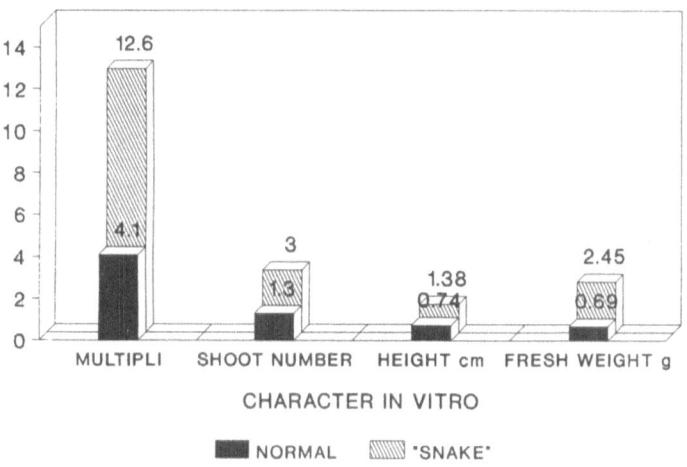

Fig. 3. In vitro characters of snake birch compared with the normal plants of the same genotype (JR1/1).

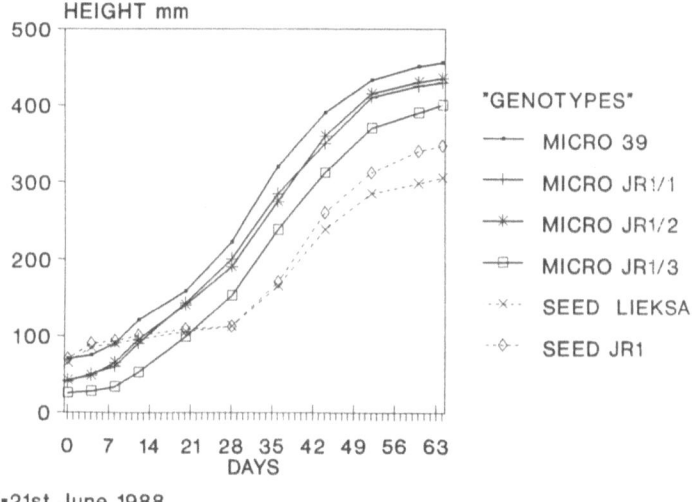

Fig. 4. The growth of birch plantlets and seedlings in Enso Ukonniemi nursery (SE Finland) in 1988

CONCLUSIONS

The micropropagation system of birch based on axillary branching in vitro developed is an effective method for mass cloning of elite mature silver birches. The project has proceeded to commercial scale in 1990. The probability of altered phenotypes seems to be low. The actual growth benefits and economics of producing birch planting material by micropropagation can be reliably assessed as data accumulated from the large scale field trials underway.

REFERENCES

1. Chu, I.Y.E., and Kurtz, S.L., 1990, Commercialization of plant tissue culture, in: "Handbook of Plant Cell Culture, Vol 5," P.V. Ammirato, D.A. Evans, W.R. Sharp and Y.P.S. Bajaj, eds., McGraw Hill, New York.

2. George, E. F., and Sherrington, P. D., 1984, "Plant Propagation by Tissue Culture," Exegetics Ltd. Hants.

3. McCown, B.H., 1985, From gene manipulation to forest establishment:shoot cultures of woody plants can be a central tool. Tappi J., 68:116.

4. Meier-Dinkel, A., 1990, Micropropagation of birches (Betula spp.). Manuscript for: "Biotechnology in Agriculture and Forestry. Trees III," Y.P.S. Bajaj, ed.

5. Smith, M.A.L., and McCown, B.H., 1982/83, A comparison of source tissue for protoplast isolation from three woody plant species, Plant Sci. Lett., 28:149.

6. Särkilahti, E., 1988, Micropropagation of a mature colchicine-polyploid and irradiation-mutant of Betula pendula Roth. Tree Physiol., 4:173.

IN VITRO STUDIES WITH WHITE ASH (FRAXINUS AMERICANA)

NODULES

John E. Preece and Sharon Bates

Department of Plant and Soil Science
Southern Illinois University
Carbondale, IL 62901 USA

ABSTRACT

Calli were obtained from cotyledonary tissue of cut
zygotic seeds that had been cultured in vitro on agar-
solidified MS medium with 5 μM 2,4-D and 5 μM BA for four
weeks. Seedlings were subcultured onto MS with 0.5 μM NAA and
0.5 μM BA. After two to four months, callus tissue was
excised and placed into liquid MS with 1 μM NAA and 1 μM BA on
a shaker at 100 rpm. Within four weeks, spherical clusters
(nodules) formed that regularly sloughed-off and produced
additional clusters. To speed multiplication, nodule clusters
could be physically broken-up with forceps and the pieces
subcultured. Nodules had a distinct morphology compared to
callus clumps. Scanning electron microscopy (SEM) studies of
nodule cross sections revealed a cortical region composed of
parenchyma cells surrounded by a layer of closely packed cells
(possibly epidermis). Areas of vascularization containing
xylem and a cambium-like tissue were evident within the
cortical region. Nodules exposed to moderately high levels of
thidiazuron produced large amounts of callus. Nodule clusters
could be removed aseptically from liquid medium and dehydrated
for up to 14 days. After desiccation for various times, the
nodules were placed on agar-solidified medium where they
rehydrated and grew as callus. Thidiazuron in the rehydration
medium stimulated callus growth. Nodules showed sensitivity
to glyphosate herbicide on agar-solidified and liquid media.
On solidified medium, glyphosate at 0.1 and 1 mM was not
lethal, and nodules remained green; 10 mM glyphosate resulted
in nodules with areas of green and brown, and 0.1 and 1 M
glyphosate were lethal. In liquid medium, nodules exposed to
1 mM herbicide became brown and died, at lower concentrations,
both green and brown sectors were evident.

INTRODUCTION

Nodules are dense, spherical structures or clusters that have
been reported to form in vitro from a variety of plant
species, including radiata pine (1), Citrus (2), carrot (4),
daylily (5), and poplar and white spruce (6). Because nodules

are uniformly shaped, can be selected for standardized size
(6), and can be multiplied easily and rapidly, we felt that
they might be suitable structures for physiological studies
including response to plant growth regulators, desiccation,
and herbicides.

White ash was selected because of its importance as a fine
hardwood and its relative ease of handling in vitro (8,9). We
have considerable experience with clonal micropropagation, and
adventitious regeneration of this species. We felt that
subjecting a uniform structure to various plant growth
regulator treatments and stresses would help to provide
additional background for the use of in vitro techniques for
the ultimate goal of tree improvement.

We therefore conducted histological investigations of nodules
to further understand their anatomical features and determine
if they were similar to nodules reported for other species.
Plant growth regulator treatments were applied to determine if
cell division and production of callus could be stimulated
from nodules. Dehydration and herbicide treatments were
selected to determine nodule response to stresses that
commonly occur during plantation establishment.

MATERIALS AND METHODS

Initial explants were mature seeds purchased from F.W.
Schumaker, Sandwich, Massachusetts, USA and stored in darkness
at 4 ± 2°C until used. Fruit were dewinged, surface
sterilized, and 1/3 of the seed opposite the embryonic axis
was transversely dissected and discarded, following the
procedure of Preece et al. (8,9). The remaining portion of
the seed was placed in vitro onto Murashige and Skoog (7)
salts plus organics supplemented with 3% sucrose (w/v). The
pH of the medium was adjusted to 5.8 with 1 N KOH or HCl prior
to the addition of 7 $g \cdot l^{-1}$ Difco Bacto agar, if used.

Cultures were incubated under cool white fluorescent
lamps that provided a 16 hr photoperiod and a PPF of 33 to 45
$\mu mol \cdot m^{-2} \cdot s^{-1}$. Explants on agar were transferred to fresh medium
monthly and liquid suspensions were subcultured biweekly.

Nodules used in these studies were initiated by excising
calli that had formed on seedlings that had been on media with
5.0 μM 2,4-dichlorophenoxyacetic acid (2,4-D) and 5 μM
benzyladenine (BA) for four weeks, then transferred to
secondary medium with 0.5 μM naphthaleneacetic acid (NAA) and
0.5 μM BA. The calli were placed into liquid medium
containing 1.0 μM NAA and 1.0 μM BA. Cultures were placed on
a platform shaker at 100 rpm. Within one month, small
spherical nodules formed. These nodules multiplied relatively
rapidly by sloughing-off new nodules or by the physical
breakup (using forceps) of larger nodules. Observations of
initial cultures indicated that the number of nodules tripled
and the size of smaller nodules doubled within one month.

Nodules were multiplied continuously throughout the study
in agitated liquid medium. Unless otherwise noted, the
uniform nodules (60-70 mm^3) used in the experiments in this
paper were removed from liquid medium and placed onto agar-
solidified medium.

An exception was the dehydration experiment, where the nodules were placed into sterile petri dishes that contained no medium or water. These petri dishes were then placed into a sealed desiccator for 6 and 14 days and stored in darkness. Relative humidity in the desiccator was maintained at 70% by placing a saturated solution of NH_3Cl and KNO_3 at the bottom of the desiccator. Following dehydration, nodules were placed on agar-solidified medium containing 10 μM TDZ and 1 μM IBA. After two weeks, ½ of the nodules were transferred to a medium containing no plant growth regulators and the other half remained on the medium containing TDZ. Nodules were transferred monthly to fresh vessels containing the same medium formulation.

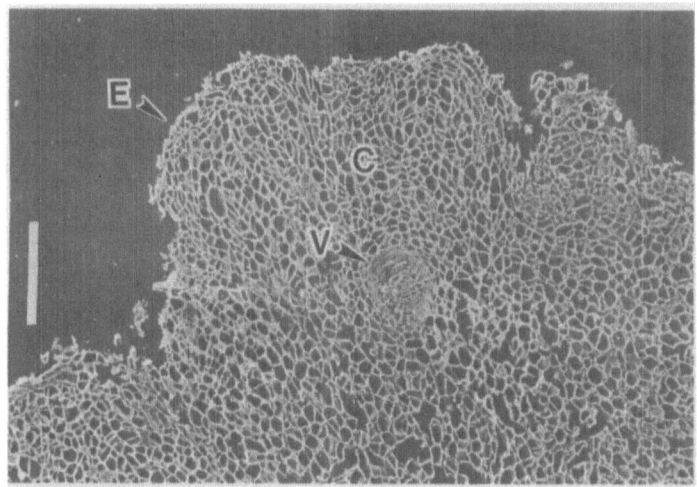

Figure 1. Cross section of a sphere of a white ash nodule cluster with the external region of closely packed cells (E), possibly epidermis; ground or cortical region (C); and vascular center (V) labelled. Scale bar = 0.30 mm.

Nodules were prepared by two methods for scanning electron microscopy observations. 1. Nodules were fixed in FAA, dehydrated through a tertiary butanol series and embedded in paraffin. Sections 12 μm thick were obtained with a microtome and placed on slides. Paraffin was removed with xylene and slides were placed in absolute alcohol. Slides were either air dried or critical point dried and coated with 250 Å AuPd. 2. Nodule slices were also prefixed in glutaraldehyde and fixed in osmium tetroxide followed by dehydration through an alcohol series. Nodules were critical point dried and coated with 250 Å AuPd.

RESULTS AND DISCUSSION

White ash nodules had distinct anatomical features and were similar, but not identical, to the <u>Populus</u> "unicenter nodules" described by McCown et al. (6). In cross section (Fig. 1), each sphere within a white ash nodule cluster largely was composed of a region of parenchymous cells that appeared analogous to cortical or ground tissue. Large intercellular spaces were present towards the inner portions of this ground tissue. Externally a layer of closely-packed cells (possibly epidermis) covered the nodules. Towards the

(A)

(B)

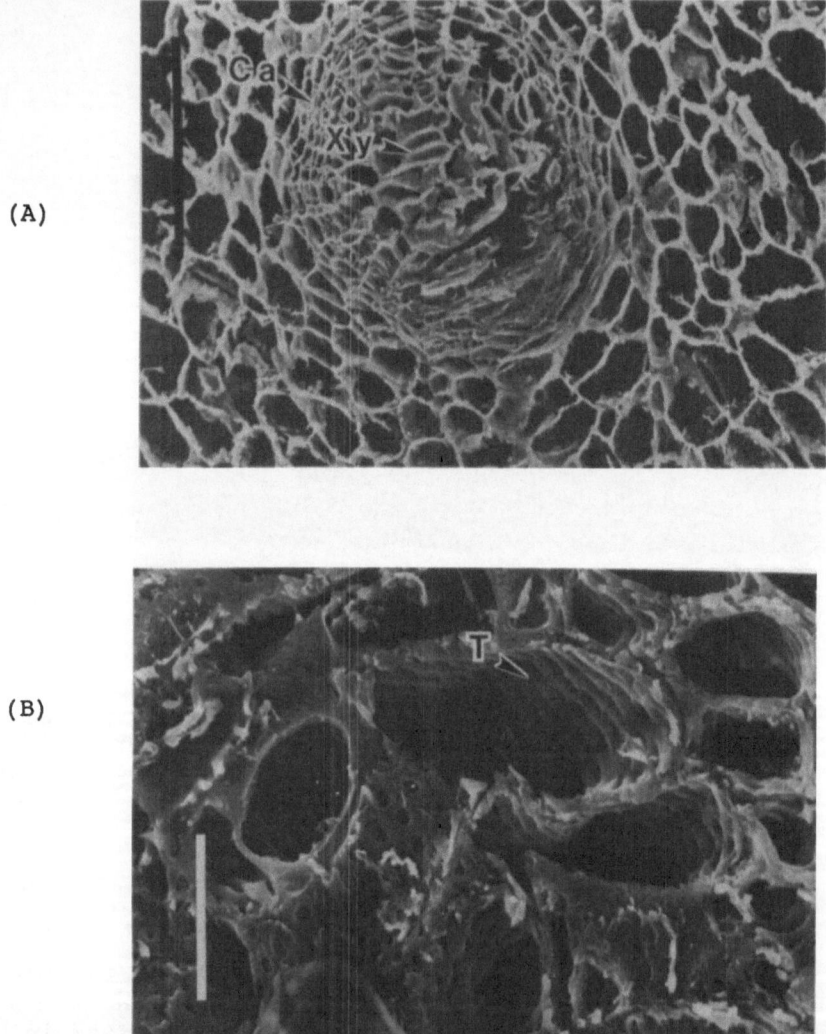

Figure 2. (A) Cross section of a vascular center and (B) close-up of xylem region showing tracheary elements with an alternate pitting pattern. (Ca) represents a vascular cambium-like tissue surrounding the xylem. (Xy) represents the xylem region. (T) represents tracheary elements with an alternate pitting pattern. Scale bars represent 150 μm in (A) and 30 μm in (B).

Table 1. Effect of thidiazuron concentration on nodule volume^z after two months in vitro.

Thidiazuron μM	Nodule volume mm^3
0.01	84.6
0.1	60.2
1	79.3
10	1333.1
100	2685.9
1000	63.2
	**
LSD	
5%	570.96
1%	761.28
Contrast	
Linear	**
Quadratic	**
Cubic	**

^{z}Each number represents the mean of 9-10 cultures.
**Significant at the 1% level.

center of each sphere within a nodule cluster was a region of vascular tissue (Fig. 2). These tissues consisted of xylem tracheary elements with an alternate pitting pattern. Closely-packed rows of cells surrounded the vascular tissue that appeared cambium-like.

Depending on treatment, nodules that were removed from liquid nodule cultures and placed on agar-solidified MS medium with various thidiazuron concentrations appeared different from each other within one week (data not presented). After eight weeks on treatment media, nodules responded in a cubic manner to increasing doses of TDZ (Table 1). Nodules exposed to <10 μM TDZ grew very little, and did not produce any callus. Much friable callus grew from nodules exposed to 10 or 100 μM TDZ with the softest callus from nodules that had been on medium with the higher concentration. The concentration of 1000 μM TDZ was lethal to the nodules which rapidly turned black. Poplar nodules produced adventitious shoots when exposed to 0.1 μM TDZ for two weeks (5); however the white ash nodules were not organogenic but produced callus.

Nodules that were removed from liquid nodule cultures and placed into empty, sterile petri dishes for desiccation, appeared shrunken after one day. Shrinkage continued throughout the 14 day desiccation cycle. When removed from the desiccation treatments and placed on agar-solidified medium, nodules began to swell and rehydrate immediately. Nodules that had been dehydrated for six days quickly rehydrated and looked similar to those that had not been desiccated; whereas nodules that had been desiccated for 14 days had necrotic-looking areas that did not fully re-expand.

Table 2. Effect of days of dehydration and thidiazuron on nodule volume[z] and dry weight[y] after three months in vitro.

| | Nodule volume (mm^3) | | Dry weight |
	Transformed[x]	Actual	(mg)
Days			
0	4.5	135.4	21.4
6	5.7	733.2	53.8
14	3.7	538.1	17.2
	**	NS	**
LSD			
5%	0.50		24.83
1%	0.66		34.11
Growth regulator			
none	4.0	148.5	
TDZ	5.2	798.3	
	**		
LSD			
5%	0.41		
1%	0.54		

[z]Each number represents the mean of 48-50 cultures for days of dehydration and 74 cultures for growth regulator.
[y]Each number represents the mean of 7-8 cultures.
[x]Log(n) transformation.
**Significant at the 1% level according to F-test.
NS Not significant.

Nodules that had been desiccated for 6 days were significantly heavier and larger than non-desiccated controls and those dried for 14 days (Table 2). It is interesting that upon rehydration, nodules responded to moderate desiccation stress by growing. Similarly, desiccation has been shown to stimulate germination and thus growth of somatic embryos of grape (3). Nodules therefore may have utilization for studies on plant water stress.

Following two weeks on medium with 10 μM TDZ and 1 μM IBA, ½ of the nodules from all desiccation treatments were placed on media containing 1 μM NAA and 1 μM BA and the other half remained on the TDZ medium. The main effect of medium and not the medium x dehydration interaction was significant. Nodule clusters continuously exposed to thidiazuron were significantly larger than those receiving the two week pulse.

Visual differences were apparent among nodules on agar-solidified medium within the first week of exposure to glyphosate. After four weeks there were significant differences in nodule appearance and growth (Table 3). Nodule rating, volume, and dry weight were affected in a cubic manner with increasing glyphosate doses. Control nodules had a higher rating (were green and growing) and were significantly larger and heavier than any exposed to glyphosate. As the

Table 3. Effect of glyphosate concentration on nodule volume, dry weight, and rating[z] after one month in vitro.

Glyphosate concn. (mM)	Rating		Volume mm^3	Dry weight (mg)
	Transformed[y]	Actual		
0	0.8	5.0	258.1	27.3
0.1	0.7	4.3	103.9	14.4
1	0.7	3.9	58.9	9.2
10	0.6	3.1	45.9	6.1
100	0.5	2.0	68.7	7.9
1000	0.5	2.0	38.5	13.7
	**		**	**
LSD				
5%	0.03		51.36	4.92
1%	0.04		68.48	6.55
Contrast				
Linear	**		**	**
Quadratic	NS		**	**
Cubic	**		**	**

Each number represents the mean of 9-10 cultures.
[z]Rating (nodule appearance)
1 = black (dead)
2 = brown (loss of color)
3 = brown and green areas
4 = green with stunted growth
5 = green with growth
[y]Arcsine $(n)^{0.5}$ transformation
NS,*,** Not significant or significant at the 5% or 1% levels, respectively.

glyphosate concentration increased, nodules became less green and were brown in appearance when exposed to 100 or 1000 mM of the herbicide. At concentrations > 0.1 mM glyphosate, nodule volume and dry weight were low compared to control nodules. The higher dry weight of nodules exposed to 1000 mM glyphosate may have been a result of very rapid death, as indicated by the brown color. By dying rapidly, nodules could not metabolize-away metabolites as at lower concentrations of herbicide, thus dry weight was higher.

Judging by color and growth, on agar-solidified medium, concentrations of ≤10 mM glyphosate were sublethal. The maximum sublethal dose is useful for in vitro selection of herbicide-resistant or tolerant lines. When nodules were placed into liquid medium with 1 and 10 mM glyphosate, those exposed to 10 mM turned brown and appeared dead, whereas those exposed to the lower concentration had areas of green and brown on their callus (data not presented). In a separate study, nodules that were exposed to 5 mM glyphosate in liquid medium had a response that was intermediate to 1 and 10 mM glyphosate. When continuously exposed to 5 mM glyphosate for more than one year, nodules eventually became brown with small green areas. These nodules grew very little and did not multiply.

Nodules are uniform, grow and multiply rapidly, and are thus convenient to use for in vitro studies. White ash nodule clusters are responsive to the plant growth regulator thidiazuron, and to various stresses, including desiccation and herbicide exposure. It was possible for us to obtain somatic embryogenesis from a nodule, but efficiency was extremely low for ash. A nodule produced two somatic embryos when placed on medium containing 0.5 μM BA and 0.5 μM NAA. It would be desirable to increase efficiency of regeneration of white ash nodules to a level comparable to that of other species. However, there is great potential for utilizing woody plant nodules for various in vitro studies.

ACKNOWLEDGEMENTS

We thank Nor-Am Chemical, Wilmington, DE for providing the thidiazuron and J.W. Van Sambeek, U.S.D.A. Forest Service, N.C.F.E.S. for reviewing the manuscript.

REFERENCES

1. Aitken-Christie, J., Singh, A. P., and Davies, H., 1988, Multiplication of meristematic tissue: A new tissue culture system for radiata pine, In "Genetic Manipulation of Woody Plants," Hanover, J. W. and Keathley, D. E., eds. Plenum Press, New York. pp 413-432.
2. Chaturvedi, H. C. and Mitra, G. C., 1975, A shift in morphogenetic pattern in Citrus callus tissue during prolonged culture, Ann. Bot. 39: 683-687.
3. Gray, D. J., 1987, Quiescence in monocotyledonous and dicotyledonous somatic embryos induced by dehydration. HortScience. 22: 810-814.
4. Jones, L. H., 1974, Factors influencing embryogenesis in carrot cultures (Daucus carota L.). Ann. Bot. 38: 1077-1088.
5. Krikorian, A. D., Staicu, S. A. and Kann, R. P., 1981, Karyotype analysis of a daylily clone reared from aseptically cultured tissues. Ann. Bot. 47: 121-131.
6. McCown, B. H., Zeldin, E. L., Pinkalla, H. A., and Dedolph, R. R., 1988, Nodule culture: A developmental pathway with high potential for regeneration, automated micropropagation, and plant metabolite production from woody plants, In "Genetic Manipulation of Woody Plants," Hanover, J. W. and Keathley, D. E., eds. Plenum Press, New York. pp 149-166.
7. Murashige, T. and Skoog, F. 1962, A revised medium for rapid growth and bioassays with tobacco tissue cultures. Physiol. Plant. 15: 473-497.
8. Preece, J. E., Christ, P. H., Ensenberger, L., and Zhao, J. 1987, Micropropagation of ash (Fraxinus). Comb. Proc. Intl. Plant Prop. Soc. 37: 366-372.
9. Preece, J. E., Zhao, J., and Kung, F. H., 1989, Callus production and somatic embryogenesis from white ash (Fraxinus americana L.). HortScience 24: 377-380.

OPTIMIZING POTENTIAL FOR ADVENTITIOUS SHOOT ORGANOGENESIS IN HYBRID POPULUS EXPLANTS IN VITRO WITH WOUND TREATMENT AND MICRO-CROSS SECTIONS

Ok Young Lee-Stadelmann, Seungwoo Lee[1], Haejoon Chung[2], Quansheng Guo[3], Myungwon Kim[4], Chunho Pak[5], and Wesley P. Hackett

Department of Horticultural Sciences, University of Minnesota
St. Paul, MN. 55108, USA

ABSTRACT

In vitro adventitious bud formation was investigated in 3 hybrid Populus clones with varying degrees of regeneration potential. Longitudinal and transverse wounding treatment of 1.0 cm leaf midvein and petiole explants both increased the shoot regneration capacity, more effectively with transverse wounding. Use of micro-cross sections (400 um in thickness) of leaf midvein or stem internode greatly improved regeneration capacity in all 3 clones. Supplementary calcium nitrate (7.05 to 9.4 mM) with optimal BA and NAA concentrations enhanced regeneration potential of micro- cross sections of 2 clones, but not of the most recalcitrant clone. Micro-cross section are excellent explant for obtaining large numbers of uniform adventitious shoots from minimum amount of material in a short regeneration time.

INTRODUCTION

Adventitious shoot formation in vitro from various explant sources has been practiced for the purpose of micropropagation and recently as a means for the selection of somaclonal variants. In either case, one of the most important tasks is to obtain as many regenerants as possible. Populus species have a relatively high capacity to form adventitious shoots in vitro from excised leaves, stems, callus, and roots. The number of shoots per unit explant, however, is still low even under optimal auxin-cytokinin ratios and concentrations (1,2,3,10,11).

We have investigated means of maximizing shoot regeneration potential of Populus explants with emphasis on factors other than those involving the culture medium.

Adventitious shoots have been induced in vitro on callus derived from leaves and stems of hybrid Populus clones. This technique took a long time (5 to 13 mo) before shoot proliferation was induced, and the number of shoots formed varied markedly from callus to callus within a clone and also between clones (8). Some clones were more recalcitrant than others under identical culture conditions. These observations led us to search for more efficient methods for regenerating plantlets via adventitious bud formation. We started by manipulating explant wounded surface using wound treatments (4) and explant size by use of micro-cross sectioning (5,6) to maximize regeneration potential. In this communication, we highlight the results of wounding experiments, and of the use of micro-cross sectioning with 3 hybrid Populus clones with varying degrees of recalcitrancy for adventitious bud formation. We will also present results

Permanent Addresses: [1]Dept. of Hort., Kyung Hee Univ., Suwon, Korea; [2]Dept. of Hort., Pai-Chai Univ., Daejon, Korea; [3]Div. of Basic Courses, Zhejiang Agr. Univ., Hangzhou, Zhejiang, P.R. China; [4]Dept. of Biol., Yonsei Univ., Kangwondo, Korea; [5]Dept. of Hort. Breeding, Mokpo National Univ., Chonnam, Korea.

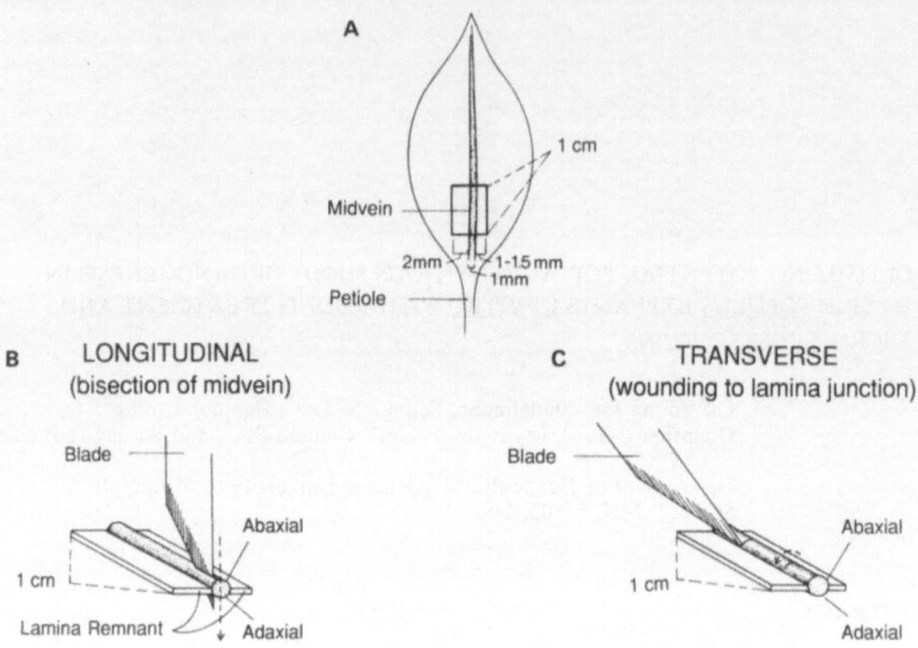

Fig. 1. Diagram of leaf midvein explant excision and wounding treatments.

of experiments which manipulate tissue orientation and Ca(NO$_3$)$_2$ levels to maximize regeneration potential of micro-cross sections.

WOUND TREATMENT

The effect of wounding was tested on 1 cm long explants from leaf midveins and petioles of clone NE 299 (Populus nigra var. betulifolia x P. trichocarpa). Midvein sections, 1 cm in length from the base of the lamina were selected as explant material (Fig. 1). A 1 to 2 mm wide strip of the lamina was left on each side of the midvein. Petiole sections, 1 cm in length, from just below the lamina attachment site were used. In preliminary studies, it was found that midvein and petiole explants from near the lamina base gave the greatest number of adventitious buds.

Explants from newly fully expanded leaves (usually third or fourth from the apex) on actively growing shoots had the highest shoot regeneration capacity. Therefore, throughout the experiment, the third or fourth leaves were used as the explant source. Explants were surface sterilized in 1 % commercial bleach (NaOCl) for 7 min. and rinsed in sterile water. For longitudinal wounding, midvein or petiole explants were split free-hand longitudinally through the vascular bundle into two pieces. For transverse wounding, two transverse incisions were made on the abaxial side of the midvein or petiole as shown in Fig. 1. Explants with the wounded side in contact with medium were cultured on modified Woody Plant Medium (WPM) (7) containing 0.7 % agar, 2 % sucrose, and 0.2 mg/l benzyladenine (BA) in culture tubes (10 x 2 cm) under 16 h of cool white fluorescent light with an irradiance of 40 to 80 μmole m^{-2}s^{-1} at 25 °C. Shoot numbers were counted under a dissecting microscope after 4 to 5 weeks of culture.

Both wounding treatments gave significantly greater numbers of shoots than the unwounded, control midveins and petioles. Transverse wounding of midvein and petiole explants increased shoot numbers more than longitudinal wounding (Table 1). When the wound surfaces were not in good contact with the medium, the number of shoots formed was reduced and the number of shoots varied greatly. On longitudinal sections, shoots could form any place along the entire cut surface. On transverse sections, shoots formed only on the proximal surface of each cut but only when the explant was in good contact with the medium. Under optimal conditions, the maximum number of shoots obtained from a 1 cm explant by wounding treatment was about 12 for midvein and 17 for petiole (Table 1). Unwounded explants gave 4 and 7 shoots for midvein and petiole, respectively.

Van Aartrijk and Blom-Barnhoorn (9) also obtained increased numbers of adventitious buds with wound treatments by removing the epidermal tissue from lily bulb scale. They attributed this increase to a wound related stimulus. It has been suggested that chemicals produced as a result of wounding may stimulate cell division and thereby produce more sites for bud formation.

However, wounding also increases cut surface area which may improve nutrient uptake from the medium since the surfaces of land plants are covered with cuticle and waxes that are impermeable to water and dissolved nutrients. Wound treatment increased the number of shoots occurring along the midvein segments when the wounded part was in contact with the medium. This suggests that the cut surface area influences the shoot forming capacity by exposing large numbers of intact cells to the medium which improves the transport or uptake of growth regulators and nutrients from the medium.

Although the wounding treatment was effective for increasing the number of shoots per explant, there were two difficulties. First, free-hand sectioning does not give consistent and uniform longitudinal cuts, and the depth of the transverse incisions varied within and between explants. In longitudinal wounding experiments often one of the resulting bisections from a midvein did not form shoots. Secondly, transverse wounded explants gave large differences in response from explant to explant. Because of differential growth of portions of the midvein, the explant was locally lifted from the agar medium leading to no or only poor contact of some of the cut surfaces with the medium.

These results and observations led us to conclude that wounding treatment is an efficient method for increasing shoot numbers and that large explant size may limit the capability for shoot formation. Because of the problems associated with the wounding treatment of large explants we have worked to develope a technique that gives uniform explant size and uniform wounding treatment.

MICRO-CROSS SECTIONS OF LEAF MIDVEINS

Micro-cross sections of leaf midveins can be made rapidly in uniform size with a vibrating microtome (Vibratome, Lancer, Model 1000) that cuts living tissue by a lateral vibrating motion and therefore with reduced damage to the tissue. Midvein segments, about 1.5 x 1.0 cm were cut from near the base of the lamina of young fully expanded leaves and mounted in a styrofoam block (Fig. 2). The block had up to seven incisions so that up to seven midvein segments could be cut simultaneously (Fig. 2).

Table 1. The influence of wounding on adventitious bud formation in hybrid Populus NE 299 midvein and petiole explants cultured in vitro on WPM with 0.2 mg/l BA. Culture period was 4 weeks.

Explant source	Wound Treatment		
	Unwounded control	Longitudinal wounding	Transverse wounding
Midvein	4.5 ± 0.3	7.5 ± 1.7	11.5 ± 1.2
Petiole	7.1 ± 0.7	11.3 ± 2.6	16.5 ± 2.4

± Standard error

Transverse sections 100, 200, 300, 400 and 500 μm thick were cut from midvein explants covered with liquid WPM. The freshly cut micro-cross sections were floated on WPM until transferred to agar solidified medium. All procedures are done under sterile conditions. Each sectioning stroke takes about one minute and a 1 cm segment gives 25 -400 μm micro-cross sections. Therefore, from 7 midvein segments in one block, about 400 micro-cross sections could be made within one hour.

One to 3 micro-cross sections were transferred to each culture tube containing agar (0.7%) solidified WPM supplemented with 2 % sucrose, 0.2 mg/l BA and 0.01 mg/l naphthalene acetic acid (NAA) at pH 5.6. The material was cultured under the same environmental conditions as used for large explants. The micro- cross sections were oriented with a cut surface towards the medium. The total number of shoots were counted 4 weeks after culture. For micro-cross sections of recalcitrant clones, only the concentrations of BA and NAA were modified.

One midvein micro-cross section, 300, 400 or 500 μm thick from clone NE 299 gave an average of 3 to 4 shoots (Table 2), nearly as many as a 1 cm midvein explant. Thus the 400 μm micro-cross sections are at least 25 times more efficient for bud formation than the 1 cm explant based on the number of micro-cross sections that can be obtained from a 1 cm explant. Micro-cross sections 200 μm thick gave an average of 1 shoot and it took longer for shoot regeneration than in the 400 μm sections. Sections 100 μm thick did not form any shoots at all. The maximum thickness of the micro-cross section tested was 800 μm. The result with this thickness was similar to that for the 400 μm micro-cross sections.

The regeneration time from micro-cross sections was shortened by 7 to 10 days as compared with the 1 cm sections. Nearly 100 % of 400 μm midvein MCS formed buds when randomly oriented with regard to whether the proximal or distal cut surface was in contact with the medium. This result suggested that either they did not display a polarity for bud formation or they reoriented themselves on the medium during culture. In either case, these results indicate that it is not necessary to place midvein MCS on the medium in a particular orientation.

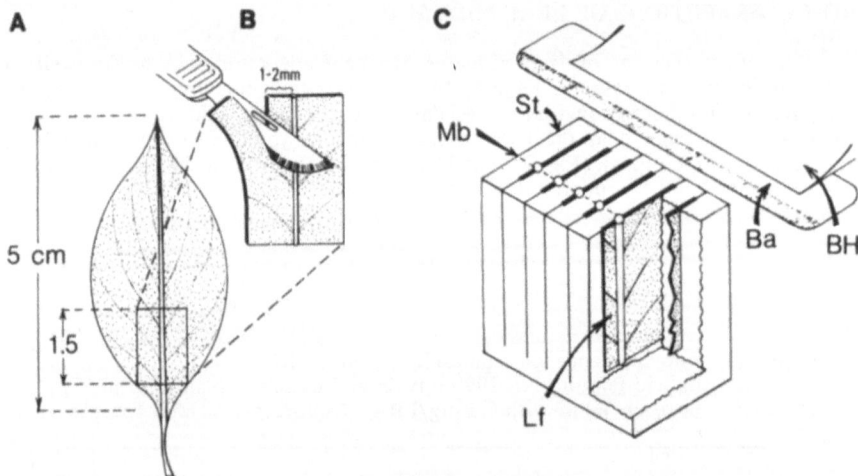

Fig. 2.   Procedures for microsectioning using a vibratome: (A) Leaf segments (1.5 x 1.0 cm) from the basal portion of the leaf, 1 cm above the petiole attachment point with the lamina; (B) lamina on both sides of the midvein removed except for 1-2 mm remnant; (C) leaf segments placed into incisions of the styrofoam block (2 x 2 cm), with the alignment of the midveins (Mb) in the center of the block. Total of six leaf sections in the block in sectional view near the midveins (Mb) showing the placement of the leaf segments inside the styrofoam block. Lf, leaf segments; St, styrofoam block; Mb, midvein; BH, blade holder; Ba, blade (from Lee-Stadelmann et al. 1989).

Shoots from micro-cross sections elongated well and were uniform in growth and morphology. So far, no abnormalities in shoot morphology have been observed in the micro-cross section derived plantlets. Roots were easily formed on shoot regenerants in 4 weeks on WPM without BA and NAA, or with of NAA at 0.005 mg/l and were ready for transplanting to the greenhouse in 8 weeks. Detailed procedures for preparing midvein micro-cross sections and details of morphogenetic observations have been described elsewhere (5,6).

Micro-cross sections of leaf petioles were also tested. Even though 400 μm petiole sections formed more shoots than midvein micro-cross sections, the handling of petiole microsections was difficult, because of their small size and lack of a convenient site to grasp them with forceps. Therefore, their use was impractical. In midvein micro-sections the lamina remnants serves as sites for grasping with forceps.

STEM INTERNODE MICRO-CROSS SECTIONS

Stem internodes below the 3rd or 4th unfolded leaves, counting from the shoot apex, are about 4 to 6 cm in length and 3 to 4 mm in diameter, 3 to 4 times larger than petiole diameters. Therefore, a large number of uniform size micro-cross sections can be obtained from an internode (about 300 to 500 sections) as an explant source. Micro-cross sections from the internodes were prepared by procedures similar to those described for midvein micro-cross sections with slight modifications for mounting the stem internode.

The micro-cross sections were oriented randomly with one or the other cut surface in contact with the medium. Shoots formed either directly from the periphery of the stem, probably from hypodermal tissue cells which contain chloroplasts (Fig. 5B), or sometimes indirectly from callus cells apparently of cortical cell origin. The average number of shoots formed from stem internode 400 μm thick stem micro-cross section was 3 and similar to that with 400 μm midvein sections on WPM (Table 3). This number of buds is less than expected considering the larger size of the stem micro-cross sections compared with the leaf midvein cross sections. Only 60 % of stem micro-cross sections formed shoots (Fig. 3A) using 2.3 mM $Ca(NO_3)_2$. This low percentage of explants forming shoots is the reason for the overall low mean for shoot numbers shown in Table 3. Most of non-shoot forming sections did not form callus and eventually died.

Table 2. Effect of midvein MCS thickness on bud formation, BA 0.2 mg/l, NAA 0.01 mg/l, 20 replications. Tissue from clone NE 299 and cultulre period was 4 weeks (from Lee-Stadelmann et al. 1989).

Thickness of MCS (μM)	No. of intact cell layers (A)	(B)	No. of shoots per MCS (mean ± SE)	Time (weeks) for macroscopically visible buds
100	2-3	0-1	0.0	-------
200	4-5	1-2	1.0 ± 0.24	3.0-3.5
300	6-7	2-3	3.1 ± 0.47	2.5-3.0
400	8-9	3-4	4.1 ± 0.38	2.0-2.5
500	10-11	4-5	4.3 ± 0.29	2.0-2.5

(A) = Junction area cells, chlorophyllous
(B) = Cortical parenchyma cells, non-chlorophyllous
MCS = Micro-cross section
SE = Standard error

In an effort to understand this relatively low percentage of micro-cross sections forming shoots, the effect of orientation of stem micro-cross sections on the medium was investigated. Either the proximal or the distal cut surface of the micro-cross section was placed in contact with medium. It was found that only micro-cross sections with the distal side in contact with medium formed shoots on the proximal side (data not shown), indicating a strong polarity for shoot morphogenesis. Because of the circular form of the stem micro-cross sections their growth in culture is symmetrical, they maintain their original explant orientation throughout the culture period. Because of this stability and the polarity of bud formation, orientation of the explant on the medium is critical for maximum shoot regeneration.

ENHANCEMENT OF SHOOT FORMATION AND VIABILITY OF EXPLANTS BY $Ca(NO_3)_2$

The optimal concentration of $Ca(NO_3)_2$ for maximum bud formation in 1 cm long midvein explants of NE 299 was 9.4 mM with about 14 shoots per explant. With midvein micro-cross sections, shoot formation was continually stimulated up to 11.75 mM (Table 3, Fig. 5E). Optimal concentration for stem micro-cross sections of NE 299 was 9.4 mM (Table 3). The percentage of explants forming buds and number of buds formed increased while callus formation decreased with increased $Ca(NO_3)_2$ level (Fig. 3A). The most striking effect of increased $Ca(NO_3)_2$ was on reduction of necrosis and increase of explant viability (Fig. 3B). Almost 100 % of the explants survived at 9.4 mM $Ca(NO_3)_2$. Without supplementary $Ca(NO_3)_2$ in WPM, death was much higher in large explants than in micro-cross sections (Fig. 3B), suggesting that micro-cross sections may be more efficient in uptake and utilization of $Ca(NO_3)_2$ and perhaps other nutrients.

We have evidence that the promotive effect of $Ca(NO_3)_2$ on bud formation is the result of interaction of Ca++ with NO_3-. Other divalent cations used in conjunction with NO_3- did not enhance shoot formation. However, the effect of $Ca(NO_3)_2$ on prevention of necrosis of explants is likely due to the effect of Ca rather than NO_3 since Ca++ is known to protect membrane structure. Thus, leakage of vital cellular substances will be prevented and the membrane transport functions such as uptake of nutrients and hormones that are vital for morphogenesis would remain intact.

Table 3.   Adventitious bud formation from midvein and stem microcross sections of NE 299 as influenced by increasing concentrations of calcium nitrate in WPM. Culture period was 8 weeks.

$Ca(NO_3)_2$ (mM)	No. of Shoots/MCS	
	Midvein MCS (400 μm)	Stem MCS (400 μm)
0	3.92 ± 0.16	2.38 ± 1.0
2.35*	2.93 ± 0.15	2.97 ± 0.51
4.7	5.09 ± 0.13	3.54 ± 0.79
7.05	5.23 ± 0.12	5.10 ± 0.99
9.4	5.29 ± 0.30	8.02 ± 0.44
11.75	6.98 ± 0.29	3.97 ± 0.17

± = Standard error
* = Concentration in WPM
MCS = Micro-cross sections

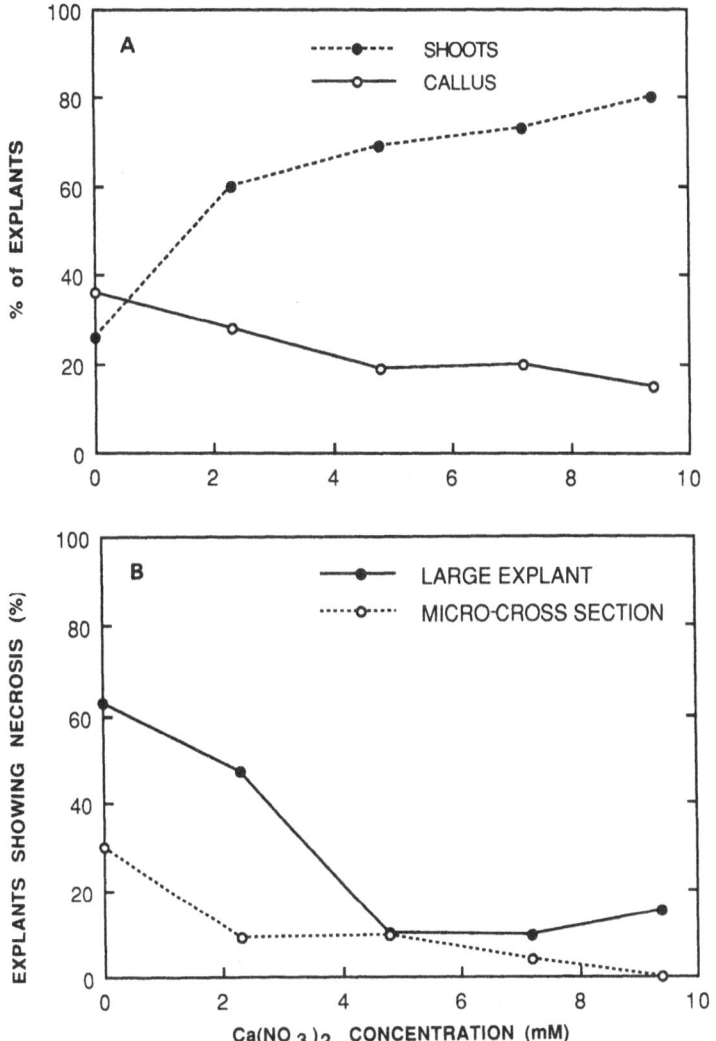

Figure 3. Influence of supplementary calcium nitrate in
WPM on (A) adventitious shoot and callus
formation from stem micro-cross sections (400 μm)
and (B) explant necrosis of 0.5 cm midvein
segments and stem micro-cross sections. Tissues
from hybrid Populus NE 299.

The positive results with micro-cross sections from clone NE 299 prompted us to test other clones for which it has been relatively difficult to induce adventitious shoots in vitro. In this work, only micro-cross sections 400 μm thick were used.

From several recalcitrant hybrid Populus clones, we have selected to test clones NE 41 ('Androscoggin', P. maximowiczii x P. trichocarpa) and DN 34 (NC 5326, P. x euramericana 'Eugenei') that are moderately and very recalcitrant, respectively. To find out the optimal concentration of NAA and BA, large midvein explants (0.5 cm) were used.

For NE 41, the optimal concentrations of NAA and BA were found to be 0.1 mg/l and 0.5 mg/l respectively, based on the growth and quality of shoots. Large midvein explants (0.5 cm) of clone NE 41 formed on average 1 shoot per explant using optimal plant growth regulator (PGR) concentrations in WPM (data not shown). Stem micro-cross sections of NE 41 formed an average of 4.5 shoots per explant using optimal PGR concentrations in WPM (Table 4) and supplementation with $Ca(NO_3)_2$ (9.4 mM) increased shoot numbers dramatically to 35 shoots.

The optimal PGR concentrations for 1 cm explants of DN 34 were 0.5 mg/l both for NAA and BA (data not shown). These concentrations were selected in the presence of supplementary $Ca(NO_3)_2$ (7.05 mM), based on callus growth and the percent of explants forming shoots. Even though callus growth was best under these optimized conditions, the number of shoots per explant was very low and the shoot vigor was poor. In WPM without supplementary $Ca(NO_3)_2$ no shoots were formed and many explants showed necrosis even in the presence of optimal PGR concentrations.

Stem micro-cross sections of DN 34 with optimal NAA and BA concentrations in WPM formed an average of 3 shoots. A higher $Ca(NO_3)_2$ concentration (9.4 mM) increased shoot numbers only slightly or not at all (Table 4). This lack of a bud formation response to $Ca(NO_3)_2$ concentration by DN 34 stem micro-cross sections is in marked contrast to the very high response by NE 41 and moderate response by NE 299. Thus, there seem to be clonal difference in the influences of $Ca(NO_3)_2$. It is not known why $Ca(NO_3)_2$ is so much less effective with DN 34 than with NE 299 or with NE 41.

The shoot regeneration capacities for the 3 clones using different conditions are summarized in Table 5. This summary shows that stem micro-cross sections are an effective means of increasing the shoot regeneration capacity of recalcitrant clones and that increased $Ca(NO_3)_2$ levels will increase the bud forming response of micro-cross sections of one recalcitrant clone, NE 41, but not the other (DN 34). With use of stem micro-cross sections efficient shoot regeneration can be obtained from even a very recalcitrant clone (DN 34).

Table 4. The influence of supplementary $Ca(NO_3)_2$ on adventitious bud formation on micro-cross sections (400 μm) of hybrid Populus NE41 and DN 34 on WPM medium with optimal BA and NAA concentrations.

Clone	Optimal conc. (mg/1) NAA	BA	$Ca(NO_3)_2$ (mM)	No. of shoots per 400 μm section
NE 41	0.1	0.5	2.35*	4.49 ± 0.06
	0.1	0.5	9.4	35.74 ± 3.65
DN 34	0.5	0.5	2.35*	3.03 ± 0.98
	0.5	0.5	9.4	4.27 ± 0.25

* = Concentration in WPM

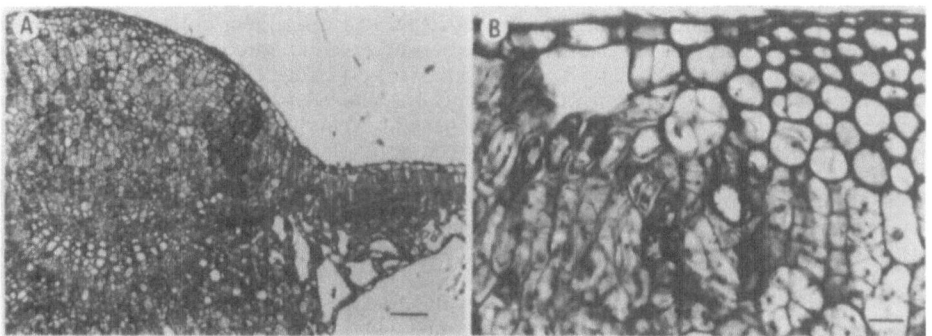

MICRO-CROSS SECTION

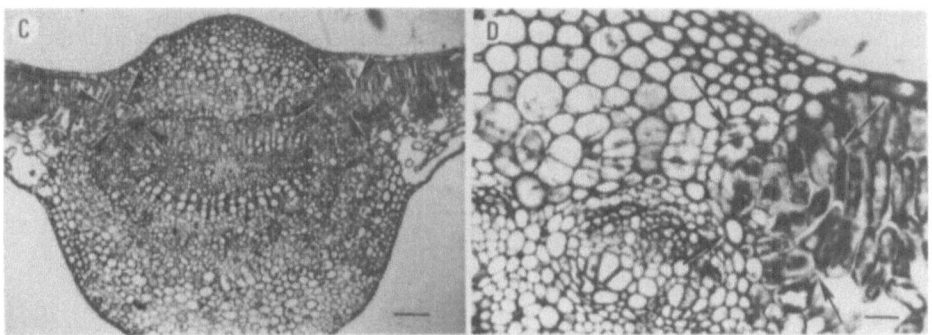

Figure 4. Histological comparison of cell division in a large explant and a micro-cross section of midveins (from NE 299) during an early stage of organogenesis. Tissues were fixed, embedded, nuclear stained and microtome sectioned (10 μm thick). A, B: large explant. Cell division occurred in most midvein cells including junction cells and those in the cortex. The midvein volume increased within 3 days after initiation of culture. Localized cell division was not observed. C, D: Micro-cross section. Localized cell division occurred only in the junction area between the vascular bundle and lamina tissue during first 3 days of culture. No cortical cells were dividing at this stage. Bars indicate 60 μm in A and C. 25 μm in B and D. Junction area: areas inside the arrow marks.

MICRO-CROSS SECTIONS FOR THE STUDY OF EARLY EVENTS IN BUD ORGANOGENESIS

Early morphogenetic responses of 400 μm midvein micro-cross sections were observed either directly in vivo using a stereo microscope (10 to 15 x) or by microscopic observation of microtome sections (10 μm thick) from fixed and embedded and nuclear-stained micro-cross sections. Controls were 1 cm explants.

In midvein micro-cross sections, the first cell divisions, localized in the junction area between vascular bundle and lamina tissue, were observed within 2 to 3 days after initiation of culture. A green, organized center was clearly visible in 3 to 5 days (Fig. 4C, 4D). In the case of large explants, junction cells and the cortical cells adjacent to them in the midvein divided concurrently. Therefore, the green organized centers observed in micro-cross sections were not discernible in 0.5 cm explants (Fig. 4A, 4B). Whether the organized center occurs at a later stage of culture in the junction area of large explants has not been determined. It is not known whether these differences in the early morphogenetic events between the large explant and micro- cross sections have any relation to the earliness of shoot organogenesis.

Direct microscopic observations of micro-cross sections suggested that the organogenetically active cells are small chlorophyllous cells located between the vascular and lamina tissues (junction area) in the case of midveins or in the hypodermal area of stem (Fig. 5B,C,D) and petiole micro-cross sections (Fig. 5B). The midvein micro-cross sections became nearly V-shaped within a week as a result of asymmetrical growth involving midvein cells. The micro-cross sections which became V-shaped formed shoots mostly from midvein junction cells. Occasionally buds formed from small veins in the lamina, but no shoot formation was observed from lamina tissue itself.

With large midvein explants, organogenetic centers could not be detected macroscopically or microscopically because the surface was covered by callus cells before the shoot emerged. Thus, because of ease of observation, micro-cross sections are a better experimental tool than large explants to study control of early events in bud organogenesis using external chemical inducers or inhibitors to manipulate them.

Table 5. Summary of differences in shoot formation response of 3 clones of 3 hybrid Populus clones to stem explant size and $CA(NO_3)_2$.

Clone	Optimal PGR conc. mg/l		Mean no. of shoots			Calculated max. no. of shoots per 1 cm explant using MCS
	BA	NAA	large explant*	Micro-cross sections (A)	(B)	
NE 299	0.2	0.01	2-6	2	8	200
NE 41	0.2	0.1	1	5	35	875
DN 34	0.5	0.5	0	3	4	100

(A) = WPM containing 2.35 mM $Ca(NO_3)_2$
(B) = WPM with supplementary $Ca(NO_3)_2$. 9.4 mM for NE 299 and NE 41 and 7.05 mM for DN 34
PGR = plant growth regulator
* = 0.5 - 1.0 explant on basal WPM (2.35 mM $Ca(NO_3)_2$)
MCS = Micro-cross section

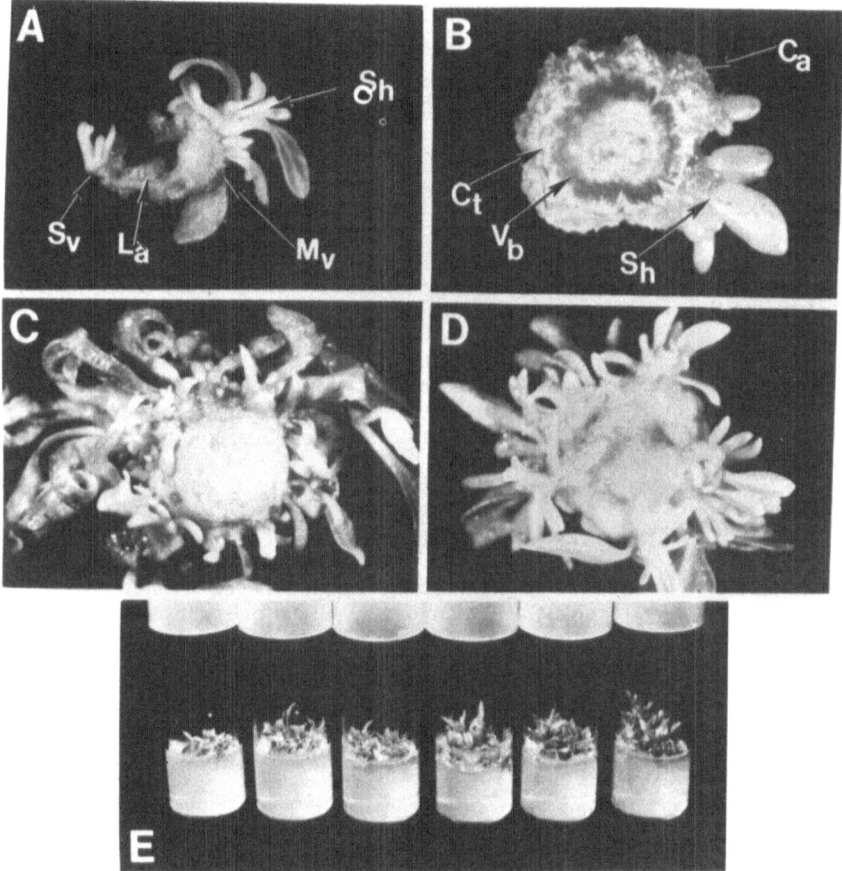

Figure 5. Adventitious shoot formation from micro-cross sections, two to three weeks after initiation of culture in WPM with optimal BA and NAA concentrations (cf. Table 4). (A) two shoots emerging from midvein and one shoot from a small vein of a midvein micro-cross section from NE 299. (B) shoots emerging from hypodermal layer of petiole micro-cross section from NE 299. (C) numerous shoots from the periphery of a stem internode micro-cross section of NE 41 with supplementary $Ca(NO_3)_2$ (9.4 mM). (D) several shoots formed from the periphery of a stem micro-cross section of DN 34 with supplementary $Ca(NO_3)_2$ (9.4 mM). (E) stimulation of shoot formation in midvein micro-cross sections of NE 299 by increasing concentrations of $Ca(NO_3)_2$ in the medium. Concentrations of $Ca(NO_3)_2$ from left to right: 0, 2.35, 4.7, 7.05, 9.4, and 11.75 mM. Sh: adventitious shoot. Mv: midvein. LA: lamina tissue, Sv: small vein. Ca: callus. Vb: vascular bundle. Ct: cortical cells.

When large midvein explants (0.5 to 1.0 cm) were cultured in a horizontal orientation, shoots were formed predominantly on the proximal (basal) cut surface of the explant with only 10 to 20 % of the explants forming shoots on the distal (apical) cut surface as well. Addition of concentrations of $Ca(NO_3)_2$ higher than in the basal WPM decreased polarity by increasing the number of shoots to a greater extent on the distal than the proximal ends of large midvein explants (data not shown).

With 400 μm midvein micro-cross sections, it was difficult to determine on which surface shoots were formed because the original orientation of the sections changed during the culture. However, using differential width of the lamina remnants on the two sides of the midvein and the asymmetric shape of abaxial/adaxial sides of the midvein, it was possible to identify the site of formation of callus and buds. Using this technique it was determined that callus and shoot formation occurred mainly on the proximal cut surface.

In comparison with midvein micro-cross sections, the site where shoots form is relatively easy to determine with stem micro-cross sections. Stem micro-cross sections maintain their original orientation because of their symmetrical growth and larger surface/volume ratio. Observations indicated that shoot formation from stem micro-cross sections was also strongly polar. Buds were formed only on the proximal cut surface, but only when the distal cut surface was in contact with the medium or when the micro-cross sections were on their edge (equivalent to a horizontal position, data not shown). These results show that there is a very strong polarity for bud formation in midvein and stem tissue and that this polarity is expressed even in micro-cross sections 400 μm thick.

MICROCROSS SECTIONS MINIMIZE MICROBIAL CONTAMINATIONS

Microbial contamination originating from explants is one of the serious problems for aseptic cultures of large explants of leaf midvein, petiole and stem internode from hybrid Populus trees. The contamination rate was variable from season to season and plant to plant, sometimes resulting in a 50 % loss of explants. With microcross sections, however, almost no explants were lost due to endogenous contaminations during culture. This was consistent and independent of season and mother plant source. This low contamination rate with micro-cross sections may be due to the small population of microorganisms existing in such a small explant.

SUMMARY AND CONCLUSIONS

1 - Both longitudinal and transverse wounding of midvein and petiole explants from NE 299 increased adventitious shoot numbers as compared with unwounded control explants. Transverse wounding was more effective than the longitudinal wounding and the maximum shoot number obtained from wounding treatment was 11 shoots for midvein and 17 for petiole. These results demonstrate that merely increasing cut surface area can increase shoot numbers without requiring changes of medium composition and culture conditions.

2 - Clone NE 299 midvein micro-cross sections (400 μm in thickness) gave an average of 3 to 4 shoots per section. The calculated number of shoots that can be obtained from 1 cm of explant material is 100. Based on this calculation micro-cross sections are 25 times more efficient for shoot formation than a 1 cm explant. Almost all micro-cross sections formed shoots and the orientation of the section did not influence the regeneration rate.

3 - Micro-cross sections from stem internodes of clone NE 299 gave similar mean shoot numbers as the midvein micro-cross sections but only about half of the micro-cross sections formed shoots. Orientation experiments indicated that orientation of the stem micro-cross sections greatly influenced organogenetic capacity.

4 - $Ca(NO_3)_2$ supplementation of Woody Plant Medium further improved the regeneration potential of large midvein explants and micro-cross sections of NE 299. Maximal number of shoots was obtained at 7.05 mM for large explants and 9.4 mM to 11.75 mM for micro-cross sections. Supplementation with $Ca(NO_3)_2$ significantly reduced explant necrosis and death.

5 - Shoot formation was mainly on the proximal cut surface in large midvein explants. This polarity of bud formation was also observed in the midvein and stem micro-cross sections.

6 - Micro-cross sectioning greatly improved adventitious shoot formation in recalcitrant clones NE 41 and DN 34. Supplementary $Ca(NO_3)_2$ greatly enhanced shoot formation in NE 41 but enhanced shoot formation only a little for DN 34.

7 - Other advantages of using micro-cross sections are faster formation of adventitious shoots, better use of the explant material, less need for culture medium and space, and greatly reduced explant-derived microbial contamination.

8 - Micro-cross section culture is a rapid and efficient method for in vitro regeneration of adventitious plantlets. This technique may be effective for clonal propagation and/or for recovery for somaclonal variants and these possibilities are under investigation.

9 - The results with wounding treatments and micro-cross sections suggest that good contact of the explant with medium during culture is an important factor for efficient shoot formation and that micro-cross sections of 300 to 800 μm thick serve this purpose effectively.

ACKNOWLEDGEMENT

This work was supported by a grant (USDA-85-FSTY-0145) from the United States Department of Agriculture, Forestry Competitive Grant Program and Cooperative Research Agreement 23-84-10 under Project FS-NC-4502 (88-02) of the USDA North Central Forest Experiment Station, St. Paul, MN 55108. Drawings were made by Kris Kirkeby. Paper of the Scientific Journal Series, Agricultural Experiment Station, University of Minnesota, St. Paul, MN. 55108, U.S.A.

REFERENCES

1. Ahuja, M. R., 1983, Somatic cell differentiation and rapid clonal propagation of aspen, Silvae Genet. 32:131.

2. Badia, N., 1981, Obtention de plantules à partir de bourgeons adventils sur feuilles de peuplier ('Serotina de Champagne' - '1214') cultivés in vitro. Proc. IUFRO-AFOCEL Colloque International Sur La Culture in vitro Des Essences Forestieres. Aug. 31-Sept. 4, 1981, p. 236.

3. Chalupa, V., 1974, Control of root and shoot formation and production of trees from poplar callus. Biologia Plant. 16:316.

4. Lee, S. W., Hackett, W. P., and Read, P.E., 1986, Adventitious bud formation on hybrid Populus midrib and petiole segments cultures in vitro, VI. Internat. Congr. Plant Tissue Culture, Minnesota, Abstract 458.

5. Lee-Stadelmann, O. Y., Hackett, W. P., Lee S. W., and Read P.E., 1987, Microthin-cross section culture for studying bud morphogenesis in vitro XIV, International Botanical Congress, Berlin, Abstract 2-02-2.

6. Lee-Stadelmann, O. Y., Lee, S. W., Hackett, W. P., and Read, P. E., 1989, The formation of adventitious buds in vitro on micro-cross sections of hybrid Populus leaf midvein, Plant Sci. 61:263.

7. Lloyd, G. and McCown, B., 1980, Commercially-feasible micropropagation of mountain laurel, Kalmia latifolia, by use of shoot tip culture, Comb. Proc. Inter. Plant Propagators. Soc., 30:421.

8. Ostry, M.E. and D.D. Skilling, 1988, Somatic Variation in Resistance of Populus to Septoria musiva, Plant Disease 72(8):724-727.

9. Van Aartrijk, J. and G. J. Blom-Barnhoorn, 1983, Adventitious bud formation from bulb scale explants of Lilium speciosum in vitro: 2,3,5-triiodobenzoic acid and temperature, Z. Pflanzenphysiol., 110 (4):355-363.

10. Venverloo, C., 1973, Formation of adventitious organs. I Cytokinin-induced formation of leaves and shoots in callus cultures of Populus nigra L. Italica. Acta Bot. Neerl., 22:390.

11. Winton, L. L., 1970, Shoot and tree production from aspen tissue cultures. Amer. J. Bot., 57:904.

SOMACLONAL VARIATION IN *POPULUS* HYBRIDS REGENERATED FROM PROTOPLAST CULTURE

Rod Serres, Mike Ostry[1], Brent McCown, and Darroll Skilling[1]

Department of Horticulture, University of Wisconsin-Madison, and [1]North Central Forest Experiment Station, USDA-Forest Service St. Paul, MN

ABSTRACT

Protoclones were regenerated from the poplar hybrid NC5339 (*Populus alba* x *P. grandidentata* 'Crandon'). Morphologic variation primarily in leaf characteristics has been observed in four vegetative generations under both greenhouse and field conditions. Three variant tree types have been identified: (1) Normal growth rate with long leaves that have reduced pubescence on the underside of the leaves. Leaves show sectoring of pubescence levels. (2) Slower growing with thick, dark green, heavily pubescent leaves in highly convoluted, brittle leaves. (3) Dwarf fast-growing tree with ovate, pale green leaves and a tendency for heavy branching. Leaves show sectoring of pubescence levels. Possible explanations for the variation observed are offered.

INTRODUCTION

Poplar species and their interspecific hybrids are important trees for short rotation intensive culture but their usefulness can be limited by insufficient resistance to important poplar diseases (4). In addition to conventional breeding programs, somaclonal variation may provide a means to obtain superior, disease-resistant trees. Reported here are the observations of somaclonal variants in a population of protoclones, clones derived from individual protoplasts (3), regenerated from the poplar hybrid NC5339 (*Populus alba* x *P. grandidentata* 'Crandon'), a fast-growing tree with good disease and pest resistance.

METHODS AND OBSERVATIONS

Protoplasts of the hybrid poplar NC5339 were isolated and cultured to regenerate shoots as described by Russell and McCown (2). Protoclones were grown in the greenhouse, selected ones transplanted to the field, grown for two years, and

Woody Plant Biotechnology, Edited by M.R. Ahuja
Plenum Press, New York, 1991

propagated by softwood cuttings to bring back into the greenhouse where they were propagated two more times by softwood cuttings. At each stage, the protoclones were evaluated for morphological characteristics in comparison to control clones derived from multiplication shoot cultures through axillary shoot development. The morphological characters investigated were leaf length (l), leaf width (w), leaf length to width ratio (l/w), leaf color, leaf pubescence, leaf shape, leaf edge, leaf tip, leaf base, leaf distortion, leaf clasping, stipule size, and stem indentation.

Variation from the control was observed at all stages (greenhouse, field, and second greenhouse). In some cases, variant protoclones noted in the field had not been noted in the greenhouse. Some protoclones variant in the greenhouse were non-variant in the field.

The presence of less pubescent sectors within normally pubescent leaves on one protoclone (#227) and the presence of normally pubescent sectors within the less pubescent leaves of another protoclone (#69) was observed (Fig. 1). These sectors were found in both greenhouse and field grown leaves. This sectoring may be due to the action of transposable elements interrupting a coding region responsible for pubescence. Activation of transposable elements has been implicated to be enhanced in the tissue culture environment (1) and may be responsible for other insertion or deletion mutations as well.

Figure 1. Sectoring of pubescence levels in protoclones #227 and #69. Background pubescence level of #227 is similar to the control level.

CONCLUSIONS

Three variant tree types have been identified that have morphologic variation in primary leaf characteristics observed through four vegetative generations under both greenhouse and field conditions:

1. A normally growing tree with long leaves (high length to width ratio) that have reduced pubescence on the abaxial leaf surface. This tree shows sectoring of pubescence levels. (ex. #69).
2. A slower growing tree with thick, dark green, heavily pubescent, and wavy edged leaves in the field. In the greenhouse this tree has a creeping form with highly convoluted, brittle, and dark green leaves. It has a poorly developed root system. This protoclone may be a polyploid variant. (ex. #223).
3. A dwarf fast-growing tree with ovate, pale green leaves and a tendency for heavy branching. This protoclone may be aneuploid with a partial or whole chromosome deletion. This tree shows sectoring of pubescence levels. (ex. #227).

Cytological observations of protoclones #227 and #223 may allow verification of possible chromosomal aberrations. Detection and verification of a transposable element in poplar will be difficult without a like-probe or information on the area of insert. Variant protoclones will be analyzed in the field again to investigate stability of those variant characters.

This project was funded in part by a cooperative research agreement with the North Central Forest Experiment Station, USDA-Forest Service, St. Paul, MN.

REFERENCES

1. Larkin, P.J. and Scowcroft, W.R., 1981, Somaclonal variation - a novel source of variability from cell cultures for plant improvement. Theor. Appl. Genet. 60:197-214.
2. Russell, J.A. and McCown, B.H., 1988, Recovery of plants from leaf protoplasts of hybrid-poplar and aspen clones. Plant Cell Reports 7:59-62.
3. Sheppard, J.F., Bidney, D. and Shahin, E., 1980, Potato protoplasts in crop improvement. Science 208:17-24.
4. USDA (ed) Forest Service, 1980, Energy and wood from intensively cultured plantations; research and development program. General Technical Report NC-28.

IN VITRO DISEASE RESISTANCE FOR EXPRESSION

OF SOMACLONAL VARIATION IN *LARIX*

Alex M. Diner

School of Forestry and Wood Products
Michigan Technological University
Houghton, MI 49931

ABSTRACT

Somaclonal variation expressed by scleroderris canker resistance or susceptibility was examined using *Larix decidua* micropropagules generated from caulogenic calli. Callus had been initiated from adventitious shoots showing persistent resistance to the pathogen. Inoculation responses by these propagules were compared to those by adventitious propagules initiated from juvenile tissues, and presumed to be canker-susceptible. Inoculation of those hosts developed on callus from one ostensibly resistant plantlet showed no resistance. Colonization by the pathogen was complete on all shoots.

INTRODUCTION

Opportunities to circumvent economic losses due to microbial diseases of important plants have been primarily found through development of disease-resistant plant genotypes. Genotype "development" implies selection of apparently resistant individuals from an otherwise infected, naturally or artificially inoculated population.

In the case of the tree host, traditional tree improvement strategies require years and large acreages, both to generate new host genotypes and to challenge them with the pathogen. In contrast, the advantages of tissue culture micropropagative systems and *in vitro* challenge with the pathogen have been shown for the larch/scleroderris canker system (1) and with others (3).

Current address of author: U.S. Forest Service, Southern Forest Experiment Station, Dept. of Plant and Soil Science, Alabama A&M University, P. O. Box 1208, Normal, AL 35762.

Woody Plant Biotechnology, Edited by M.R. Ahuja
Plenum Press, New York, 1991

However, the most promising source of novel genotypes of plants appears to be via somaclonal variation from cell culture. Not only may variants be generated at high frequency, but since genotype variation is random, the potential multiplicity of genotypes may be limited only by the size of the genome. Such multiplicity has obvious advantages for subsequent natural propagation, over monoclones or systems generating only a small number of clones.

Methods have been developed for shoot organogenesis using callus generated from short shoot buds of *Larix* (5). Calli/cell cultures are those in which genetic variation may be expected to occur most frequently. We may thus anticipate such variation in calli grown from short shoot buds of plantlets expressing either resistance or susceptibility to inoculation with *Gremmeniella abietina* (Lagerb.) Morelet, the causative agent of scleroderris canker disease. This would result in resistant shoots from an otherwise genetically susceptible cell population, and vice versa. The objective of this study was to examine for somaclonal variants among populations of callus-initiated *L. decidua* Mill. propagules.

MATERIALS AND METHODS

Short shoot buds were excised from two-year-old *Larix decidua* rooted tissue culture plantlets in a greenhouse. These plantlets included nine elongated from propagules which earlier and repeatedly (4x) had resisted *in vitro* inoculation with conidia of *G. abietina* isolate 18-46 from northern Wisconsin (unpublished). Buds were surface-sterilized, scales removed, and shoot primordia placed on a modified Schenk and Hildebrandt medium (5). Tissue was subcultured monthly. A second group of propagules was initiated from juvenile tissues as described (1). Propagules from both groups were elongated to shoots of approximately 15 mm stem height, on a growth regulator-free Litvay medium containing glutamine as the sole source of amino nitrogen (4). To each shoot was then applied a single 0.01 ml aqueous drop suspension of 1750 viable conidia of isolate 18-43 from northern Wisconsin prepared as described by Abdul Rahman et al. (1). Nine 15 mm adventitious shoots from juvenile tissues were also inoculated, as were non-caulogenic callus cultures, grown on a modified Brown and Lawrence medium (1) from cambial explants of each member of the two groups of rooted plantlets. Two replicates of each callus genotype were inoculated. Shoots and calli were examined during 30 days for colonization by the pathogen.

RESULTS

Callus developed slowly from only 6 of 211 (3%) cultured shoot primordia. Isolated, small needles and a few bud-like structures eventually appeared on several calli. However, only three buds on callus from one canker-resistant plantlet elongated to shoots. Within two weeks following inoculation with conidia, all three showed extensive colonization by the pathogen, as did the nine inoculated adventitious shoots. All non-caulogenic callus cultures were similarly colonized.

DISCUSSION

The low frequency of *L. decidua* caulogenic cultures reported here, is not unique (personal communication: Jan Bonga, Forestry Canada; Johanne Bruhn, Michigan Technological University). And, the apparent complete loss in callus culture of presumed genetic resistance to scleroderris canker reported here, cannot be considered a manifestation of somaclonal variation. Susceptibility to pathogen expressed by hosts developed from a resistant "parent" may simply be a result of the use of a different fungal isolate, though from the same area. The fact that plantlets from which caulogenic calli were derived had repeatedly resisted inoculation, suggests the need for continued study of caulogenic calli as regards reproducible phenotype selection. Such calli, however, constitute only one of several opportunities for clonal micropropagation of *Larix* (and other conifers), which may then be challenged by a pathogen or other stress. Indeed, adventitious buds are inducible from needle meristems on seedlings, plantlets and mature trees (2). And, although opportunities may ultimately prove unsatisfactory for the development of adventitious somaclonal conifer variants, phenotype heritability may be examined among micropropagules generated using various ontogenic procedures.

ACKNOWLEDGEMENT

The author gratefully acknowledges project funding by the USFS North Central Forest Experiment Station Cooperative Agreement AG-88-20.

REFERENCES

1. Abdul Rahman, N. N., A. M. Diner, D. D. Skilling, and D. F. Karnosky. 1987. *In vitro* responses of conifer adventitious shoots and calli inoculated with *Gremmeniella abietina*, For. Sci., 33:1047-1053.

2. Diner, A. M. 1990. Clonal micropropagation of mature *Larix*, New For. (in press).

3. Diner, A. M. and D. Karnosky. 1987. Tissue culture application to forest pathology and pest control, P. 351-373 *in*: "Cell and Tissue Culture in Forestry", Volume 2, J. Bonga and D. Durzan, eds., Martinus Nijhoff Publishers, Boston.

4. Diner, A. M., A. Strickler, and D. Karnosky. 1986. Initiation, elongation, and remultiplication of *Larix decidua* micropropagules, New Zealand J. For. Sci., 16:306-318.

5. Laliberte, S. and M. Lalonde. 1988. Sustained caulogenesis in callus cultures of *Larix* x *eurolepis* initiated from short shoot buds of 12-year-old trees, Am. J. Bot. 75:767-777.

JUVENILITY, MATURATION AND REJUVENATION

GENE EXPRESSION DURING GROWTH AND MATURATION

Keith W. Hutchison[+], Patricia B. Singer[+], and
Michael S. Greenwood[*]

[+]Department of Biochemistry and [*]Department of
Forest Biology, University of Maine, Orono, Maine 04469 U.S.A.

ABSTRACT

The expression of *cab* and *rbc*S gene families were measured in RNA extracted from needles from larch trees ranging in age from 1 y to 75 y. Steady state *cab* mRNA levels are relatively higher (~40%) in newly expanding short shoot foliage from juvenile plants compared with mature plants. Later in the season no consistent difference in *cab* expression between juvenile and mature plants was detected. Unlike *cab* gene expression, the expression of the *rbc*S gene family did not seem to vary with age. These data show that the maturation-related changes in morphological and physiological phenotypes are associated with changes in gene expression. No causal relationship has been established, however. Indeed, we conclude that the faster growth of juvenile scions is not due to increased net photosynthesis or *cab* expression.

INTRODUCTION

Maturation in the conifers is a developmental process which is, at present, poorly understood. It is exemplified by such characteristics as a reduced rate of growth, reduced efficiency in the rooting of cuttings, changes in foliar morphology and the onset of flowering. Maturation in woody plants has recently received much attention because of the maturation-related decrease in the ability to clone selected individuals using explants from mature plants (10,11). Maturation and its reversal are of interest for several reasons. The rapid growth of the juvenile phase is essential for the young tree to compete with other vegetation. Prolonging the juvenile phase may be of use in the production of biomass from conifer forests. Shortening the juvenile phase would be of use in accelerated breeding programs for the production of genetically improved trees. The ability to reverse maturation will also be important for the propagation of specific mature trees via either rooted cuttings or tissue culture. Finally, an understanding of the maturation process is necessary to assure that any rejuvenation achieved is followed by a normal maturation sequence (10).

The events which may regulate maturation are not known. The fact that grafted scions from juvenile and mature plants maintain their respective phenotype (9,11) suggests that maturation is a result of permanent genetic changes within the meristem of the developing conifer, rather than a consequence of the increased size and/or structural complexity of the plant, as suggested by others (3,5).

Woody Plant Biotechnology, Edited by M.R. Ahuja
Plenum Press, New York, 1991

It has been our working hypothesis that maturation is the result of genetic changes in the meristem of the tree, and that the different phenotypes of the mature and juvenile phases will be reflected in the RNA transcripts found in the respective plants. We have initiated a study to detect genes which are differentially expressed in juvenile and mature trees, using the conifer *Larix laricina* (eastern larch) as a model system. We are using larch for several reasons: 1) its juvenile and mature characteristics have been described and are typical of those for other conifers (11), 2) it is very responsive to all forms of vegetative propagation relative to most other conifers, 3) its rapid growth rate and ease of propagation make it attractive as a species for plantation establishment, and 4) its genome is among the smallest of the extraordinarily large conifer genomes (6). Complicating the study is the fact that conifers, including larch, are highly outbred, with diverse genetic backgrounds which may affect the expression of individual genes (4,8). An understanding of maturation at the molecular level would not only help elucidate the mechanism of maturation, but the feasibility of reversing it as well.

MATERIALS AND METHODS

Plant Material

The material used for these studies has been previously described (11). Briefly, there were 40 trees separated into 4 distinct age classes of ten trees each. The average ages were 1 year (age class 1), 5 years (age class 2), 17 years (age class 3) and 45 years (age class 4). For analysis of gene expression, newly expanding short shoot needles were harvested from each of the 10 trees in each of the four age classes. The samples within an age class were pooled, quick frozen in liquid N_2 and stored at -70°C until the RNA was extracted. Sampling began 4 weeks after bud-break and was repeated approximately every two weeks until 10 weeks post bud-break, at which time the needles were fully expanded. Sampling was carried out on clear days, at 7 hours after sunrise to eliminate diurnal effects on measuring gene expression.

Extraction of RNA and Slot Blots

RNA was extracted from larch needles, roots and seedlings by the method of Whitmore and Kreibel (18), with the exception that 0.1% Triton X-100 (Boehringer Mannheim) was substituted for the SDS in all phases of the procedure. Ten grams wet-weight of tissue gave approximately 5 mg of total RNA. For slot blots serial dilutions of RNA were denatured by heating at 65°C for 10 min in 50% deionized formamide, 20 mM MOPS buffer (pH 7.0). The samples were cooled on ice and loaded onto Zetaprobe nylon filters using a BRL Slot Blot apparatus. The RNA was fixed to the filter using UV light.

The construction of cDNA libraries will be described elsewhere (Hutchison, et al. submitted). Differential hybridization used for screening these libraries was as previously described (17). Lambda-ZAP (Stratagene) libraries were also screened by subtractive hybridization using the Subtractor™ kit of Invitrogen. Screening was done as described by the supplier.

The slot blots were prehybridized at 42°C for 3-4 h in 50% formamide, 5X SSC, 50 mM $NaPO_4$ (pH 7.2), 1 X Denhardt's solution (5), 1% SDS, 50 μg/ml each of poly A and poly C (Pharmacia), and sonicated salmon sperm DNA (50 μg/ml). The filters were then hybridized overnight at 42°C in fresh hybridization buffer and ^{32}P-labelled probes (1 x 10^6 cpm/ml). After hybridization the filters were washed 4x in 0.2X SSC, 1% SDS at 50°C. The filters were then blotted dry an exposed to either X-Omat AR5 or X-Omat K X-ray film as described above. After exposure to the film the blots were erased by putting them in 500 ml of 0.2X SSC, 0.2% SDS at 95°C and shaking while the solution cooled to room temperature. This procedure was repeated a second time after which the filters were blotted dry and stored until used in the next hybridization.

RESULTS AND DISCUSSION

We have previously reported that chlorophyll content of the long and short shoot foliage of eastern larch increases linearly with the $\log_{10}$ of age (11). Net photosynthetic rate in larch long shoot needles also increases with age (Hutchison et al., submitted for publication). The increase in NPS appears to be a consequence of the increased chlorophyll content since stomatal conductance showed no change with increasing age (Hutchison et al. submitted for publication). Needle length of long shoot foliage declines slightly but significantly with age, but needle thickness and cross sectional area increase. The specific leaf weight also increases with age suggesting that long shoot needles become more massive with increasing age. The chlorophyll content of short shoot foliage varies in the same way with age as the long shoot foliage (11). Although we have not measured the photosynthetic capacity of short shoot needles, there is no reason to expect that short shoot foliage will behave any differently than that of long shoots. Therefore, maturation results in both morphological and physiological changes in the long shoot foliage.

Larch cDNA libraries have been constructed from short shoot buds of juvenile and mature trees in both lambda-GT10 (15) and lambda-ZAP (Stratagene) vectors. These libraries represent genes that are expressed in meristems and in the surrounding leaf primordia and developing needles. The libraries have been extensively screened by differential hybridization with cDNA probes made from juvenile and mature plants. We have, to date, not been able to detect either juvenile- or mature-specific clones by this procedure.

Differential hybridization is useful for detecting genes that are both highly expressed and show significant differences in expression between juvenile and mature plants. Transcripts that are expressed at low levels would not be detected by this method. Therefore we have screened lambda-ZAP cDNA libraries by subtractive hybridization using a the Subtractor™ Kit of InVitrogen. Using such a procedure we selected for juvenile and mature specific clones. From cDNA libraries of 50,000-100,000 clones the subtraction protocol produced approximately 700 clones from the juvenile cDNA library and 70 clones from the mature cDNA library. We do not know the reason for the 10-fold difference in recovery. None of the clones from either library have proven to be differentially expressed between juvenile and mature plants upon further investigation, suggesting they represent clones that escaped the selection process.

One conclusion from this effort is that any uniquely expressed transcripts must be present at extremely low levels, below the sensitivity of the subtractive hybridization protocol. Alternatively, we may conclude that maturation does not arise from the induction of new, maturation-specific genes or from the repression of juvenile-specific genes. We do find that some sequences are differentially expressed but the difference in expression in juvenile vs. mature plants is small (see below). We propose that few, if any genes will show major differences between juvenile and mature plants, and that the differences in phenotypes may be due to small changes in the expression of a number of gene families.

Since juvenile and mature plants showed significant differences in both chlorophyll content and CO_2 fixation rates we were interested in the expression of genes related to the photosynthetic apparatus. For analysis of gene expression newly expanding short shoot needles were harvested from each of the 10 trees in each of four age classes. The average ages were 1 year (age class 1), 5 years (age class 2), 17 years (age class 3) and 45 years (age class 4). RNA samples were collected at two week intervals during the period of needle expansion, starting 4 weeks post-bud break. These RNAs (which represent pooled samples from 10 trees in each age class) were applied to a slot blot and probe in succession with a larch *rbc*S cDNA probe, a larch *cab* cDNA probe and a larch 18S rRNA probe. The latter probe was to control for equal loading and/or binding of RNA to the nylon membrane.

71

In newly expanding short shoots *cab* was expressed at higher levels in juvenile plants than in any of the older age classes (Fig. 1A). A similar trend was seen with the second time point. As the season progressed the amount of *cab* mRNA present in a total RNA sample generally decreased. In addition, the later time points showed no consistent differences between the level of expression in juvenile plant and that in mature plants. For example, at 8 weeks post-bud break age class three was the only sample showing a level of expression substantially different (in this case higher) from age class one. At 10 weeks post-bud break there was a great deal of variation among the four age classes with regard to *cab* expression.

We have now analyzed *cab* expression in a large number of samples of newly expanding short shoots from different trees and during different years. In all cases, juvenile plants have higher levels of *cab* mRNA than do mature plants (Hutchison et al., submitted), consistent with what we report here (Fig. 1A). It is our interpretation of the later time points that there is no consistent difference in *cab* expression between juvenile and mature plants. However, we are not as confident in the measurements made on later time points because those RNA samples contain a significant amount of carbohydrate contamination which may be interfering with the hybridization reactions. Yields of total RNA were also lower in the later time points.

The higher level of *cab* expression in newly expanding needle from juvenile plants is inconsistent with our observation that there is more chlorophyll in mature plants than in juvenile plants (11). At least part of the explanation may reside in the fact that early in the growing season time we do not detect any difference in chlorophyll content between juvenile and mature trees for either long or short shoot foliage (11). Furthermore, in angiosperms it has been shown that the control of pigment production, and the steady state level of the chlorophyll *a/b*-binding proteins are not directly related to the level of *cab* gene expression (1).

Unlike *cab* RNA expression, the *rbc*S sequences showed no consistent differences among the four age classes, even at the earliest time point (Fig. 1B). This conclusion is corroborated by a more in-depth analysis of *rbc*S expression in newly expanding short shoot foliage (Hutchison, et al. submitted). Only age class 3 appeared to have less *rbc*S RNA than age class one. However, samples from other trees of the same range of ages does not show the same trend (Hutchison et al. submitted). The variability of expression at the later time points was even greater in our measurements of *rbc*S expression than for *cab* expression. We do not know the reason for this. However, one possibility is that *rbc*S expression is much more sensitive to genetic background or environmental factors, which coupled with the carbohydrate contamination of the later RNA samples gives a wide range of apparent levels of expression. We are currently operating with the model that unlike *cab* expression, *rbc*S expression does not vary in a maturation-related fashion. We are designing alternative experimental approaches to confirm this conclusion.

Clearly, maturation affects the morphology and physiology of larch needles so that the mature needles appear to be relatively more massive, contain more chlorophyll, and exhibit more net photosynthesis. These anatomical and physiological changes are associated with a change in the expression of genes coding for the chlorophyll *a/b*-binding proteins. Unique gene products associated with the juvenile or mature state were not detected. At present, we do not know whether the differential expression of the *cab* gene family is regulated at the transcriptional or post-transcriptional level. Maturation also affects the needle anatomy of a number of other woody species (12), but the effects of these anatomical changes on photosynthesis have only been investigated in English ivy (2) (*Hedera helix*) and red spruce (16) (*Picea rubens*). In *Hedera*, mature leaves also appear more massive (they are thicker and have a higher specific leaf weight), and net photosynthesis per unit surface area is also higher, which appears to be due to a higher stomatal frequency and conductance. In contrast to larch, total chlorophyll content is actually greater in juvenile than mature leaves (2). The mature foliage of red spruce is also thicker and more massive than juvenile foliage, but

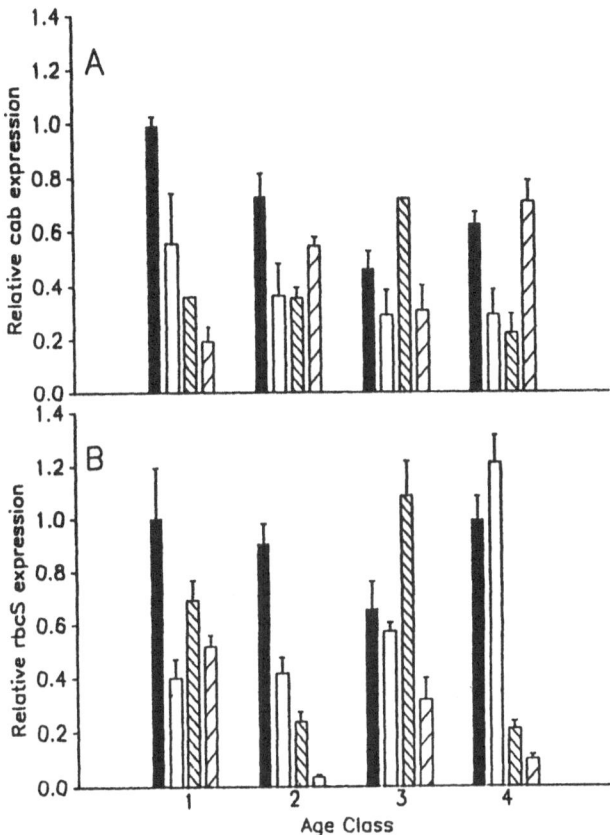

Figure 1. Expression of *cab* and *rbc*S gene families in short shoots from juvenile and mature larch. RNA was extracted from pools of needle samples from 10 trees in each of four age classes. The RNAs were applied to slot blots and probed with a larch *rbc*S probe. After exposure of the autoradiogram, the filters were erased and reprobed with a larch *cab* probe. Finally, they were probed with a larch 18S rRNA probe. The expression of the *rbc*S and *cab* sequences were normalized to the amount of rRNA detected on the filter and are expressed as relative to the amount of RNA in age class 1 at 4 weeks post-bud break. A) *Cab* gene expression. B) *Rbc*S gene expression. Age class 1 = 1y, age class 2 = 5 y; age class 3 = 17y; age class 4 = 45y. Bars = standard error. ■■■ = 4 weeks post-budbreak; ▭ = 6 weeks post-budbreak; ▨ = 8 weeks post-budbreak; ◣ = 10 weeks post-budbreak.

the juvenile foliage exhibits greater net photosynthesis, which appears to be due to increased stomatal conductance. Chlorophyll content is similar in both types of foliage. With the exception of a tendency for mature foliage to be more massive, the physiological differences between mature and juvenile foliage do not follow a consistent pattern (12). Clearly whatever is causing the observable differences between juvenile and mature foliage in these species does not elicit totally similar physiological responses. Therefore, one would not expect to see the same patterns of differential expression of *cab* genes among these species. However, since the mature foliage does share common morphological traits, there may be a common maturational process which can have varied effects on physiological processes. At present we have not found any obvious evidence that patterns of gene expression are anything but a consequence of whatever controls maturation. Hackett et al. (14) also report little differential gene expression during root formation by petioles of juvenile and mature *Hedera*. Cross comparison of cDNA libraries from both types of petioles, with and without IAA treatment, reveals only 1 or 2 unique clones, found in mature petioles treated with IAA. Since decreased rooting seems to always be associated with maturation, one cannot conclude at this time that the increased rooting capacity of juvenile tissue is due to the expression of particular genes which is subsequently lost with maturation. Instead, there may be gene products specific to the mature state which inhibit rooting.

The role of gene expression in maturational changes involving rooting or photosynthesis, not to mention other maturational characteristics would appear to be quite subtle. Tracing a common underlying maturational cause by examining gene products associated with changes in rooting or photosynthetic capacity would, in our opinion, be very difficult. But the maturational time courses for change by the suite of morphological and physiological characteristics in larch are similar, and the rate of change is most rapid in the first 5 years (11). Consequently, there may be a single (or small number) of events, several steps prior those events we report here, which affect all maturational processes.

The phenomenon of maturation is a complex developmental problem and the genetic mechanisms of control are likely to be similarly complex at the molecular level. We believe, however, that the genetic mechanisms that ultimately control phase change are likely to be simpler, at the conceptual level, than might be suggested by the variety of phenotypic changes associated with maturation.

ACKNOWLEDGEMENTS

This work was supported USDA Grant 85-FSTY-9-0140, and funds administered through the Maine Agricultural Experiment Station. Maine Agricultural Experimental Station publication #1446.

REFERENCES

1. Anderson, J. M., 1986, Photoregulation of the composition, function, and structure of thylakoid membranes. *Annu. Rev. Plant Physiol.* 37: 93-136.
2. Bauer H. and Bauer, U., 1980, Photosynthesis in leaves of the juvenile and adult phase of ivy (*Hedera helix*), *Physiol. Plant.* 49: 366-372.
3. Borchert, R., 1976, Differences in shoot growth patterns between juvenile and adult trees and their interpretation based on systems analysis of trees. *Acta Hortic.* 56: 123-130.
4. Cheliak W. M. and Pitel, J. A., 1985, Inheritance and linkage of allozymes in *Larix laricina*, *Silvae Gen.* 34: 142-148.
5. Denhardt, D. T., 1966, A membrane-filter technique for the detection of complementary DNA, *Biochem. Biophys. Res. Commun.* 23:641-646.
6. Dhillon S. S., 1987, DNA in tree species. *in*: "Cell and Tissue Culture in Forestry: General Principles and Biotechnology," J. M. Bonga and D. J. Durzan. eds. Martinous Nijhoff, Boston. pp. 293-313.
7. Durzan, D. J., 1984, Special problems: adult vs. juvenile explants, *in*: "Handbook of plant cell culture" v2, Sharp, Ammirato, Yanada, eds., MacMillan Pub. Co..
8. Fins L. and Seeb, L. W., 1986, Genetic variation in allozymes of Western larch, *Can. J. For. Res.* 16: 1013-1018.
9. Greenwood, M. S. Phase change in loblolly pine: shoot development as a function of age, *Physiol. Plant.* 61: 518-522.
10. Greenwood, M. S., 1987, Rejuvenation of forest trees, *Plant Growth Reg.* 6: 1-12.
11. Greenwood, M. S., Hopper, C. A. and Hutchison, K. W., 1989, Maturation in larch. I. Effect of age on shoot growth, foliar characteristics, and DNA methylation. *Plant Physiol.* 90: 406-412.
12. Greenwood M. S. and Hutchison, K. W., Maturation as a developmental process, *in*: "Clonal Forestry: Genetics, Biotechnology and Application." M. R. Ahuja and W. J. Libby, eds., Springer Verlag, New York (*in press*).

13. Hackett, W. P., 1985, Juvenility, maturation and rejuvenation in woody plants, *Hort. Reviews* 7: 109-155.

14. Hackett, W. P., Murray, J. and Woo, H., 1991, Biochemical and molecular analysis of maturation related characteristics in *Hedera helix. in*: "Woody Plant Biotechnology," M. R. Ahuja, ed. Plenum, NY.

15. Hunyh, T. V., Young, R. A. and Davis, R. W., 1985, Constructing and screening cDNA libraries in lambda-GT10 and lambda-GT11. *In*: "DNA Cloning: A Practical Approach," D. Glover, ed., IRL Press, Oxford, pp.49-78.

16. Hutchison, K., Greenwood, M., Sherman, C., Rebbeck, J. and Singer, P., 1990, The molecular genetics of maturation in eastern larch (*Larix laricina* [Du Roi] K. Koch), *in*: "Molecular basis of plant ageing." R. Rodriguez, ed., Pergamon Press pp. 141-145.

17. Hutchison, K. W., Singer, P. B. and Greenwood, M. S., 1988, Molecular analysis of gene expression during the development and maturation of larch, *in*: W. M. Cheliak and A. C. Yapa (eds), "Molecular Genetics of Forest Trees. IUFRO Molecular Genetics Workshop." Petawawa National Forestry Institute, Chalk River, Ont. Information Report PI-X-80, pp. 26-33.

18. Whitmore F. W. and Kreibel, H. B., 1987, Expression of a gene in *Pinus strobus* ovules associated with fertilization and early embryo development, *Can. J. For. Res.* 17: 408-412.

CELLULAR, BIOCHEMICAL, AND MOLECULAR ANALYSIS OF

MATURATION RELATED CHARACTERISTICS IN HEDERA HELIX

W.P. Hackett, J. Murray, and H. Woo

Department of Horticultural Science
University of Minnesota, St. Paul, MN 55108

ABSTRACT

Regulation of maturation-related root initiation potential and anthocyanin accumulation in Hedera helix are discussed. Evidence is presented for regulation of these characteristics at the level of gene expression. In the case of anthocyanin, m-RNA for the enzyme dihydroquercetin-4-reductase accumulates in juvenile leaf discs during sucrose induced anthocyanin accumulation, but not in sucrose-treated mature leaf discs that do not accumulate anthocyanin. These results suggests that regulation of gene expression for this maturation-related enzyme is at the transcriptional level. A wound inducible cDNA clone is a highly expressed in mature but not juvenile petioles and may be related to maturation-related reduced rootability.

INTRODUCTION

English ivy (Hedera helix L.), a woody perennial with a juvenile phase lasting 10 or more years, exhibits a number of distinct differences in phenotypic characters between the juvenile and mature phases (Table 1). This species is a classical example of stable dimorphism. The juvenile and mature forms can be maintained as separate plants through use of cuttage propagation or they can co-exist on the same plant. Through use of asexual propagation, the same genotype can be maintained in a greenhouse as separate individuals with very different phenotypes for long periods of time (years). Treatment of the shoot of a mature plant with GA_3 will induce development of juvenile characteristics. These characteristics make it very useful for doing an experimental analysis of the basis for differential phenotypic expression.

For the last 10 years, we have been doing such an experimental analysis of the control of individual maturation-related characteristics using H. helix. The characteristics that we've studied are: 1) shoot apical and subapical meristem size, configuration and activity in relation to internode length and phyllotaxis; 2) root initiation potential; and 3) anthocyanin accumulation. In this paper, I will limit discussion to root initiation potential and anthocyanin accumulation.

Maturation-Related Rooting Potential

Juvenile phase H. helix plants have stem aerial roots while mature phase plants do not. We have discovered that the rooting potential of detached leaves and in vitro cultured delaminated petioles of juvenile and mature ivy is very similar to that of stem cuttings (1,2). Detached juvenile leaves and delaminated petioles respond to auxin by initiating roots from small, discrete areas of cells in the exterior part of the phloem of each vascular bundle and the adjacent cortex. In contrast, mature leaves and delaminated petioles respond to auxin by forming only callus as a result of cell division in these same areas as well as throughout the cortex.

Table 1. Phenotypic characteristics that differ between clonal juvenile and mature phase English ivy (Hedera helix L.)

Characteristic	Juvenile	Mature
flowering ability	absent	present
stem orientation	plagiotropic (horizontal)	orthotropic (upright)
leaf morphology	palmately lobed	ovate and entire
phyllotaxy	distichous (alternate)	spiral
stem and petiole pigmentation	anthocyanins present	anthocyanins absent
stem adventitious roots	present	absent

Detached leaves and delaminated petioles have the following experimental advantages for studying rooting potential: 1) no possibility of pre-formed root initials; 2) variable rooting potential in tissues easily selected for similar physiological age and anatomical organization and having identical genetic makeup; 3) simple, uniform, fully differentiated tissue system in which source of plant growth substances can be endogenous (with lamina) or exogenous (without lamina); 4) easy in vitro manipulation of environment and precise provision of nutrients and plant growth substances as pulses or continuously fed from proximal or distal ends (delaminated petioles); 5) relatively rapid, specific and morphogenetically distinct responses to auxin; 6) ease of grafting to obtain composite cuttings; 7) tissues readily available in large amounts on a year-round basis.

Experiments have been performed using de-laminated petioles of juvenile and mature leaves and leaf cuttings composed of the four reciprocal combinations of juvenile and mature petioles with juvenile and mature lamina to study potential for root initiation. Procedures for using delaminated petioles and composite juvenile and mature leaf cuttings for studying rooting potential are described in Hackett et al (2). Figure 1 illustrates the use of grafting to make composite juvenile and mature leaf cuttings. The results of experiments with these systems lead to the following conclusions: (a) the morphogenetic process of root initiation is very different in easy- and difficult-to-root tissues; (b) auxin is required for root initiation in debladed juvenile petioles; (c) exogenously applied auxin and its metabolites have similar distribution patterns in juvenile and mature debladed petioles; (d) differences in ethylene metabolism do not appear to be causally related to differences in rooting potential in juvenile and mature debladed petioles; (e) root initiation is mainly a function of the rooting potential of cells localized in the petiole; (f) there is no evidence of a rooting inhibitor being transported from mature lamina; and (g) a translocatable substance(s) formed in juvenile lamina can either induce rooting initiation in mature petioles or increase rooting potential of new cells formed as a result of auxin treatment.

To investigate the molecular difference between juvenile and mature H. helix plants, the poly A(+) RNAs were isolated from juvenile and mature delaminated petioles at different time points after excision and translated in rabbit reticulocyte lysates. These translation products were analyzed by two-dimensional gel electrophoresis. At day 0 of in vitro culture, differences in only two polypeptides were detected. One polypeptide was specific in juvenile petioles and had a pI of ~9 and a size of ~25 kDa. The other polypeptide was specific in mature petioles and had a pI of ~5.3 and a size of ~28 kDa. A histological study showed that the differentiation of root initials in juvenile petioles starts at day 7 or 8. Because a primary interest was in messages related to root initiation potential, we tried to detect changes in translation products occurring before that time point. After day five of in vitro culture, we could detect only one polypeptide which was increased two or three times more in juvenile petioles than in mature petioles. Its pI was ~4 and its size was 40 kDa. These results show that

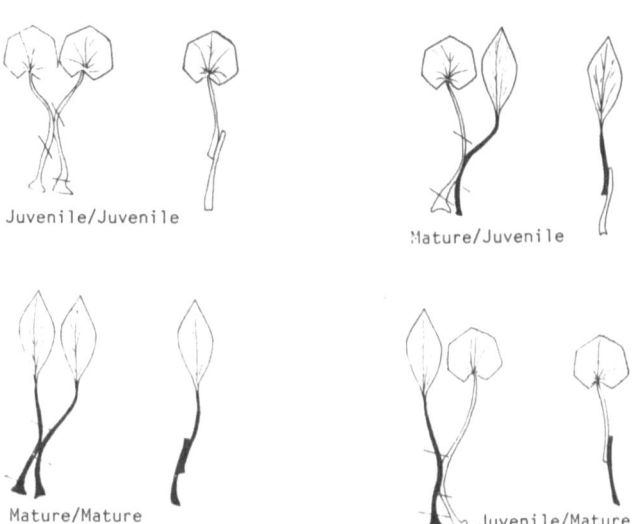

Juvenile/Juvenile

Mature/Juvenile

Mature/Mature

Juvenile/Mature

Figure 1.   Schematic representation of detached leaf approach grafting procedure used to obtain cuttings with reciprocal combinations of juvenile and mature petiole with juvenile and mature lamina.

the abundant mRNA species detectable via in vitro translation products are very similar in juvenile and mature petioles, both before and 5 days after auxin treatment. However, less abundant mRNAs most likely involved in synthesis of regulatory proteins wouldn't be detected by in vitro translation techniques.

To get more information on maturation-related reduced rootability, we made a cDNA library of poly A(+) RNA from auxin treated, 5 day cultured juvenile petioles and did differential screening by comparing the messages from auxin treated and non-treated juvenile and mature petioles. The initial screening of ~20,000 recombinants from the juvenile library [JA-5(S)] gave one cDNA clone which was specific in juvenile petioles. This clone (pHW101) represents a mRNA that is constitutively expressed at a higher level in juvenile than in mature petioles but is expressed at a lower level in auxin treated than non-treated juvenile petioles. It is also expressed at a higher level in juvenile than in mature leaf lamina and stem.

The disadvantage of differential screening is in its low sensitivity for the detection of low abundant messages. In our research we are also looking for low abundant messages which are also cell specific for differentiation. Therefore we have done subtraction hybridization screening to isolate low abundant messages. By using hydoxylapatite column chromatography, we could subtract about 85% of common messages. After subtraction hybridization screening of ~20,000 recombinants, we isolated a second differentially expressed clone. The second clone (pHW103) represents a mRNA that is expressed at a higher level in mature than juvenile petioles after 3 days of in vitro culture. But this clone was not expressed at all in juvenile or mature lamina, stem, or root tissues. Northern analysis shows that the mRNA size of pHW101 is ~1.4 kb and pHW103 is ~1.0 kb. Also, northern analysis shows more than one band which suggests these two clones may be in gene families. Neither of these cDNA clones seems to be closely related to the root initiation process. However, clone pHW103 which is expressed at a much higher level in mature petioles than juvenile ones, appears to be wound inducible since it isn't expressed at day 0 in any tissue tested but is expressed after excision and culture for 3 days. The wound inducibility of the pHW103 clone has been confirmed in wounded petioles of otherwise intact mature plants. Wounded petioles of otherwise intact juvenile plants do not express this clone. The function of this cDNA clone is being investigated by sequencing it and determining its histological site of expression by in situ hybridization.

Recently, Keller and Lamb (3) have identified a gene for a novel cell wall hydroxyproline-rich glycoprotein which is specifically expressed in tobacco in a sub-set of endodermis and pericycle cells at the inception of lateral root initiation. We have histological evidence that the thickness and blue light fluorescence of cell walls of juvenile and mature petioles is different, particularly for those cells involved in root initiation in juvenile petioles. Based on these findings, we will be investigating the importance of expression of genes for cell wall constituents as a basis for maturation related reduced rootability. The molecular analysis of excised juvenile and mature petioles during root induction demonstrates that there are differences in gene expression in juvenile and mature ivy but the number of genes differentially expressed is small.

Phenylpropanoid and Flavonoid Metabolism

Because of the observed accumulation of anthocyanin in juvenile but not mature leaves and stems of H. helix, we have studied phenylpropanoid and flavonoid metabolism at the cellular and biochemical level using stem and petiole tissue and leaf discs. Biochemical and enzymological studies show that the specific activity of phenylalanine ammonia-lyase is twice as high in mature as juvenile tissue and this is reflected in 50% higher extractable phenylpropanoids in mature than juvenile tissue. So it is unlikely that early steps in the phenylpropanoid-flavonoid pathway limit accumulation of anthocyanin in mature tissues.

The above conclusion led us to concentrate on flavonoid metabolism. Juvenile phase H. helix accumulates two classes of flavonoid glycosides in the dermal tissue of stems and petioles. The flavonoids, flavonols and anthocyanin are derived from dihydroflavonols late in the flavonoid biosynthetic pathway. The anthocyanin accumulates in the hypodermal tissue consisting of 3 to 4 layers of collencyma cells. The flavonols also accumulate in the dermal tissue, however, it has not been possible to determine if they are strictly localized in the hypodermis or epidermis. Mature phase ivy synthesizes and accumulates flavonols but not anthocyanin in its anatomically similar dermal tissue. The lack of synthesis of anthocyanin in mature phase hypodermal tissue is due to the lack of activity of dihydroquercetin-4-reductase (DQR), the enzyme that catalyzes the initial step of the conversion of dihydroflavonols to anthocyanin. Leaf laminae of neither phase accumulate anthocyanin when grown at temperatures above 15°C and there is no detectable activity of DQR. In addition, DQR and anthocyanin accumulation can be induced by sucrose in leaf discs from juvenile but not mature plants. These results showing that accumulation of anthocyanin in juvenile leaves and lack of accumulation in mature leaves of H. helix is due to the differential activity of DQR, is the only example we know of in which expression of a phase-related characteristic is due to the activity of a single polypeptide. Furthermore, by use of a cDNA probe for DQR from snapdragon, it has been demonstrated that mRNA for DQR accumulates in sucrose treated juvenile leaf discs but not mature leaf discs. This suggests that the phase specific characteristic of anthocyanin accumulation is due to differential gene expression at the transcriptional level. This hypothesis is being tested.

Histological studies of juvenile and mature petioles using blue light epifluoresence of fixed and stained tissues as a measure of wall bound phenylpropanoids show that fluorescence is lower in specific cells destined for root morphogenesis in juvenile petioles as compared to cells in the same location in mature petioles destined for fiber differentiation. This latter observation suggests that the two maturation characteristics anthocyanin accumulation and root potential might be metabolic related through the phenylpropanoid-flavonoid biosynthetic pathway.

Conclusions

These results provide a strong rationale for: (a) identification of a translocatable factor(s) that influences rooting potential; (b) study of enzymes specifically involved in cell wall lignin metabolism as they relate to rooting potential; (c) concluding that maturation-related differences in phenotypic characteristics are the result of differential gene expression; (d) use of the differential expression of DQR activity in juvenile and mature H. helix as a basis for investigating the molecular control of gene expression for a maturation-related characteristic.

REFERENCES

1. Geneve, R.L., Hackett, W.P., and Swanson, B.T., 1988, Adventitious root initiation in debladed petioles from juvenile and mature phases of English ivy, J. Amer. Soc. Hort.
   Sci., 113:630-635.

2. Hackett, W.P., Geneve, R.L., and Mokhatari, M., 1988, Use of leaf petioles of juvenile and mature _Hedera helix_ to study control of adventitious root initiation, _Acta Hortic._, 227:141-144.

3. Keller, B. and C.J. Lamb. 1989. Specific expression of a novel cell wall hydroxyproline-rich glycoprotein gene in lateral root initiation. _Genes and Development_, 3:1639-1646.

ADVENTITIOUS BUD PRODUCTION FROM MATURE PICEA ABIES: REJUVENATION

ASSOCIATED WITH FEMALE STROBILI FORMATION

Kathryn X. Wang, David F. Karnosky and Roger Timmis

Michigan Technological University   Weyerhaeuser Company
School of Forestry and Wood Products   Tacoma,
Houghton, Michigan 49931 USA   Washington 98477 USA

ABSTRACT

Developing female strobili of mature Picea abies were induced to form adventitious bud primordia when cultured on half-strength LP medium, containing 10^{-5} M 6-benzyl amino purine. Further development of buds required transfer to media free of growth regulators. A high frequency of strobili discs forming bud primordia was obtained only when strobili were collected at about the time of meiosis. Adventitious buds arose mainly from the outside edges of strobili discs; some also arose from the cut surface of the ovuliferous scales. The frequency of adventitious bud formation varied from tree to tree, indicating intraspecific variation in bud formation.

INTRODUCTION

Picea abies (L.) Karst is one of the world's most important timber and ornamental conifers. Native to northern and central Europe, it has been widely planted in Europe, the Caucasus, Siberia, China, Japan and North America. The wood is light in weight, long fibered, elastic and slightly resinous so that it is good for timber and for paper production.

Natural propagation in Picea abies is by seed, but the long vegetative phase before flowering makes breeding in this species a very slow process. During the past ten years, remarkable progress has been achieved in the induction of organogenesis (24,26) and embryogenesis (2,3,15,16,18,25,28) in Picea abies cell and tissue culture with most work being done with embryos or young seedlings. However, cloning of mature trees is generally preferred over cloning of embryos or seedlings because it is not possible to determine if these embryos or seedlings have the desired qualities and traits of the mature selection such as vigor, height growth, stem taper, branching habit, crown shape and disease resistance (10,19).

Successful in vitro multiplication using explants from mature trees has been achieved for some hardwoods (1,12,14), but clonal propagation of most mature hardwoods and conifers in vitro is still very difficult or in some cases impossible. Successful multiplication and production of good quality plants via micropropagation from mature tissues of conifers has been reported in only a few conifers, including Pinus radiata (17), and Sequoia sempervirens (11), and Larix decidua (20). Rejuvenation of the mature tissues appears to be important in the process of propagation from mature trees. While most studies have focused on newly developing vegetative shoots, Bonga and co-workers have shown that somatic tissues of female strobili collected at about the time of meiosis also have the capacity to form adventitious shoots (5,8,9).

The purpose of this study was to see if the somatic tissues of female Picea abies strobili might be rejuvenated at the time of meiosis and thereby serve as useful source tissue for the induction of adventitious buds which would be useful for micropropagation.

MATERIAL AND METHODS

Two large Picea abies trees (greater than 40 feet in height) on the Michigan Technological University campus (#1 and #2), five of the same size at the University golf course (#4, #5, #6, #7 and #8) and eight smaller flowering trees (approximately 25 feet in height) in a row of some 40 Picea abies trees located about five miles from campus (#3, #3C, #3D, #3E, #3F, #3G, #3H, #3I) were chosen for this study. All chosen trees produced cones in 1987 and had flower buds in 1988 and 1989. Female strobili were collected during the period from April 15, 1988 to June 6, 1988 and from May 9, 1989 as shown in Table 1.

Table 1.   Collection dates and tree numbers of Picea abies female strobili collected in 1988 and 1989.

Collection Number	1988					1989	
	I	II	III	IV	V	VI	VII
Sampling Date	4/15	4/28	5/12	5/23	6/6	5/9	5/18
Tree Number	1	1	2	2	3C	3H	3H
	2	2	3	3C	3D	3G	3I
	3	3	4	3D	4	3I	5
	4	4	5	4	5		6
	5		6	5	6		
	6		7	6	7		
			8	7	8		

Female strobili, ranging in length from 0.8 to 12 cm and in width from 0.5 to 2.9 cm, as shown in Table 2, were stored in plastic bags at $4^{o}C$. Care was taken to have the strobili surface dry before they were refrigerated to minimize subsequent fungal contamination.

The female strobili were surface sterilized as described for Larix decidua by Bonga (7). After removing the outside bud scales, the strobili were

Table 2.   Picea abies female strobili sizes at the various collection dates in 1988 and 1989.

Collection	Sampling Date	Length (cm)	Width (cm)
I	4/15/88	0.8-1.0	0.5-0.6
II	4/28/88	1.0-1.2	0.5-0.6
III	5/12/88	1.5-1.7	0.6-0.8
IV	5/23/88	5.8-6.2	1.7-1.9
V	6/6/88	8.0-12.0	2.5-2.9
VI	5/9/89	0.7-1.0	0.5-0.6
VII	5/18/89	2.2-3.0	0.7-1.0

placed in 6% sodium hypochlorite (commercial bleach) and stirred for ten minutes and then placed in 70% ethanol-HCL (one drop of HCL per 100 ml of 70% ethanol) for two minutes, followed by three rinses in sterile distilled water.

Surface sterilized female strobili were placed in sterile Petri dishes in an autoclaved solution of 1 000 mg/l malonic acid. The strobili were sliced sliced into 1.5 mm thick transverse discs and these were submerged in malonic acid solution for ten to 15 minutes before they were placed flatly with the acropetal side up on the nutrient medium. In some experiments, the scales and the central core of the strobili were cultured separately.

Different basal media were tested to determine the best one for in vitro culture of P. abies strobili explants. The basal media tested were SH (23), LP (27) WPM (22), DCR (13) and LM (21). After initial trials suggested LP 1/2 and SH 1/2 as the best basal media, additional salt concentrations (1/4X, 1/2X and 1X LP and 1/2X and 1X SH) were tested for disc cultures. One-half strength LP medium in 1988 was found to be the best medium for the culture of discs and scales from P. abies female strobili based on explant survival and so was used for the 1989 collections.

After initiation on the LP 1/2 (LP with all minerals reduced to one half) basal medium with 10^{-5} M BAP for the first three weeks of culture, the explants were subcultured every three weeks to LP 1/2 media free of growth regulators. The concentration of sucrose was 1% in all media. Initially various media solidifying concentrations were tried, including 0.5% gelrite, 0.7% agar and 0.8% agar. The 0.7% agar appeared to give best results in early trials based on explant survival and was used in all subsequent studies. The pH of the media was adjusted to 5.6 before autoclaving, which was done at 121°C for 15 minutes. All cultures were maintained in 10 cm X 1.5 cm Petri dishes at 20°C and with a 24-hour photoperiod.

In an attempt to minimize the browning of disc explants, several antioxidants were tested by adding them to the initial media in 1988. They were ascorbic acid (1.0 X 10^{-4} M, 4.0 X 10^{-4}M, 8.0 X 10^{-4}M, 1.6 X 10^{-3}M, and 3.2 X 10^{-3} M) tyrosine (0.05 mg/l and 0.1 mg/l), and DMSO (dimethyl sulfoxide) (0.75%).

RESULTS

In the 1988 experiments, adventitious bud formation was observed only with the discs and scales derived from strobili of collection II and III (Table 3). For the first two weeks in culture, most discs remained green and they swelled to about two times their original size. But the central core of most of the discs gradually turned brown before the cultures were transferred to medium without BAP. Only those discs with some green tissue evident were transferred. Even when considerable browning of the disc explants occurred, the peripheral tissue often remained green and vigorous for about eight to ten weeks.

Adventitious buds developed either directly around the edge of the discs or from individual scales after about six weeks in cultures. In total, adventitious buds developed on 15 discs and five scales. Tree-to-tree variation in culturability was noticed as adventitious buds were only produced in two of four trees in collection II and four of seven trees in collection III in 1988. In total, 23 adventitious buds were produced from two trees for collection II and 26 were produced from four trees for collection III. These data are summarized in Table 4. All adventitious buds eventually died before elongating more than 1 to 2 mm. Their death appeared to be related to a general browning of the discs and/or scales that generally occurred within two to three months of culture initiation.

Table 3. The number of Picea abies female strobili explants plated initially, the number of explants subcultured to two stages and the number of number of explants that produced adventitious buds in 1988.

Collection	I	II	III		IV		V	
Explant Type	d+s*	d+s	d	s	d	s	d	s
Original Number of Explants	296	92	293	265	203	289	118	10
Number of Explants Subcultured to Hormone-Free Media	49	62	241	187	150	170	104	6
Number of Explants Remaining Green for Six Weeks	7	11	151	103	0	0	0	0
Number of Explants Producing Adventitious Buds	0	3	12	5	0	0	0	0
Number of Adventitious Buds Produced	0	23	17	9	0	0	0	0

* d=1.5 mm thick cross-section discs of female strobili with scales removed; s=scales cultured separately from the central core of the female strobili; d+s=discs with scales still attached.

Table 4. Number of Picea abies female strobili disc and scale explants that produced adventitious buds for each tree in collection II and III in 1988.

	Collection II (4/28)					Collection III (5/12)							
Tree Number	1	2	3	4	Total	2	3	4	5	6	7	8	Total
Initial Discs	8	21	22	41	92	30	68	64	23	22	39	47	239
Initial Scales			Not Tested			0	61	43	39	24	57	41	265
Discs Forming Adventitious Buds	2	0	1	0	3	0	6	2	1	3	0	0	12
Scales Forming Adventitious Buds			Not Tested			0	3	1	1	0	0	0	5

Explants of collection I remained green for a few months but did not develop. Explants of collections IV and V browned quickly (about one month or less) and did not develop.

The 1988 results suggested that a developmental "window" exists in which the adventitious buds can be induced from the mature tissues. The experiment was repeated with material that was collected the next year at the stage of strobili development close to that where the best response occurred in 1988. For the two collections of 1989, good response was obtained only on discs collected on May 18 (Table 5). With explants from this date, adventitious buds developed after transferring discs to hormone-free medium. After six weeks of incubation, 47 of the 173 discs formed a total of 83 adventitious buds from the edge of discs (Table 5). In some adventitious buds, needles were readily distinguishable. All the adventitious buds were obtained from discs of tree #3H and #6 (Table 6).

Table 5. The number of Picea abies female strobili disc explants plated initially, the number of explants subcultured to two stages and the number of explants that produced adventitious buds in 1989.

Collection Number	Date	Original Number of Explants	Number of Subculturable Explants	Number of Explants Producing Buds	Number of Buds Produced
I	5/9	71	0	0	0
II	5/18	310	169	47	83

Table 6. The number of *Picea abies* female strobili disc explants for May 18, 1989 and the number of adventitious buds produced for each of four trees.

Tree Number	Initial Discs	Discs Forming Buds	Number of Buds
3H	72	19	47
3I	56	0	0
5	81	0	0
6	101	28	36

LP 1/2 medium was the best medium for retaining explant vigor and for having more discs with adventitious buds formed than the other media (Table 7).

Table 7. Effects of different nutrient media on the relative health (green=healthy as opposed to brown=dead) and adventitious bud induction on *Picea abies* female strobili discs.

Media*	1/2SH	1/2LP	1/2LM	DCR	WPM
Initial Disc Number	23	19	21	38	27
Number of Discs Still Green after Three Weeks	16	18	7	25	11
Discs Forming Buds	5	11	0	2	0

* SH=Schenk and Hildebrandt medium (23); LP=LP medium (27); LM=Litvay's medium (21); DCR=DCR medium (13); WPM=Woody plant medium (22).

Because the most adventitious bud initiation occurred on the 1/2 SH or 1/2 LP media, additional strengths of these two media were tested. The test was conducted with collection III. 1X and 1/2X of SH medium and 1X, 1/2X and 1/4X of LP medium were compared. Both 1/2 SH and 1/2 LP were good for initiation of adventitious buds on female strobili discs (Table 8).

Table 8. Comparison of different strength SH and LP media for relative health (green=health as opposed to brown=dead) and adventitious bud production on *Picea abies* female strobili discs.

Media*	SH	1/2SH	LP	1/2LP	1/4LP
Discs Green After Three (3) Weeks	-**	+	+	+	+
Bud Formation	-	+	-	+	-

* SH=Schenk and Hildenbrandt medium (23); LP=LP medium (27).
** --=no; +=yes.

The results of the ascorbic acid studies are shown in Table 9. The addition of 4.0×10^{-4} M ascorbic acid gave the best results in keeping the strobili discs green. Additions of 0.05 mg/l and 0.1 mg/l of tyrosine and 0.75% of DMSO had no significant effects in improving disc survival. It was observed that in spite of the browning of the central part of the discs, adventitious buds still initiated and developed on the discs whose margins remained green.

Table 9. The effect of various concentrations of ascorbic acid on *Picea abies* female strobili disc culture.

Ascorbic Acid Concentration (M)	0	1.0×10^{-4}	4.0×10^{-4}	8.0×10^{-4}	1.6×10^{-3}	3.2×10^{-3}
Original Number of Discs	11	18	19	19	18	13
Number of Discs Green after Three (3) Weeks	0	1	7	2	0	0

DISCUSSION

Bonga and co workers (7 8 9) have been able to induce organogenesis on tissues of mature <u>Larix decidua</u> trees by slicing female strobili into cross-sectional pieces and plating them on tissue-culture medium supplemented with BA. There appears to be a short (one to two week) window of time when this occurs. During this window (around the time of meiosis), adventitious buds can be induced from diploid tissues surrounding the ovule. Thus, there appears to be a rejuvenation stimulus that is diffusible from the ovule out into the surrounding tissues. Our results suggest that a similar rejuvenation stimuli occurs in <u>Picea abies</u> and that adventitious bud production from mature trees can be obtained during that time.

The main factors that determined the extent of organogenesis in <u>Picea abies</u> female strobili were (1) collection date (about the time of meiosis was best) (2) genotypes (only about one third of our trees produced adventitious buds in culture); (3) media strength (half strength media appeared to be better than full strength or quarter strength); and (4) media type (LP and SH media were better than LM, WPM, or DCR).

Most of the adventitious buds were produced in discs from strobili collected about the time of meiosis. The possible relationship between meiosis and morphogenetic activation of the nearby somatic tissues is far from understood. It has been speculated that meiosis may temporarily stimulate the organogenetic ability of some somatic tissues near the site of meiosis (4). During the lifetime of the tree, a relatively stable alteration in physiology occurs when the tree changes from juvenile to mature is reversed in the sexual process. This reversal probably occurs mostly in the meiocytes (6). During meiosis, some somatic tissues of the female strobili become morphogenetically active, suggesting that they are activated by the same factors that initiate rejuvenation in the meiocytes. But whether or not this explains the formation of the adventitious buds obtained in this or other studies still needs more research. A better understanding of the mechanisms of rejuvenation during meiosis is important because it may suggest means of removing the mature determination from cells of mature trees, thus re-establishing their capacity for organogenesis and true-to-type propagation.

The methods used for the <u>in vitro</u> culture of embryo or juvenile tissues are generally not applicable to tissues from mature trees. In contrast to <u>in vitro</u> culture of juvenile tissues, a low-salt medium is generally necessary for culture of mature trees. In general, it seems that slow-growing organs should be cultured on more diluted media than fast growing ones and that the organized development of organs requires more diluted medium than does unorganized cell growth. In comparison of different media (SH, LP, WPM, DCR, LM) for mature <u>Picea abies</u> culture, it was found that lower total ionic strength media like 1/2SH or 1/2LP were more effective in organogenesis of female strobili of <u>Picea abies</u> than were higher ionic strength media.

Shoot elongation did not occur in any of the buds, probably because our subculture medium was inadequate. To establish an optimum nutrient composition and concentration is a prerequisite for shoot elongation and rooting and should be the main focus of future experiments.

LITERATURE CITED

1. Ahuja, M.R. 1987. <u>In vitro</u> propagation of poplar and aspen. In Bonga, J.M. and D.J. Durzan (Eds.) Cell and Tissue Culture in Forestry. Vol. 3. Martinus Nijhoff Pub., Boston. P. 207-223.
2. Becwar, M.R., T.L. Noland and S.R. Wann. 1986. Somatic embryo development and plant regeneration from embryogenic Norway spruce callus.

Tappi Research and Development Conf. Tappi Press, Atlanta, Georgia. P. 125-130.

3. Becwar, M.R., S.A. Verhange, and S.R. Wann. 1987. The frequency of plant regeneration from Norway spruce somatic embryos. Proceeding 19th Southern Forest Tree Improvement Conf. College Station, Texas. The National Technical Information Service, Springfield, Virginia. P. 92-100.

4. Bonga, J.M. 1980. Plant propagation through tissue culture, emphasizing woody species. In F. Sala, B. Parisi, R. Cella and O. Cifferi (Eds.). Plant Cell Cultures: Results and Perspectives. Elsevier/ North Holland Biomedical Press. New York. P. 253-264.

5. Bonga, J.M. 1981. Organogenesis in vitro of tissues from mature conifers. In Vitro 17:511-518.

6. Bonga, J.M. 1982a. Vegetative propagation in relation to juvenility, maturity and rejuvenation. In J.M. Bonga and D.J. Durzan (Eds.). Tissue Culture in Forestry. Martinus Nijhoff. Dr. W. Junk Pub. Boston. P. 387-412.

7. Bonga, J.M. 1982b. Shoot formation in callus from the stalks of young female strobili of Larix decidua. Can. J. Bot. 60:1357-1359.

8. Bonga, J.M. 1984. Adventitious shoot formation in cultures of immature female strobili of Larix decidua. Physiol. Plant. 62:416-421.

9. Bonga, J.M. and P. von Aderkas. 1988. Attempts to micropropagate mature Larix decidua Mill. IUFRO Workshop on somatic cell genetics of woody plants. Grosshansdorf, West Germany.

10. Bornman, C.H. 1983. Possibilities and constraints in the regeneration of trees from cotyledonary needles of Picea abies in vitro. Physiol. Plant. 57:5-16.

11. Boulay, M. 1987. Conifer micropropation: Applied research and commercial aspects. In J.M. Bonga and D.J. Durzan (Eds.). Cell and Tissue Culture in Forestry. Vol. 3. Martinus Nijhoff. Dr. W. Junk Pub. Boston. P. 185-206.

12. Chalupa, V. 1987. European hardwoods. In J.M. Bonga and D.J. Durzan (Eds.). Cell and Tissue Culture in Forestry. Vol. 3. Martinus Nijhoff. Dr. W. Junk Pub. Boston. P. 224-246.

13. Gupta, P.K. and D.J. Durzan. 1985. Shoot multiplication from mature trees of Douglas fir (Pseudotsuga menziesii) and sugar pine (Pinus lambertiana). Plant Cell Rep. 4:177-179.

14. Gupta, P.K., A.F. Mascarenhas, and V. Jagannathan. 1981. Tissue culture of forest trees: Propagation of mature trees of Eucalyptus citriodora by tissue culture. Plant Sci. Lett. 20:195-201.

15. Hakman, I. and S. von Arnold. 1985. Plantlet regeneration through somatic embryogenesis in Picea abies (Norway spruce). J. Plant Physiol. 121:149-158.

16. Hakman, I., L.C. Fowke, S. von Arnold and T. Eriksson. 1985. The development of somatic embryos in the tissue cultures initiated from immature embryos of Picea abies (Norway spruce). Plant Sci. 88:53-59.

17. Horgan, K. 1987. Pinus radiata. In J.M. Bonga and D.J. Durzan (Eds.). Cell and Tissue Culture in Forestry. Vol. 3. Martinus Nijhoff Pub. Boston. P. 128-145.

18. Jain, S.M., R.J. Newton and E.J. Soltes. 1988. Enhancement of somatic embryogenesis in Norway spruce. Theor. Appl. Genet. 76:501-506.

19. Karnosky, D.F. 1981. Potential for forest tree improvement via tissue culture. Bio. Sci. 31:114-120.

20. Laliberte, S. and M. LaLonde. 1988. Sustained caulogenesis in callus cultures of Larix x eurolepis initiated from short-shoot buds of a 12-year-old tree. Amer. J. Bot. 75:767-777.

21. Litvay, J.D., M.A. Johnson, D. Verma, D. Einsphahr, and K. Weyrauch. 1981. Conifer suspension culture medium development using analytical data from developing seeds. Inst. Paper Chemistry, Appleton, Wis. Tech. Paper No. 115.

22. Lloyd, G.B. and B.H. McCown. 1980. Commercially feasible micropropaga-
    tion of mountain laurel (Kalmia latifolia) by use of shoot-tip
    culture. Proc. Inter. Plant Propagators Soc. 30:421-437.

23. Schenk, R.U. and A.C. Hildenbrandt. 1972. Medium and techniques for
    induction and growth of monocotyledonous and dicotyledonous plant-
    cell cultures. Can. J. Bot. 50:199-204.

24. von Arnold, S. 1982. Factors influencing formation, development and
    rooting of adventitious shoots from embryos of Picea abies (L.)
    Karst. Plant Sci. Lett. 27:275-287.

25. von Arnold, S. 1987. Improved efficiency of somatic embryogenesis
    mature embryos of Picea abies. J. Plant Physiol. 128:233-244.

26. von Arnold, S. and T. Eriksson. 1979. Bud induction on isolated needles
    of Norway spruce grown in vitro. Plant Sci. Lett. 15:363-372.

27. von Arnold, S. and T. Eriksson. 1981. In vitro studies of adventitious
    shoot formation in Pinus contorta. Can. J. Bot. 59:870-874.

28. von Arnold, S. and I. Hakman. 1986. Effect of sucrose on initiation of
    embryogenic callus from mature zygotic embryos of Picea abies
    (Norway spruce). J. Plant Physiol. 122:261-265.

EUCALYPT JUVENILITY

D. M. Paton

Forestry Department, Australian National University
GPO Box 4, Canberra, ACT 2601 Australia

ABSTRACT

Rejuvenation of adult tissue of E.grandis is possible
after grafting or multiple bud culture. Seedling rooting
capacity and low G content are both regained during these
rejuvenation treatments. Provided that some possible
limitations are recognized, such rejuvenated tissue from elite
large trees could be useful for biotechnology, including
genetic transformation. This approach has advantages over the
use of interspecific hybrids where lethal genotypes can arise
in adult tissue.

INTRODUCTION

Current methods in woody plant biotechnology almost
invariably rely on embryonic or juvenile seedling tissue for
experimental material. Just how far the results obtained can
be extrapolated to later stages in tree ontogeny is a
challenging question that is likely to become increasingly
relevant as some of the promising advances are tested in the
field. Another question arising from these current methods, is
the often equivocal terminology used in tree development where
stages in ontogeny and life-span need definition at the plant
level as well as for organs such as leaves. Both these
questions are addressed in the present paper especially as
they apply to Eucalyptus.

Most results refer to E.grandis which has developmental
features that are often intermediate for other forestry
species of the genus. For example, the ontogenetic decrease in
rooting capacity of E.grandis cuttings (cf.Fig.1) occupies an
intermediate position when compared with four other selected
species[20]. A root-shoot gradient in the plant growth
regulator, G[12], has been postulated to be causally related to
the decrease in rooting capacity during seedling ontogeny in
this species[23]. While G is so far restricted to E.grandis
(from which it gets its name), other plant growth regulators
are likely to control comparable but not identical root-shoot
gradients for rooting capacity in other eucalypts[20].

Woody Plant Biotechnology, Edited by M.R. Ahuja
Plenum Press, New York, 1991

In any one eucalypt species, the root-shoot gradient for rooting capacity can differ markedly from other ontogenetic gradients affecting for example, leaf trends in shape, colour and phyllotaxis on the young expanded shoot. This difference is reported for the five species in Fig.1, to emphasise the need in the present paper to identify rooting capacity alone as the ontogenetic gradient under consideration. The question of terminology is especially relevant in this regard.

MATERIALS AND METHODS

Shoot, leaf and bud terminology

For the purposes of eucalypt physiology and perhaps more generally for tree biotechnology, two mutually exclusive sets of terms can be used to specify stages in the seedling - tree sequence. One set is related to the life-span of the tree, or individual shoots or individual leaves, where ageing involves the progression from young to old. The other set of terms refers to ontogeny of the individual tree (rather than individual shoots or leaves) as it passes through the juvenile - adult sequences involved in the different ontogenetic gradients in flowering, phyllotaxis and rooting capacity. It should be re-emphasised that the main experimental results in the present paper are restricted to juvenile - adult trends in rooting capacity, and particularly the relationship of these trends to root-shoot gradients in G content.

The term, basal, refers to the shoot position that is closest to roots as occurs at the base of a seedling stem (juvenile) or at the base of adult scions grafted close to the cotyledonary node of seedling stocks. Distal refers to stem positions at least 1m above the basal position.

As intermediate stages usually occur between juvenile and adult, perhaps adolescence is an appropriate term to indicate the often unpredictable behaviour between the more stable extremes[16]. Heteroblastic leaves in many eucalypts also involve intermediate morphology between the seedling (juvenile) and adult leaves, but while these intermediate leaves are often noted as juvenile in species descriptions[24], the present study does not use the term, juvenile, in this sense. Neither are the terms, maturation, mature and senescence used as they frequently do not discriminate between ontogeny and life-span. For example, there is a need in the present study to refer quite specifically to young adult shoots for grafting on young juvenile stock. Any reference to young mature (adult) shoots or mature (old) juvenile shoots could be a source of confusion.

Eucalypts commonly abort most axillary buds on young adult shoots[24] leaving internal accessory buds[8] to produce the prolific epicormic shoots so typical of fire damaged trees. Basal epicormic shoots may arise from basal accessory buds either at the very base of the stump of felled trees equivalent in height to the first few seedling nodes, or at the base of adult scions close to the graft union and cotyledonary node of juvenile stocks.

Vegetative propagation

Leafy stem cuttings of most eucalypts root readily provided that the base of the cutting is near the cotyledonary node[22]. The trends in Fig. 1. illustrate how the ontogenetic decreases in rooting capacity differ for five eucalypts when the cutting base is taken from successively higher nodes.

Stumps of felled E.grandis trees can produce basal epicormics that vary in rooting capacity from 10-80% depending on genotype[23]. It is not uncommon, however, for some stumps to produce no basal epicormics or sparse epicormics that rarely root. Such unpredictable outcomes are typical of intermediate (adolescent) ontogeny, and indicate the limitations of this method for propagating elite large trees.

Grafting of young or old adult scions on juvenile seedling stock provides very suitable flowering shoots for manipulated crosses[26]. Despite some graft incompatibility, basal accessory buds from the adult scion of these grafts can produce epicormic shoots capable of rooting provided that the accessory buds have developed within the secondary phloem for 6-18 months[23].

Multiple bud (MB) culture for in vitro culture of nodal explants of eucalypts[2,11], has proved to be very effective for micropropagation of old trees[3,14] especially when the nodal explants are taken from previously adult scions grown under glasshouse conditions[3,13,14]. Even in relatively pest-free glasshouses, microbial infections pose a serious problem for MB culture[2,11] but the sterilisation protocol described below, offers one solution for young adult eucalypt shoots[15].

In all these various propagation methods, rooting capacity is best recorded as % rooting[22] rather than root weight as in mung bean bioassays[12]. When the source data for MB cultures, grafts and stump epicormics[22] are given as the range in % rooting, the mid-point of this range approximates to mean % rooting as derived for all non-seedling points in Fig.2.

Sterilization for MB culture of adult shoots

Material selected visually for insect-free growth, was washed in running tap water for one hour. Nodal cuttings or explants were obtained from a position three to four nodes from the apex. Each explant consisted of one node with about 0.5cm of internode either side. About 0.5cm of the petiole was retained but all leaf lamina was removed. The explants were chosen to exclude any with either growing axillary shoots or rough abscission scars from the shed axillary shoots.

These explants were immersed in a saturated solution of calcium hypochlorite for a period of 75 minutes which is much longer than treatment times reported in previous studies[3,9,17]. The explants were then transferred to sterile, covered containers of distilled water for ten minutes. This washing procedure was repeated twice. All subsequent steps were carried out in a sterile cabinet (a Clemco Laminar Flow Work Station, Model 43S, fitted with hot cathode silica-quartz tubes: standard germicidal ultra-violet, 254 nm).

For UV sterilization, the explants were exposed to four hours of UV radiation at approximately 1m distance from the UV source while being agitated in a beaker of distilled water using a magnetic stirrer. After irradiation, the explants were cultured in tubes of autoclaved plant growth medium[2].

Following the UV irradiation, much of the external tissue of the explants turned brown to black during the first few days in culture. Contamination with bacteria or yeast was minimal but many sterile explants were killed outright by the sterilization treatment. Explants were considered to have survived sterilization if some green tissue remained or green shoots subsequently developed. On this basis, about 50% survival was obtained with a combination of 75 minutes calcium hypochlorite and four hours UV[15].

At three weeks to two months after sterilization about 5% of the surviving explants produced a weak shoot from the accessory bud positioned within the axillary tissue. When this surviving shoot produced basal shoots, the culture was subdivided as usual for MB culture[11]. Roots can form spontaneously after 8-10 transfers but an unusually high level of auxin (10^{-4}M IBA) promoted rooting at about the third transfer[15].

Location of the often sterile accessory buds in the phloem[8], offers some protection when more external tissues are killed. These external tissues include the chlorophyll containing cells of the outer cortex which judging from rapid loss of the original green colour of the explant, are especially vulnerable. As high G content in the adult explant is restricted to these green cells, their destruction reduces the G content that inhibits rooting in bioassays[12]. Thus such drastic sterilization probably has a direct role in the production of sterile MB cultures with some rooting capacity.

G content and other plant growth regulators

Methods for extraction, purification and identification of G[12] are now greatly improved by use of solvent gradient HPLC[6,20]. Although this method allows the G content of portions of one adult leaf of E.grandis to be determined with reproducibility within 5%, all G contents given in Fig.2 involved bulked samples of usually four leaves (two leaf pairs) or shoot and callus of MB cultures. It should be noted that high G contents of > 0.5mg/g fr.wt. of leaves are restricted to E.grandis and adult non-basal leaves of this species. Except for P in E.pulverulenta (Fig. 3), the chemical structure of equally potent plant growth regulators in other species is not known although their presence in some is indicated by strong bioassay activities equivalent to G[20].

All non-seedling G contents in Fig.2 were determined using standard TLC methods[12], after extraction of bulked material from about four leaves depending on leaf size. With the very small modified leaves of MB cultures, at least five large cultures were required for the small amounts of G detected (Fig.2). One limitation of all data used for Fig.2 was that G content and rooting capacity were determined on different but comparable shoot material.

RESULTS

Ontogenetic loss of rooting capacity

Ontogenetic loss in rooting capacity for E.grandis decreases at a rate less steep than in E.viminalis and E.pauciflora but more steep than E.camaldulensis and E.deglupta (Fig.1). As E.deglupta can form roots on stem cuttings taken from the tree crown 22, this tropical species exhibits little10,22 or no (Fig.1) ontogenetic loss in rooting capacity. Thus as far as rooting behaviour is concerned, E.deplupta retains seedling juvenility indefinitely, while E.grandis is intermediate for the genus.

To illustrate why this paper on eucalypt juvenility is restricted to rooting capacity, it is helpful to compare the ontogenetic trends in Fig.1 with ontogenetic trends in phyllotaxis. These different leaf trends (as indicated by the number of paired opposite leaves on the seedling stem4) are 3-4 (E.camaldulensis and E.grandis),3-5 (E.pauciflora),and indefinite (E.viminalis and E.deglupta). The ontogenetic trends for phyllotaxis and rooting capacity are thus not related for these five species. Perhaps there are chance associations in other eucalypts leading to the claim that presence of juvenile foliage is associated with the ability to form adventitious roots on the stem17 but there is no evidence of this kind in the present study. This paper accordingly, emphasises rooting capacity and related G contents.

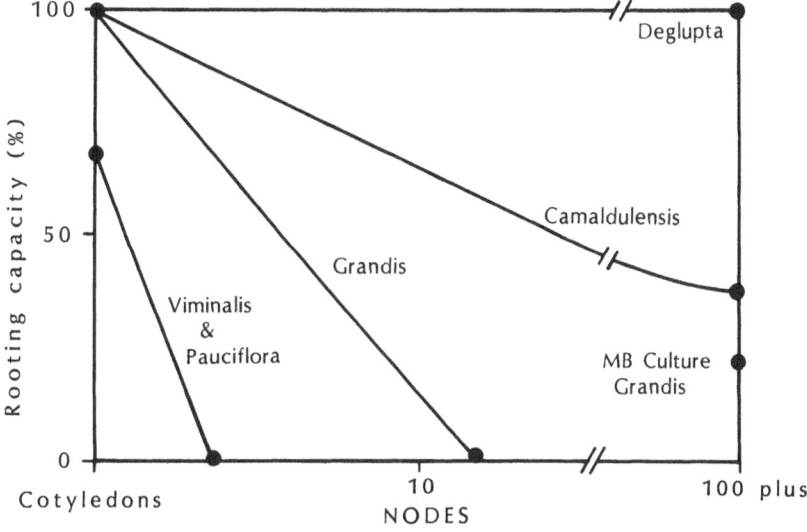

Fig.1.Root-shoot gradients for rooting capacity in five eucalypts. Species as indicated with their ontogenetic decrease in rooting capacity related to the node number of the base of the cutting. Data points represent maximum and minimum values for rooting capacity which apart from E.camaldulensis, are largely independent of the genetic variation affecting intermediate values. The point for MB culture of E.grandis (Fig.2) involves rejuvenation since the adult donor material had zero rooting capacity.

The trend line for seedling points in Fig.2, indicates that rooting capacity at successive nodes above the cotyledons is inversely related to G content of the leaves at those nodes. Both points for MB culture of adult nodal explants agree remarkably well with this ontogenetic trend. The point for adult scions on seedling stocks is slightly displaced from the ontogenetic line but perhaps other factors including graft incompatibility may have an effect. Bark thickness and growth of basal epicormics may be amongst factors involved in the displacement of the point for stump epicormics in Fig.2. Despite these unknown factors and the limitations noted in Materials and Methods, the trends in Fig.2 are clear enough for the available evidence to suggest that low G content is invariably associated with some rooting capacity, irrespective of the ontogenetic history of the rooted material.

Rejuvenation clearly occurred during MB culture of adult nodal explants when some roots were spontaneously produced. In this case as well as with growth of the adult scions, juvenility for rooting capacity has been regained. In contrast, juvenility has been retained by the stump basal epicormics.

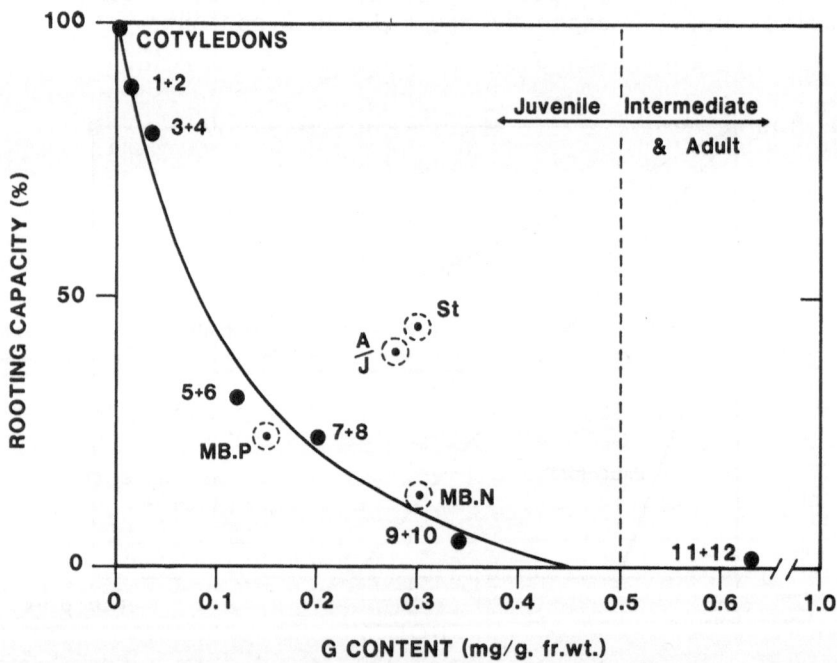

Fig.2.Relationships between rooting capacity and G content in E.grandis. Ontogenetic changes during seedling development shown by solid circles for node-leaf numbers (firm data derived from Fig.1 [22] for rooting capacity, and Fig.3 [19] for G content). Other points shown by dotted circles (preliminary data from Table 1 [23]), refer to basal stump epicormics (St), basal epicormics from adult scions on juvenile stocks (A/J), and MB cultures with donor plants grown in nursery (MB N) or phytotron (MB.P)

The trends in Fig.2 suggest that rooting is absent or
rare above a G content of 0.5 mg/g fr.wt. As the values for
both node and G content associated with 50% rooting are much
more variable than those for nil rooting, the G content of
0.5mg/g fr.wt. is accordingly taken as a convenient but
arbitrary separation between juvenile tissue (low G,high
rooting) and intermediate-adult tissue (high G,low rooting).
Genetic control of juvenility in E.grandis is implied by this
variation observed when about 50% rooting capacity is lost
during seedling ontogeny. Comparable variation occurs when
juvenility is regained (MB culture and grafts) or retained
(stump epicormics) The variation in adult scions grafted on
seedling stocks is especially relevant in this context as a
small trial using clonal adult scions on half-sib seedling
stocks, indicates that both stock and scion genotypes affect
rooting capacity and G content. In all cases, however, rooting
of scion basal epicormics appears closely associated with G
contents in the basal leaves of less than 0.5 mg/g fr.wt.

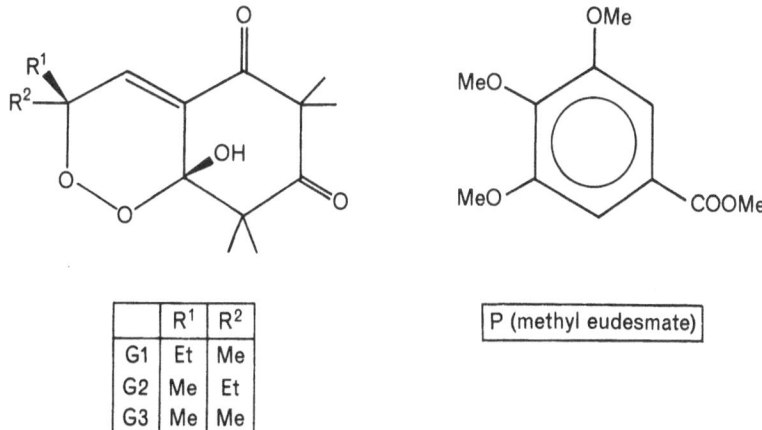

	R¹	R²
G1	Et	Me
G2	Me	Et
G3	Me	Me

Fig.3.Chemical structures of G and P. Bioassay results for
mung-bean rooting[12], indiate that there is no marked
difference in the biological activities of G1,G2 and G3.
Neither do G and P differ much in their effective
concentrations for responses in germination, rooting and
antitranspirant bioassays[7].

Despite the constancy of this juvenility cut-off in G
content, G is not a good genetic marker for juvenility in
intraspecific crosses of E.grandis. The problem arises because
a high level of phenotypic variation in G content results from
seasonal changes related to night temperature[20] and day
temperature[19]. This phenotypic variation effectively masks
much of the genotypic variation in G content expected in
E.grandis crosses. For this reason, progenies from an
interspecific cross between E.grandis and E.pulverulenta were
used for the inheritance study.

Inheritance of G and P

In E.pulverulenta, G is replaced by P, the plant growth regulator with most bioassay activity and most leaf content[5]. E.grandis has no P. The chemical structure of P differs from that of G in ways that suggest early separation of their respective biosynthetic pathways, and thus likely independent and non-competititve inheritance of P and G.

As indicated by the data points for each of the two F1 plants in Fig 4, parental levels of G and especially P,are inherited as non-dominant factors. Non-dominance is even more marked in the F2 progency that consists mainly of low G and low P phenotypes. No F2 plants have high parental levels of both.

The most likely explanation of this unusual inheritance pattern is that combined high leaf contents of G and P are lethal. Either G or P alone, is not. This interpretation presumes that the observed seedling death at about the 10 node stage and later, occurs when the ontogenetic increase in G content interacts with P. Direct evidence for such an interaction has, however, proved very elusive. One reason is that preliminary in vitro bioassays with combinations of G and P gave inconclusive results. Another problem is that the rapid increase in G content at and above node 10, allows little time to sample for high G content before the plants die.

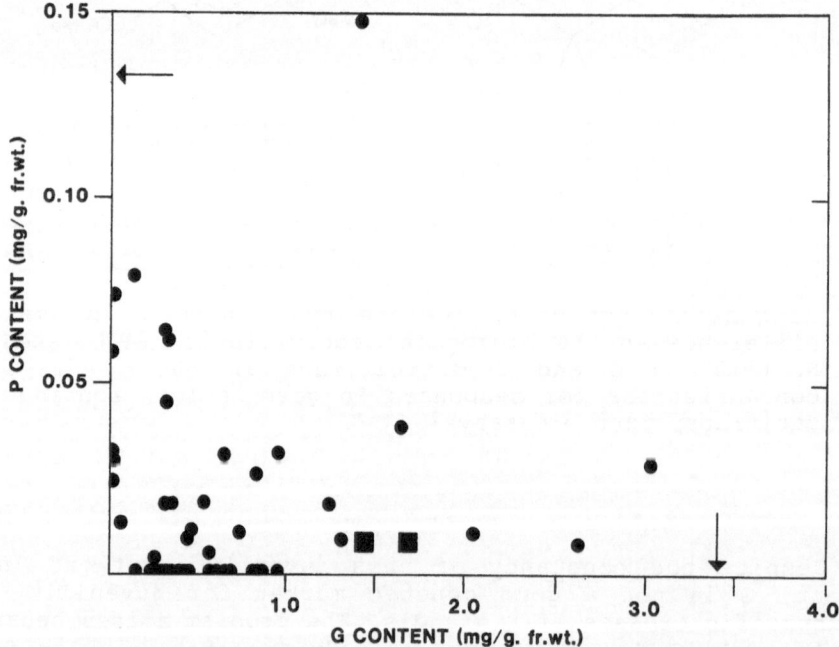

Fig.4.Relationship between G and P contents in progeny from the E.grandis (G) and E.pulverulenta (P) cross for two F1 plants(■) and numerous F2 plants (●). Parental levels of G and P indicated by arrows. All contents determined by HPLC analyses of extracts from expanded distal leaves on plants 1-2m high (F2) or 10m trees (F1).

In addition to decreased rooting capacity and increased G content during seedling ontogeny in E.grandis, the converse relationship now demonstrated between them for several cases of rejuvenation leaves little doubt that a close relationship is involved. ·The results further suggest that seedling ontogeny may be a special case of a more general root-shoot gradient in G content[23] in which callus can substitute for roots in establishing the gradient. Questions of cause and effect are notoriously difficult to answer in developmental plant physiology, but these various relationships taken together strongly suggest some causal, inverse relationship between G content and juvenile rooting capacity.

A series of checks of the rooting capacity of young basal epicormics produced successively from the same adult scions, suggests a possible optimum period of 12 months after grafting for maximum rejuvenation to occur. This is sufficient time for new functional accessory buds to develop from the scion cambium, and replace the original accessory buds that become inactive before being cut off by the cork cambium. Rejuvenation of these developing accessory buds has thus occurred under some slow-acting influence from the roots. The least complex interpretation of this root effect is that some graft transmissible factor from the roots either inhibits G synthesis or promotes G breakdown in developing accessory buds of adjacent stem tissue. In both cases there is substantial in vitro and in vivo evidence that G-Mannich bases would be involved in such synthesis or breakdown[6]. Although roots do not appear to act as a sink for G from basal stem tissue, the presence of various G breakdown products in the root needs to be carefully checked before this possibility can be excluded. Similarly, preliminary evidence that root extracts do not degrade G or affect G biosyntheses, needs confirmation.

Despite these uncertainties about the precise mechanism for rejuvenation in adult scions of E.grandis, the time requirement of about 12 months agrees generally with overall times involved in cascade grafting of stem tips[13], and MB cultures in other laboratories[2,3,9,17]. Since apical[13] and axillary[2,3,9,17] buds appear to be as effective as accessory buds for rejuvenation, the common factor is a bud meristem developing in close proximity to either roots or callus. These physiological aspects of rejuvenation combined with known chemical details of G metabolism[6], offer some real possibilities for control of G content, and thus control of juvenility in E.grandis. Such juvenility control would be very relevant for experimental treatments including genetic transformation and subsequent regeneration of transgenic plants. There is, however, at least one possible problem that could block this promising approach in eucalypts.

This is the practical difficulty experienced in this laboratory when attempts are made to maintain eucalypt tissue and especially cell cultures, at reliably high growth rates required in most techniques for genetic transformation. For example, the MB cultures in the present study did not always maintain constant growth after root formation, presumably because their juvenility status was somewhat unstable being intermediate (adolescent) between the original adult tissue and the fully rejuvenated state. Callus and cell cultures of mesophyll and hypocotyl parenchyma from E.grandis were even more short-lived under our conditions. It is possible, however, that the sucess rate could be higher elsewhere as the most recent summary of in vitro culture of eucalypts[17], lists 22 out of 38 species that produced callus cultures. Six of these formed roots and shoots from the callus but only one of these, E.marginata[3] appeared capable of root and shoot development from callus derived from adult donor material. Apart from some exceptional cases, including E.marginata, eucalypts thus appear to be a rather refractory genus as a source of adult donor material for in vitro culture.

General use of juvenile tissue in genetic transformation techniques raises other uncertainties that could have quite subtle but serious consequences. Some of these uncertainties may even provide a bonus for the biotechnologist, but the more usual case is likely to be the opposite. The direct relationship between G content and frost resistance in E.grandis[19] is a good example of the first outcome where the low G content of callus would give no indication of the potential for high G content and thus increased frost resistance in the adult tree. The attempt to increase frost resistance in E.grandis by incorporation of P from the more frost resistant E.pulverulenta, is an equally good example of the second outcome where lethal effects are expressed in adult but not in juvenile tissue. The implications arising from these ontogenetic lethal effects are, however, much wider than for E.grandis alone.

Comparable ontogenetic lethal effects are observed in classical plant breeding programs using other interspecific eucalypt crosses[25]. If the grandis x pulverulenta cross is taken as typical, then the as-yet unidentified but equally potent plant growth regulators in other eucalypts[20], could replace G and P as the executive molecules for expression of these ontogenetic lethal effects.

Although sexually produced lethal genotypes in plants are not generally regarded as being dangerous, potential dangers of new gene combinations produced asexually by modern biotechnology are of great social concern. While individual cells with new lethal or deleterious genomes are likely to be quickly excluded from regenerating callus cells used in transformation studies, the consequences can be quite different if lethal or dangerous affects are delayed until adult tissue develops. For transgenic plants, there is thus a real possibility that the juvenile phenotype is a most uncertain basis for extrapolating to the adult phenotype. Even strict quarantine of the experimental material could be quite misleading if checks were restricted to juvenile tissue but expression of a dangerous gene combination was delayed to adult tissue. The possibility of ontogenetic effects in transgenic plants is accordingly, one more area of concern in transformation studies where due caution is called for.

When viewed in this light, the need for increased understanding of plant juvenility takes on a degree of social urgency that possibly rivals the disasterous consequences arising from controversies over cereal vernalization about 50 years ago[1]. The link between vernalization and juvenility is still not clear[18,27], and in common with juvenility studies in woody plants, theoretical considerations of possible controlling mechanisms are commonly restricted to speculation. The present study has thus made a considerable advance as the concepts presented for eucalypt juvenility, and for *E.grandis* in particular, do not involve undue speculation when postulating roles for a specific molecule in ontogenetic ageing and rejuvenation. Without a firm chemical basis of this kind, both these aspects of juvenility are all too often treated as rather mystical processes where rejuvenation is the reversal of an otherwise autonomous and largely unknown process of ontogenetic ageing[23].

REFERENCES

1. Ashby, E., 1947, Scientist in Russia. Penguin Books, UK.

2. Barker, P.K., de Fossard, R.A. and Bourne, R.A. ,1978, Progress towards clonal propagation of *Eucalyptus* species by tissue culture techniques. *Proc. Int. Pl. Prop. Soc.* *27*, 546-56

3. Bennett, I.J. and McComb, J.A.,1982, Propagation of Jarrah (*Eucalyptus marginata*) by organ and tissue culture. *Aust. For. Res.*, *12*, 121-7.

4. Blakely, W.F., 1965, A key to the eucalypts. C'wealth Govt. Printer, Canberra, Third edition.

5. Bolte, M.L., Bowers, J., Crow, W.D., Paton, D.M., Sakurai, A. and Takahashi, N.,1984, Germination inhibitor from *Eucalyptus pulverulenta*. *Agric. Biol. Chem*. *48*, 373-6.

6. Bolte, M.L., Crow, W.D. and Paton, D.M, 1987, Frost hardiness in *Eucalyptus grandis*: a possible molecular mechanism. Pl. Biol., V, 129-39.

7. Bowers, J., 1983, The growth regulator in *Eucalyptus pulverulenta* (P1) and inheritance of P1 & G in crosses with *E.grandis* (G). Honours thesis, Botany Dept, Aust.Nat.University.

8. Cremer, K.W., 1972, Morphology and development of the primary and accessory buds of *Eucalyptus regnans*. *Aust. J. Bot. 20*, 175-95.

9. Cresswell, R. and Nitsch, C., 1975, Organ culture of *Eucalyptus grandis*. *Planta 125*, 87-90.

10. Davidson, J., 1974, Reproduction of *Eucalyptus deglupta* by cuttings. *N.Z.J. for. Sc. 4*,191-203.

11. De Fossard, R.A., Nitsch, C., Cresswell, R. and Lee, E.C.M., 1974, Tissue and organ culture of *Eucalyptus*. *N.Z. J. For. Sc. 4*, 276-78.

12. Dhawan, A.K., Paton, D.M. and Willing, R.R. ,1979, Occurrence and bioassay responses of G: a plant growth regulator in _Eucalyptus_ and other Myrtaceae. _Planta_ 146, 419-22.

13. Francelet, A. and Boulay, M., 1983, Micropropagation of frost resistant eucalypt clones. _Aust. For. Res._ 13, 83-9.

14. Hartney, V.J., 1981, Vegetative propagation of the eucalypts. _Aust. For. Res._ 10, 191-211.

15. Holden, P.G. and Paton, D.M.,1981, Sterilization of field-grown _Eucalyptus_ for organ culture. _Aust. Inst. Hort._ 3 (10): 5-7.

16. Libby, W.J., 1989, Comment at this workshop.

17. McComb, J.A. and Bennett, I.J., 1986, Eucalypts _in_ Biotechnology in Agriculture and Forestry, Vol.1, 340-62, (ed Y.P.S. Bajaj), Springer-Verlag, Berlin.

18. Paton, D.M., 1969, Vernalization, photoperiodic induction and flower initiation on the late pea cultivar, Greenfeast. _Aust. J. biol. Sc,_ 21, 609-17.

19. Paton, D.M., 1981, _Eucalyptus_ physiology. III. Frost resistance. _Aust.J. Bot._ 29, 675-88.

20. Paton, D.M., 1987, Frost resistance in _Eucalyptus_: are plant growth regulators involved? _Pl. Biol., V.,_ 117-27.

21. Paton, D.M., Dhawan. A.K. and Willing R.R.,1980, Effect of _Eucalyptus_ growth regulators on the water loss from plant leaves. _Plant Physiol._ 66, 254-6.

22. Paton, D.M., Willing, R.R. Nicholls, W. and Pryor, L.D.,1970, Rooting of stem cuttings of _Eucalyptus_: a rooting inhibitor in adult tissue. _Aust J. Bot._ 18, 175-183

23. Paton, D.M., Willing R.R. and Pryor, L.D.,1981, Root-shoot gradients in _Eucalyptus_ ontogeny. _Ann. Bot._ (_London_) 47, 835-8.

24. Pryor, L.D.,1976, Biology of eucalyptys. Edward Arnold, London.

25. Pryor, L.D. and Johnson, L.A.S., 1971, A classification of the eucalypts. _Aust. Nat. Univ._, Canberra.

26. Pryor, L.D. and Willing, R.R., 1963, The vegetative propagation of _Eucalyptus_ - an account of progress. _Aust. For._ 27, 52-62.

27. Reid, J.B., 1981, Flowering in _Pisum_: Evidence that vernalization does not influence the length of evocation. _Aust.J. Plant Physiol._ 8, 319-27.

SOMATIC EMBRYOGENESIS

VARIATIONS IN SOMATIC POLYEMBRYOGENESIS: INDUCTION OF ADVENTITIOUS EMBRYONAL-SUSPENSOR MASSES ON DEVELOPING DOUGLAS FIR EMBRYOS

L. Hong, M. Boulay[1], P.K. Gupta[2] and D.J. Durzan[3]

Department of Environmental Horticulture
University of California
Davis, CA 95616 USA

ABSTRACT

Embryonal-suspensor masses (ESMs) from seven commercial genotypes of Douglas fir are distinguished from embryogenic calluses by origin, cytochemical staining and developmental sequence. Zygotic ESMs with a dominant embryo can be rescued and new embryos multiplied and reconstituted in darkness by cleavage and budding polyembryony without a callus phase. New ESMs can also be induced directly on the surface of developing embryos. By contrast, the recovery of embryos derived from an intermediary callus phase follows a spatio-temporal wave of activity from the radicle towards the cotyledons starting 4 to 8 weeks after fertilization. In all embryo recovery systems, callus formation and teratogenic structures may be avoided by careful subculture, minimal exposure to low levels of plant hormones, and by selection of true-to-type ESMs grown in darkness.

Callusing of protodermal cells is promoted in light. The dedifferentiated cells can be induced to redifferentiate back into embryogenic ESMs in darkness. Callus-based regenerative processes can be followed microscopically by the increased formation of acetocarmine-reactive factors produced by nuclei and amyloplasts of redifferentiating cells. These factors are absent in nonembryogenic calluses. The alternative origins of ESMs are important to distinguish because the mechanisms by which proembryos are reconstituted and multiplied may contribute to different genotypic and phenotypic variances in regenerated products.

INTRODUCTION

Our study reveals different origins for embryonal-suspensor masses (ESMs) of Douglas fir from fertilization to the ripening of the seed. The aim is to distinguish the origin and role of an embryogenic callus in the process of somatic embryogenesis and

---

[1] Current Address: Moët-Hennessy Research, 92704 Colombes, Cedex, France.

[2] Current Address: Weyerhaeuser Technical Centre, Tacoma, Washington.

[3] To whom correspondence should be addressed.

Woody Plant Biotechnology, Edited by M.R. Ahuja
Plenum Press, New York, 1991

polyembryogenesis and to show that callus is not needed for the reconstitution and multiplication of proembryos. Moreover, the development of new ESMs, which can be induced and redifferentiated from callus, can now be followed by acetocarmine staining to show that proembryos multiply by the cleavage and budding process.[8] Here, multiplication occurs by cleavage and budding polyembryony and is not uniquely based on the derivation of one embryo from one callus cell. Dedifferentiation, induction and redifferentiation are the central processes in the recovery of multiple embryros from a callus. Observations related to the establishment of ESMs in cell suspension cultures are described in greater detail.

This study reaffirms in greater detail the earlier observations that Douglas fir, which does not normally exhibit cleavage and budding polyembryogenesis in nature,[1,4] does so *in vitro*. High levels of abscisic acid are an important inhibitory factor for the recovery of multiple embryos by polyembryogenesis.[8,2] Embryos of Douglas fir can now also be recovered from protoplasts, with the transient expression of the *luc* gene introduced by electroporation in protoplasts of the ESM.[11] Taken together, the experimental developmental biology of conifer embryogenesis now enters a new era of understanding and should assist greatly in our efforts to manipulate these processes for the scaled-up recovery of true-to-type embryos.[7]

MATERIALS AND METHODS

Two to five female cones from six genotypes (567, 568, 569 and 570 in 1988 and 171 and 168 in 1987) of Douglas fir (*Pseudotsuga menziesii*) were obtained from Weyerhaeuser (Tacoma, WA) on a weekly basis between June 6 and August 9. Cones were transported in a cold box and stored for not more than one week at 4°C until ESMs and zygotic embryos could be rescued.

Fig. 1. Development stages (1988) for female cones of Douglas fir (genotype series 500 Weyerhaeuser) used to recover embryonal-suspensor masses and the zygotic embryo. Dates represent the day of collection. ESM and embryo rescue would occur one week later. The largest cone is approximately 10 cm high.

Table 1. A basal nutrient medium (DMH) for the culture of rescued embryonal-suspensor masses of Douglas fir. Composition is formulated in mg/l and adjusted to pH 5.8 before autoclaving.

Nitrates		Fe·EDTA	
NH_4NO_3	220	$FeSO_4 \cdot 7H_2O$	13.96
$Ca(NO_3)_2 \cdot 4H_2O$	278	$Na_2EDTA \cdot 2H_2O$	18.62
KNO_3	170		
Sulfates		**Vitamins**	
		Thiamine·HCl	1.00
$MgSO_4 \cdot 7H_2O$	185	Nicotinic acid	0.5
$MnSO_4 \cdot H_2O$	11.2	Pyridoxine·HCl	0.5
$ZnSO_4 \cdot 7H_2O$	4.3		
$CuSO_4 \cdot 5H_2O$	0.013	**Carbon and Nitrogen Sources**	
Halides		Sucrose	30,000
$CaCl_2 \cdot 2H_2O$	55	myo-Inositol	1,000
KI	0.41	Casein hydrolysate	500
$CoCl_2 \cdot 6H_2O$	0.012	L-Glutamine	450
$NiCl_2$	0.012	Glycine	2
Phosphate, Borate, Molybdate			
KH_2PO_4	85		
H_3BO_3	3.1		
Na_2MoO_4	0.12		

Cones were sprayed with alcohol and flamed for surface sterilization. Bracts were peeled back to expose the female gametophytes bearing the embryos for excision and placement on agar plates. Agar plates were formulated with a DMH medium (Table 1) for growth in darkness at $23 \pm 1°C$. This rescue method showed little contamination and had the advantage that tissues were not exposed to strong surface sterilants before being placed on the DMH medium.

Recovery of Embryonal-Suspensor Masses (ESMs)

For each genotype, from 40 (June sampling dates) to 150 (July and August) developing seeds were recovered from each cone (Fig. 1). Twenty rescued ESMs with or without their female gametophytes were rescued on each plate of DMH medium. No fewer than five replicates per genotype were used in this study.

The emergence of proliferating ESMs was monitored visually during a week-long subculture cycle. When clear, mucilaginous ESMs emerged, they were removed and subcultured in darkness, always independently of the female gametophyte.

The effect of four levels of 2,4-D (5, 10, 25 and 50 μM) on ESM growth was followed for no fewer than 20 ovules per plate with four replicates. ESMs selected for this study originated together with callus directly on the hypocotyl between July 9 and July 30. Callus (white and nonmucilaginous), when produced, turned brown in darkness. When subcultured to weak light (< 2.0 μW cm^2/nm in blue, red and far-red), a green callus with a rapidly decreasing organogenetic and embryogenetic potential could be established.

Microscopy and Cytology

A few μl sample of ESM derived from hypocotyls was removed from proliferating masses, placed on a glass slide and double-stained as described by Gupta and Durzan (1987) using a Zeiss Inverted Photomicroscope. Preparations were examined under bright field, polarized light and UV light for fluorescence microscopy (exc .365 nm, em > 418 nm).

Callus, characterized by poor double staining, slight browning on DMH media, and by unorganized daughter cell growth patterns, was isolated and cultured separately for microscopic examination. Teratogenic structures, derived from the overproliferation of proembryonal cells at high 2,4-D levels, could be detected by acetocarmine staining and easily distinguished from callus in terms of the proembryonal cell not being able to form discrete individual proembryos during the cleavage process.

Cell Suspension Culture

ESMs cultured in darkness on agar plates were transferred to 1 liter nipple flasks (1 rev. min^{-1}) or to 250 ml Erlenmeyer flasks (100 rev min^{-1}) with fluted bottoms, as described by Durzan and Gupta (1987), and subcultured every seven days in darkness. Inoculation density was 8 to 10 ml packed cell volume per 100 ml medium.

Experimental Design

General. Experiments were based on how ESMs were rescued and calluses (embryogenic and nonembryogenic) induced from developing seeds. Two designs were based on the attachment to or separation from the ESM and the female gametophyte. The first set was based on the rescue of ESMs independently of the female gametophyte with genotypes collected in 1988 (500 series on a modified DMH medium with 5 μM 2,4-D, and 2 μM each of BA and Kn) by one of us (LH). The second set was performed (by MB and PKG) on genotypes collected in 1987 (168 and 171). ESMs and embryos remaining in partial contact with the female gametophyte were collected in June and early July. Developing, cracked gametophytes with the exposed ESMs or zygotic embryos were placed on a DMH basal medium with 50 μM 2,4-D, 20 μM each of BA and Kn. Once ESMs proliferated, the ESMs were rescued and subcultured on DMH with 5 μM 2,4-D and 2 μM each of BA and Kn. Cell suspension cultures were established in nipple and Erlenmeyer flasks. In some cases, filter paper bridges were also used.

Specific Aspects. For set one, the effect of cone harvest date and ESM-exposure levels to 2,4-D on agar plates were examined in terms of polyembryogenesis and callus induction. Adventitious buds were also induced on rescued zygotic embryos. For set two, the effect of variety of liquid media were examined for the rescue of the zygotic-derived ESMs during the month of June. Media tested were described by Durzan and Gupta,[8] Litvay et al.,[16] and Hakman and Von Arnold.[12]

OBSERVATIONS

Embryo and ESM Rescue Without Gametophytic Interaction

Importance of Collection Date and Developmental Stage. In earlier work, the zygotic ESMs were rescued by coculture, with their corresponding female gametophytes.[8] In a separate study by one of us (LH), ESMs were dissected and separated physically from the female gametophyte and placed on DMH medium with 5 μM 2,4-D and 2 μM each BA and Kn in 0.7% Bacto (Difco) agar. Under these conditions and between June 7 and June 14, surviving ESMs could not be rescued without attachment to their female gametophyte. However, after mid-June, proliferating ESMs were induced on the surface of embryos and recovered within 40 days with increasing ease until July 21 (Fig. 2). Zygotic embryos rescued later tended to form a callus (Fig. 3). Callusing increased till August 12. The calluses, so recovered, were embryogenic, i.e. they could revert to ESMs, provided that darkness and a 7 to 10 day subculture rate was maintained at genotype-suitable levels of 2,4-D (Fig. 4).

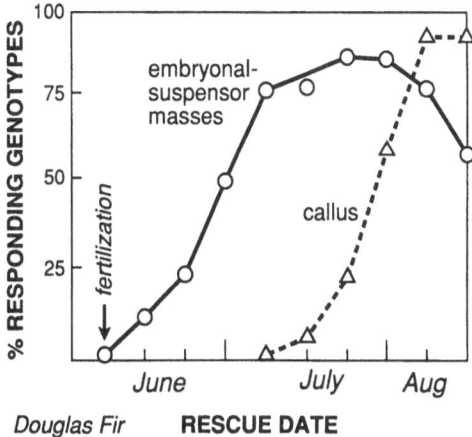

Fig. 2. Percentage of all genotypes responding (all or none) by producing ESMs and/or callus over all levels of 2,4-D (5 to 50 ppm) as a function of rescue date.

Effect of 2,4-D on ESM and Callus Formation. The time-course exposure to four levels of 2,4-D revealed that all genotypes responded poorly to 50 μM in terms of ESM recovery after 40 days. The tendency was for callus to form, especially after mid-July. The recovery of ESMs, free of their female gametophyte, was best at 10 μM, especially for genotypes 566 and 567. At 10 μM, calluses were recovered after 40 days only in late July and August when cotyledons were evident. Although cotyledon-derived ESMs were recovered at the later dates, the recoveries were genotype dependent, with 569 and 570 superior to 566, 567 and 568.

Mucilaginous ESMs, recovered at all levels of 2,4-D, were easily distinguished from the white callus, which upon subculture turned brown. Callus obtained from all levels of 2,4-D, when subcultured in weak light, turned green and formed compact cell masses.

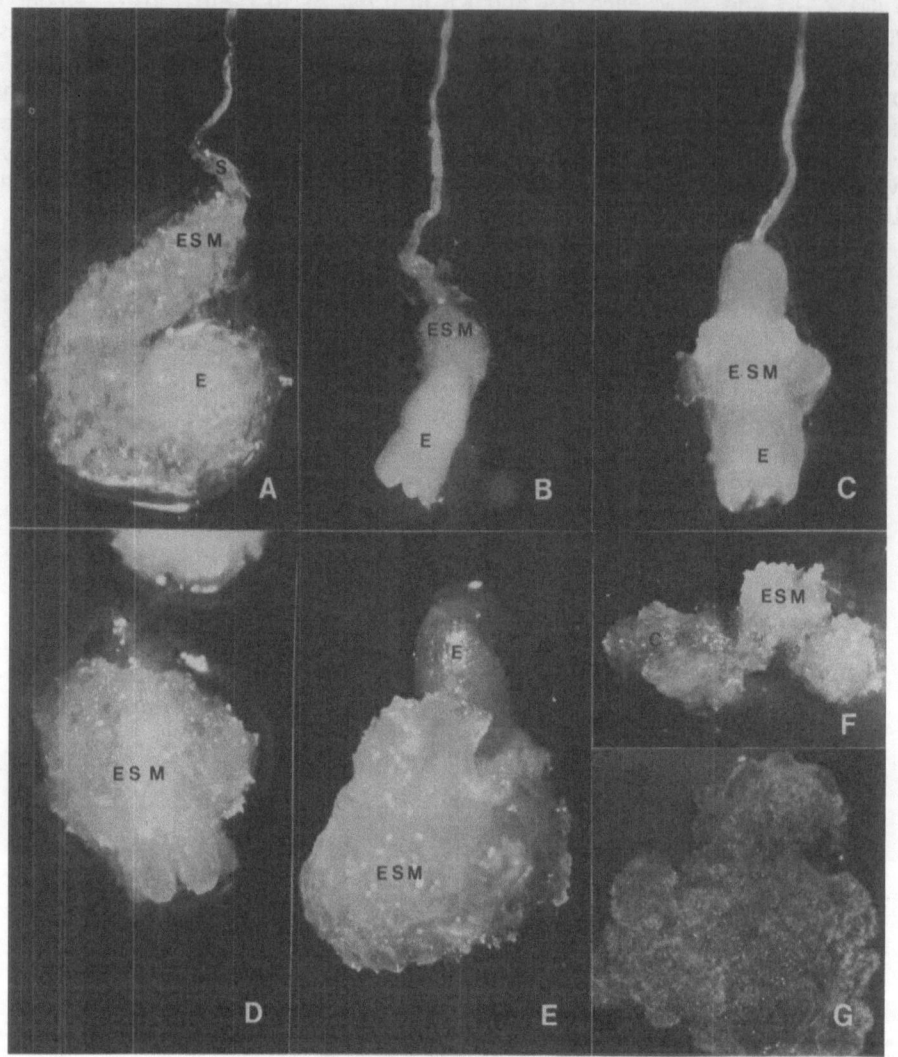

Fig. 3. Induced development of an ESM directly on isolated zygotic embryos not attached to the female gametophyte and with or without a callus phase. Zygotic embryos and the remaining ESM were rescued and cultured on the following date.

A. ESM develops from the entire rescued proembryo in late June; ESM proliferation is most prominent along the zygotic ESM that now occupies the suspensor (**S**) (Genotype 171).

B. By the end of June to July 19, the initial ESM is still recoverable after 4 to 5 days at the radicle of the embryo (**E**) (Genotype 567).

C. In July the ESM on the same medium is now found after 3 days above the radicle and predominantly in the hypocotyl (Genotype 570).

D. By August 9 the ESM continues to be derived from the hypocotyl after 15 days.

E. By August the ESM is recovered mainly from cotyledons.

F. Residual callus (**C**) and ESM from the surface of a rescued Douglas-fir embryo rescued in early August.

G. Callus initiated in mid-August tends to brown (in darkness) and green (in light). This process can be overcome by selection and induction of a new ESM from an embryogenic callus that represents a transitional state (Fig. 7).

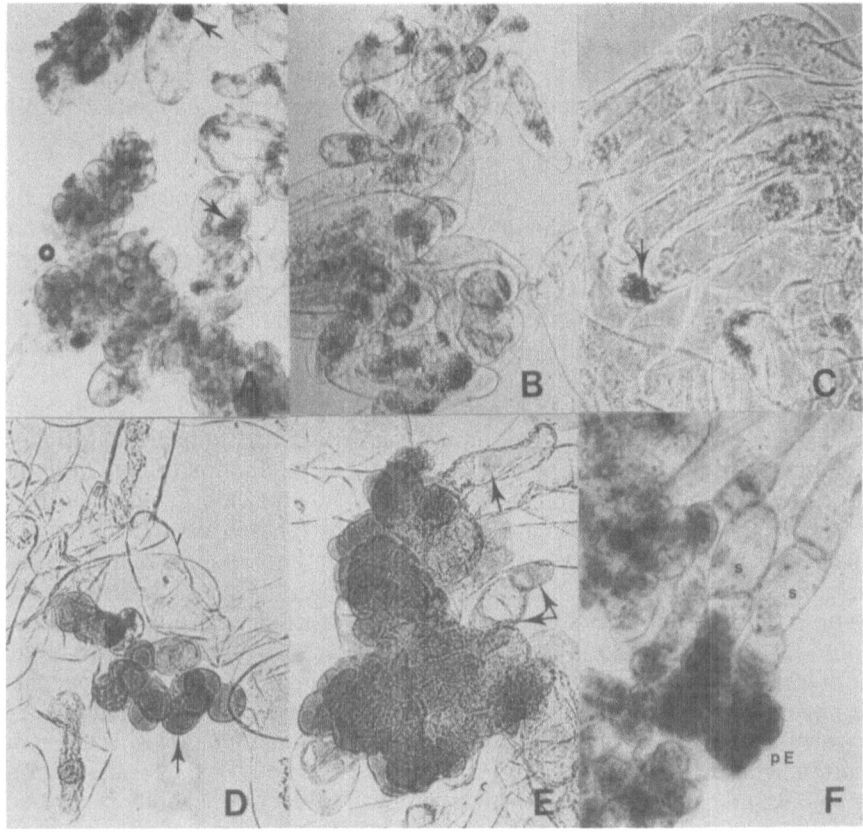

Fig. 4. Cytochemical staining characteristics of callus cells as they are transformed into proembryonal cells. Callus originates from rescued embryos not remaining attached to the female gametophyte.

A. In callus (**C**) from embryos collected on August 9 (*cf.* Fig. 1) browning occurs in residual cells below calls with the loss of acetocarmine reactivity. Note swelling of amyloplasts and their clustering around nuclei (arrows) as cells turn increasingly brown.

B. Acetocarmine reactivity of cells after 5 to 6 days and start of elongation and clustering of amyloplasts during induction of ESM (Genotype 570, collected July 15).

C. Progressive elongation of protodermal cells during induction of ESM after 7 days in culture. Arrow points to acetocarmine reactive protoplasm that establishes polarity for the formation of a proembryonal cell.

D. Acetocarmine-reactive proembryonal cells (arrow). Cluster of proembryonal cells from Genotype 570, collected July 15.

E. Increased acetocarmine staining in induced proembryonal cells in Genotype 568, collected on July 15. The group of proembryonal cells is poorly organized, somewhat teratogenic, and typical of transitional stages found in long term cultured embryogenic calluses. Arrow shows budding polyembryony.

F. The formation of proembryonal cells and suspensors on the protodermal surface of hypocotyls (Genotype 570, collected July 15).

The mean percentages of rescued zygotic embryos (July 14) initiating ESMs after 40 days at 5 μM 2,4-D and 2 μM each of BA and Kn were 23, 32, 17 and 22% for genotypes 567 to 570, respectively. Variances did not exceed 21% of the mean.

Rescued zygotic embryos or ESMs wounded during their excision were considered separately from those remaining intact. The wounded embryos and ESMs for all genotypes consistently initiated more ESMs (25 vs. 12%) than intact transplanted tissues.

Recrudescence of ESMs from Callus. The isolation and subculture of callus lines from rescued ESMs led to a gradual browning of all callus cell masses. This browning occurred at 10 to 50 μM 2,4-D, i.e. levels at which ESMs were maintained as viable, white and mucilaginous in darkness. In both sets of experiments, calluses deteriorated, and new white and mucilaginous ESMs emerged spontaneously from callus cells after four to five subcultures for all genotypes (Fig. 4).

Cytochemical Changes During Induction of ESMs from Callus. Callus and ESMs formed simultaneously from rescued embryos independent of attachment to the female gametophyte for all genotypes by mid to late July. Examination of the development of callus from the hypocotyl region showed in early July that most cells contained starch and "weak" acetocarmine staining of the cytoplasm. These cells were atypical of cells in the zygotic and induced ESMs because of their intermediate staining properties. Acetocarmine reactivity was found around amyloplasts (Fig. 4). It is these cells that were precursors for the ESMs formed by recrudescence later in July. The detection of various stages in this process coincides with the need to dedifferentiate, induce and redifferentiate cells of the zygotic embryo for the formation of an "embryogenic callus." An embryogenic callus, when obtained, combines unorganized and organized daughter-cell growth patterns. Callus varies in its structural presentation, being friable to nodular and often light brown in darkness (green in light). While cells may contain starch, very little cytoplasmic acetocarmine reactivity or Evan's blue permeability is seen. Growth rates, while rapid under the 7 to 10 day subculture routine, are not as great as observed with ESMs. Rates comprise a wide variety of daughter cell shapes and growth patterns.

Embryo and ESM Rescue With Gametophytic Interaction

Collection dates for genotypes 168 and 171 were June 28, July 6, 14 and 22. Culture of the developing ESM-gametophyte complex yielded a proliferating mucilaginous ESM (Fig. 5). After six subcultures of the ESMs on the same medium, they were examined by double staining with acetocarmine and Evan's blue. ESMs containing only proembryonal and suspensor cells were inoculated into liquid media. For both genotypes, the best ESMs were recovered from seeds collected on July 22. Callus can also be obtained from the embryo and gametophytic interactions. However, after four subcultures on 5 μM 2,4-D and 2 μM 2,4-D each BA and Kn, calluses turned brown.

Different liquid media were used to culture ESMs for at least four weekly transfers at an inoculation density of ca 1 gm fresh wt per 100 ml medium. Transfer of cells from a BM$_3$ medium[10] to a loblolly pine medium[16] led to ESM deterioration. Cells recovered when subcultured back into DCR$_3$.[8] Transfer of cells to a DCR$_3$ or to a Hakman and Von Arnold basal medium[12] (with 10 μM 2,4-D and 2 μM BA) also gave good cultures and maintained white ESMs.

Polyembryogenesis in Suspension Culture

When ESMs are established as cell suspension cultures, the origins and behavior of the polyembryonic phenomena become more accessible for study (Fig. 6). Study is facilitated by the cytochemical staining of cells in suspension culture from the start to

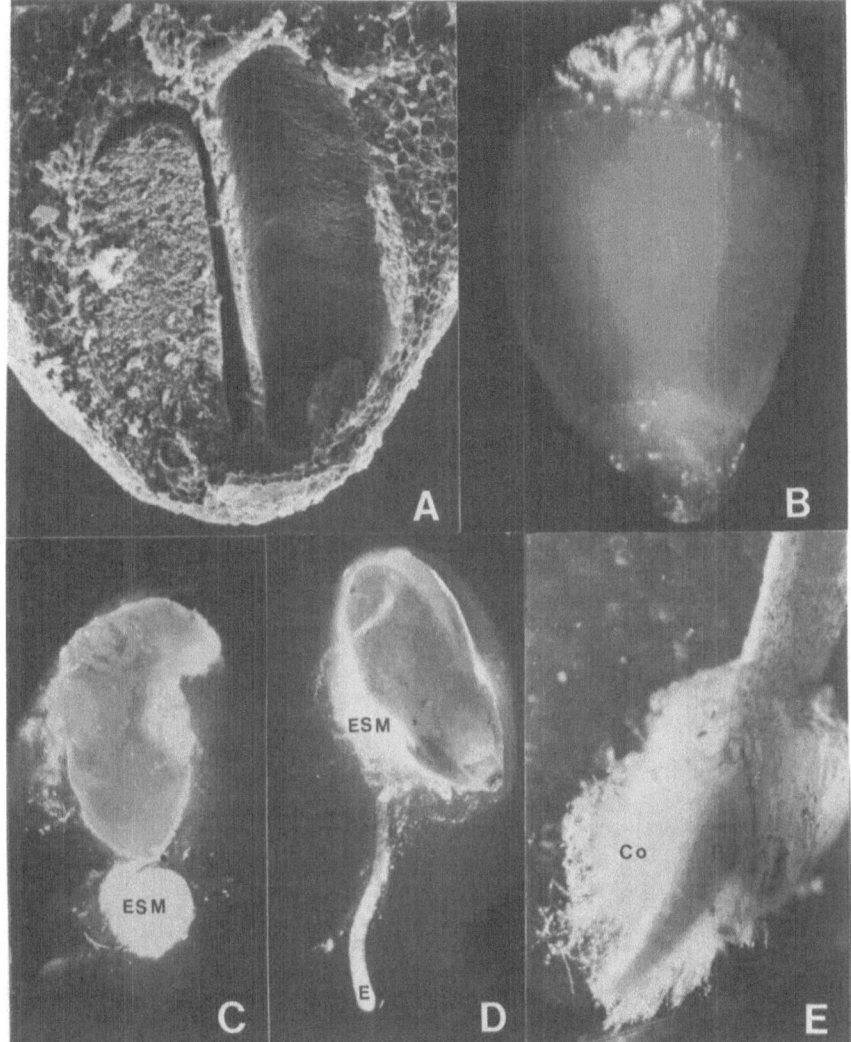

Fig. 5. Rescue of the ESM with the attached female gametophyte.

A. Longitudinal cut section showing one of two archegonia in a bract taken from cone. The multiplication of archegonia represents simple polyembryony. Female gametophytes bearing ESMs (archegonia) are removed and placed on agar media. × 850.

B. Isolated female gametophyte with an enclosed ESM at bottom from cones collected on June 7. Length ca 4 mm.

C. Emergence of ESM (Genotype 168) from cones harvested on June 22.

D. Zygotic embryo (**E**) (Genotype 168) from cones harvested on June 22.

In C and D the ESM remains attached to the female gametophyte.

E. Contents of the residual female gametophyte remaining attached to the radicle end of a germinating (5 to 6 day) Douglas fir seedling (wild type genotype, Placerville, CA). The contents are largely coleorhizae (**Co**) that obscure remnants of the original ESM. × 8.

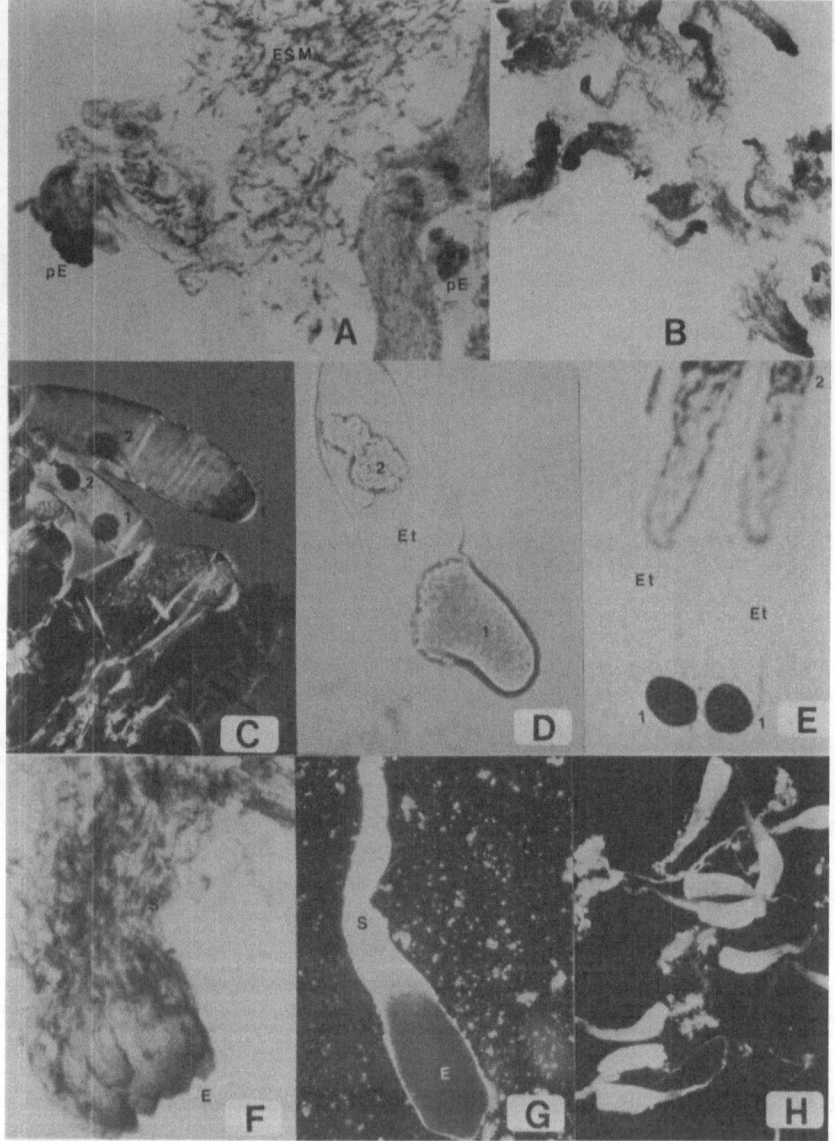

Fig. 6. Somatic polyembryogenesis in Douglas fir cell suspension cultures (*cf.* Durzan 1988 for color plates).

A. Acetocarmine-Evan's blue staining of the original zygotic suspensor (**S**) and ESM (Genotype 168) rescued on June 22. After one subculture in darkness, the elongating cells of the ESM give rise to other proembryos (**pE**). × 20.

B. Suspension culture of elongating cells of the ESM leads to the more or less synchronous production of cleaving, Douglas fir embryos (stained with acetocarmine and Evan's blue). × 15.

C. An embryonal-suspensor mass stained with orcein to reveal the free nuclear state (nuclei 1 and 2) and nuclear-cytoplasmic migration in embryonal tube cells undergoing the lobing or budding process. Arrow at lower right shows the process of necking of protoplasm in these cells. × 64.

end of each 6 to 10 day subculture. Examination of cells in suspension shows the origin of free nuclei, their migration and the establishment of cleaved "cassettes" of adhering embryos. The tendency for cleavage polyembryony is inhibited by the addition of abscisic acid (3-4 μM) to the culture medium at the proembryo stage.

Formation of Adventitious Buds

Nearly half of the zygotic embryos collected and cultured in light after July 28 produced adventitious shoot buds after 40 days on a medium with 5 μM 2,4-D and 0.5 μM each of BA and Kn (Fig. 7). In darkness, adventitious buds were rare. The tendency was the continued proliferation of callus.

DISCUSSION

In the recovery of somatic embryos from proliferating ESMs, several origins are possible. Apart from the rescue of fraternal embryos and ESMs (simple polyembryony) not examined in this study, other origins include 1) proliferation of the original zygotic ESM, 2) ESM formation on the zygotic embryo (direct formation) and 3) induction via an embryogenic callus (indirect by dedifferentiation, induction and redifferentiation). The latter two appear to originate from protodermal cells on the surface of the zygotic embryo.

In practice, the successful recovery of ESMs is a function of genotypes, isolation and culture technology, and the stabilizing and directional selection methods imposed by the investigator. Differences in origins and recovery systems will help to account for genotypic and phenotypic variances among the recovered embryos (Table 1). Figure 8 summarizes the variations in regeneration that can be obtained from ESMs and protodermal cells of the zygotic embryo. The contribution of teratogenic products of the ESM is not indicated. These aberrant structures, which occur only at high levels of 2,4-D, are not necessarily considered as "callus." The teratogenic behavior contributes further to the diversity of cell types found in cultured cell masses derived from the ESM or from callus with proembryogenic cells.

D. Lobing embryonal tube (**Et**) with migrating pE (**1**) and pU (**2**) nuclei covered with neocytoplasm reacted with calcofluor. $\times$ 64.

E. Two embryonal tube cells (**Et**) stained with acetocarmine and Evan's blue to show large reactive nucleus (**1**) with a thin layer of neocytoplasm and a second nucleus (**2**) at the opposite pole associated with an Evan's blue permeable and collapsed protoplasm. $\times$ 88.

F. Cleavage polyembryogenesis in cell suspension culture. Proembryonal cells have cleaved longitudinally to form a mass of 12 or more embryos (**E**) with their supporting suspensor system (**S**). The differential elongation of cells leads to an asynchrony in the process, i.e. not all cleavage products are at the same stage of development. $\times$ 56.

G. Cleaved embryos shown in Figure G can be grown separately under the influence of abscisic acid. Even so, abscisic acid is not always required, as indicated by this single embryo (**E**) with its suspensor (**S**) stained with acetocarmine. $\times$ 64.

H. Numerous somatic embryos after one month in cell suspension culture at the stage of cotyledon formation. All embryos have suspensors, some with remnants of the embryonal-suspensor mass remaining attached. $\times$ 25.

All seven genotypes yielded true-to-type ESMs and callus. The optimal yield and recovery time for ESMs (Fig. 2) could be related to a "wave of potential" starting from the zygotic ESM at the base of the proembryo and ending with induced ESMs found on cotyledons of the ripe embryo (Fig. 3). The propensity for the formation of ESM or callus was genotype-dependent over the 2-year study. ESMs were more easily recovered from seeds with female gametophyte remaining attached to the zygotic ESM. Embryogenic calluses with true-to-type and teratogenic ESMs were more commonly

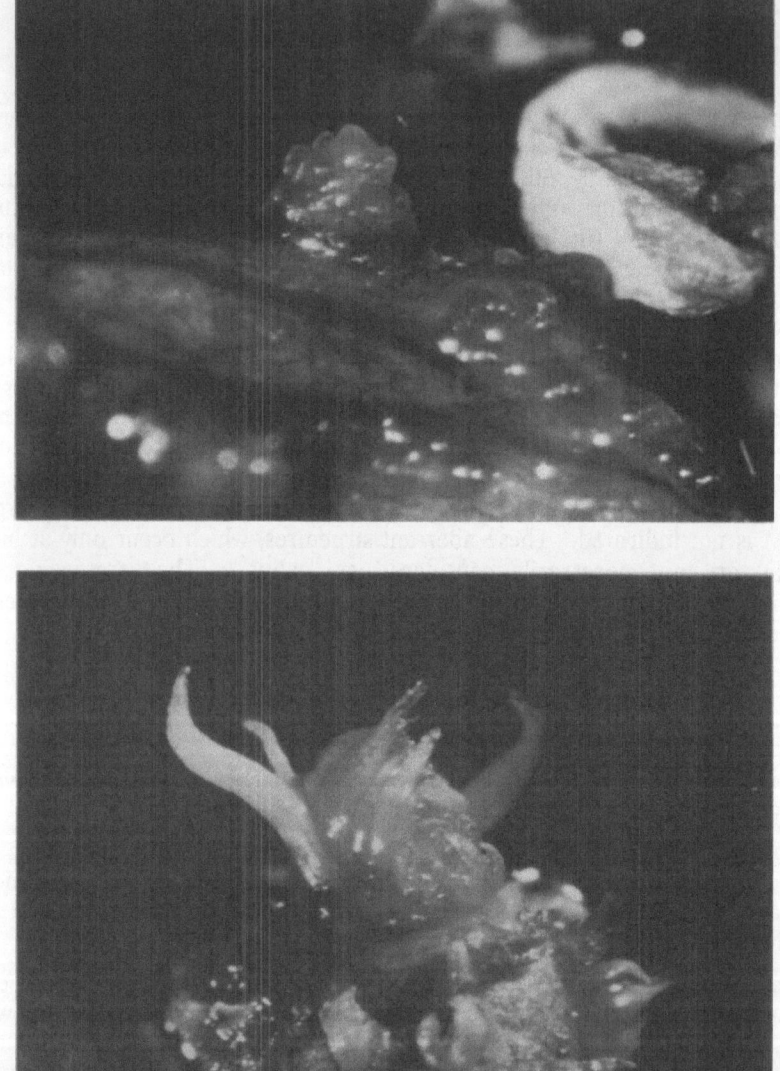

Fig. 7. The formation of adventitious buds on zygotic embryos rescued in early August (genotype 568).

A. Early stage (2 mm high) of bud formation on cotyledons after 40 days.

B. Later stage (4 to 6 mm high) after 47 days. All cultures were in weak diffuse light at 23°C on DMH medium (Table 1).

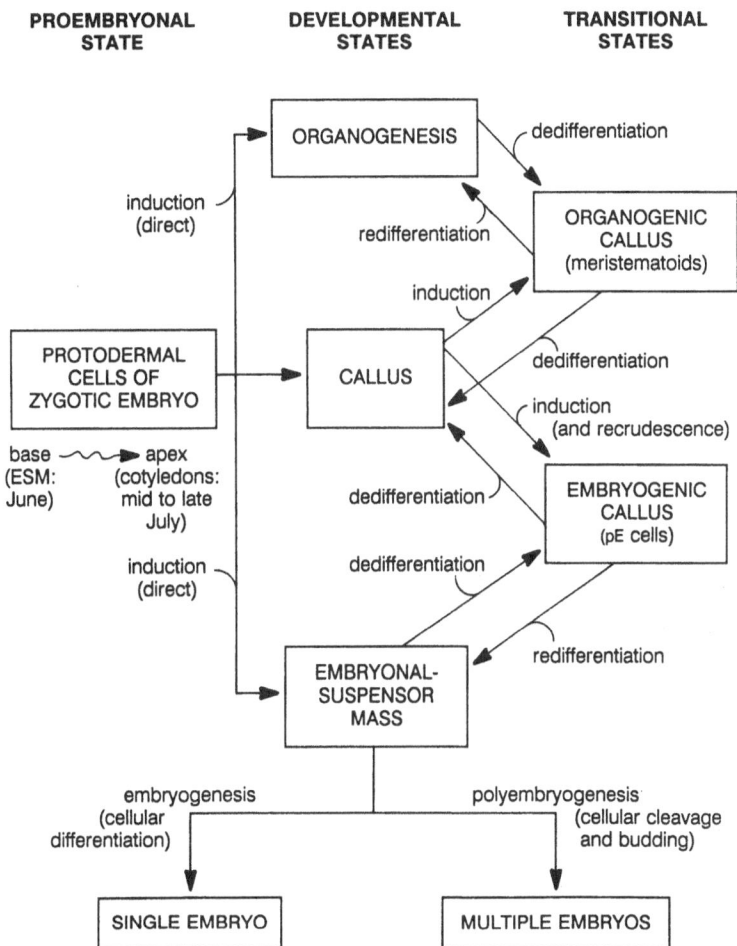

PROEMBRYONAL STATE   **DEVELOPMENTAL STATES**   **TRANSITIONAL STATES**

ORGANOGENESIS

dedifferentiation

induction (direct)

ORGANOGENIC CALLUS (meristematoids)

redifferentiation

induction

PROTODERMAL CELLS OF ZYGOTIC EMBRYO

CALLUS

dedifferentiation

base ～～▶ apex
(ESM: (cotyledons:
June)   mid to late
             July)

induction (and recrudescence)

EMBRYOGENIC CALLUS (pE cells)

dedifferentiation

dedifferentiation

induction (direct)

EMBRYONAL-SUSPENSOR MASS

redifferentiation

embryogenesis (cellular differentiation)

polyembryogenesis (cellular cleavage and budding)

SINGLE EMBRYO   MULTIPLE EMBRYOS

Fig. 8. Variations in regeneration by somatic embryogenesis and polyembryogenesis from protodermal cells. The reconstitution of proembryos from an embryonal-suspensor mass (ESM) may occur by somatic polyembryogenesis using transplanted ESM several weeks after fertilization. In somatic embryogenesis, the indirect regeneration of proembryos may occur from a callus. This process is based on the dedifferentiation of a cell of the differentiated embryo, followed by induction and redifferentiation to cells of the upper suspensor (U), suspensor (S) and embryo (E), i.e. the USE algorithm of Dogra (1984) is established. Adventive embryos arise directly from protodermal cells. If an ESM develops directly or indirectly from a protodermal cell, we may reconstitute proembryos by somatic polyembryogenesis or embryogenesis.

recovered from rescued embryos separated from their female gametophyte. Another factor, related to the genotype and local environment, was the fact that all cones at any one date were not exactly at the same stage of development, i.e. the degree of maturity varied within a tree and within the cone itself. This variation was more evident in cones collected in July than any other month.

The ESM recovery rate from zygotic embyros was greatest in mid-July. By mid-July, the zygotic embryos have already benefited from nourishment and conditions in the developing seed. These "enabling" conditions have yet to be properly and fully mimicked by composition of media, cultural conditions and current tissue culture technology. In our experience, it is the premature exposure to light and the interaction of light and nitrate in the medium, coupled with long subculture intervals and high levels of plant hormones, that encourages growth and development of embryogenic calluses.

Embryo development does not proceed in a true-to-type way in an embryogenic callus, unless ESMs are reestablished and conditions in the zygote are maintained. For this reason, we have tended to maintain as much contact of the ESM and zygotic embryo with the female gametophyte in darkness and in media lacking or lower in nitrates. Results from a separate study (in preparation) have indicated the way nitrates interact with the developing embryos and alter the embryonic protein complement (unpublished data).

Our observations indicate that strong stabilizing and directional selection pressures must be placed on cells within the ESM. These pressures tend to select for the specific true-to-type proembryonic multiplication involving proembryonal (pE) or embryonal-tube (Et) or embryonal-suspensor cells (Es). By constantly maintained cell selection pressures, the multiplication rates become specified and directed according to cell type and origin, genotypic characteristics and opportunistic histogenic algorithms. For these reasons, we have preferred to recognize the term "embryonal-suspensor mass" with reference to the zygotic model, as distinct from callus and embryogenic callus. This distinction draws attention to the mode of multiplication as being embryonic (one cell, one embryo), polyembryonic (one cell, multiple embryos) or teratogenic (aberrant). Polyembryony may occur by cleavage of pE cells in the ESM or occasionally by budding with a free nuclear stage. Whatever the origin of polyembryony, the developmental plan is of the *"Pseudotsuga* type."[18,4] While the cellular origins of achieving ontogeny in cell suspension cultures are multiple, all variations are interpreted in terms of zygotic behavior under field conditions.

Polyembryony is a function of ABA levels in the seed and medium: high levels of ABA inhibit polyembryony (e.g., ref. 2). Moreover, polyembryony in seeds is often found in pioneering species at high elevations or tree lines and may represent an opportunistic method for survival.[14] In this context, our cultural conditions could reimpose natural environmental stresses to direct and stabilize polyembryony.[15,19]

The developmental features that distinguish adventitious, mucilaginous ESMs from calluses or embryogenic calluses are evident from cytochemical studies that reveal transitions involving the recrudescence of ESMs from calluses. Intermediate developmental stages found in embryogenic calluses show the origin of newly emerging proembryonal cells (acetocarmine reactive). Here one must be careful to distinguish that the multiplicity of proembryos arises not only from the transformation of a callus to a proembryonal cell, but also from the polyembryonic phenomena associated with the adventitious ESM derived from callus. In our experience, embryogenic calluses represent a shifting population of cells. More directional and stabilizing selection needs to be imposed on cell masses of this type. For this reason, their use to identify morphogenic markers must be treated with caution and the proper controls.

Table 2. Phenotypic variances (V) possible in somatic embryos regenerated by the polyembryonic process (modified from Durzan 1988a).

PHENOTYPIC VARIANCE

V_a Average effect of genes

V_d Dominance

V_r Epistatic genetic variance

V_c Variance introduced by clonal process

      V_{sp} Simple or polyzygotic polyembryony (fraternal proembryos)

      V_{cp} Cleavage polyembryony from pE cells

      V_b Lobing or budding polyembryony of pE and its derivatives

      V_R Variance due to relic nuclei (rare)

      V_{ab} Variances arising by passage through a callus or teratogenic phase

$V_{a,d,}$ Interaction of genes in embryonal-suspensor mass (a,d,r,c)
$_{r,c,g}$ with the maternal female gametophyte or its equivalent

$V_{a,d,r,}$ Variance based on interactions among genes and environment
$_{c,gxe}$

V_e Variance due to different environments

A puzzling, but intriguing, nonspecific indicator for the differentiation of proembryonal cells is the acetocarmine reaction. This reaction emerges in the cytoplasm around amyloplasts and indicates an important role for the paternally derived organelles during the formation of a neocytoplasm. According to Camefort[3] and others, the maternal chloroplasts but not mitochondria are lost after fertilization and replaced by chloroplasts or amyloplasts from the pollen.[17] This is followed by the formation of an acetocarmine-reactive neocytoplasm that emerges from the migrating nuclei (*cf.* Fig. 6C,D). Results of a separate study will reveal the different components associated with the acetocarmine reaction (unpublished data).

The travelling wave of morphogenic potential at the protodermal surface involves a series of events. The surface of proembryos is covered by a nonacetocarmine-reactive layer (Fig. 6G). Under this layer are protodermal cells that may have several fates (Fig. 7). In such a portrayal we have to sort out functional interactions, spatio-temporal organization and fluctuations of cell populations for each item depicted. Cells, scheduled for epidermal layer development are transformed directly or indirectly to various fates, again depending on cultural conditions, response capacities, cell system sensitivities, selection differentials, and subculture rates. For each alternative fate, the degree of daughter cell adhesion and intercellular matrices that bind cells vary from mucilaginous (embryonic) to rigid (organogenic). Transitional stages involving callus reveal transitional stages in the polymerization of the intercellular matrix (unpublished data).

The variances in the quality of recovered embryos are embodied in the phenotypic and genotypic variances outlined in Table 2. When embryogenesis and polyembryo-genesis are coupled with the lethality prevalent in some genotypes and with the pathology of development *in vitro*, we are indeed faced with a complex embryo recovery situation. This complexity is being sorted out and better managed through the use of cell suspension cultures to identify and select cells that are more amenable to further process control (*cf.* ref. 7). The aim of our future work in propagation is to eliminate variants which arise by uncontrolled disturbances in the culture system and to establish a proven fitness from batch to batch for each genotype as appropriate.

REFERENCES

1. Allen, G.S. and Owens, J.N., 1972, "The Life History of Douglas-Fir." Environment Canada, Forestry Service, Ottawa.

2. Boulay, M.H., Gupta, P.K., Krogstrup, P. and Durzan, D.J., 1988. Conversion of somatic embryos from cell suspension cultures of Norway spruce (*Picea abies* Karst). Plant Cell Reports, 7:134-137.

3. Camefort, H., 1969, Fécondation et proembryogénèse chez les Abiétacées (notion de néocytoplasme). Revue Cytologie Biologie Végétales, 32:253-271.

4. Dogra, P.D., 1984, The embryology, breeding systems and seed sterility in Cupressacea -- a monograph. In: "Glimpses in Plant Research", P.K.K. Nair, ed., Vikas House Ltd., Delhi, India.

5. Durzan, D.J., 1988a, Somatic polyembryogenesis and plantlet regeneration in selected tree crops. Biotech. and Genetic Eng. Revs., 6:339-376.

6. Durzan, D.J., 1988b, Process control in somatic polyembryogenesis. In: "Molecular Genetics of Forest Trees," J.-E. Hällgren, ed. Swedish Univ of Agric. Sciences, Dept. Forest Genetics and Plant Physiol., Umeå, Sweden, Frans Kempe Symp., June 14-16th, Report No. 8, pp. 147-186.

7. Durzan, D.J. and Durzan, P.E., 1990, Future technologies: Model-reference control systems for the scale-up of embryogenesis and polyembryogenesis in cell suspension cultures. In: "Micropropagation," P. Debergh and R.H. Zimmerman, eds. Martinus Nijhoff, Dordrecht, pp. 389-426.

8. Durzan, D.J. and Gupta, P.K., 1987, Somatic embryogenesis and polyembryogenesis in Douglas fir cell suspension cultures. Plant Sci., 52:229-235.

9. Gupta, P.K. and Durzan, D.J., 1987, Biotechnology of conifer-type somatic polyembryogenesis and plantlet regeneration in loblolly pine. Bio/Tech., 5:147-151.

10. Gupta, P.K. and Durzan, D.J., 1986, Plantlet regeneration via somatic embryogenesis from subcultured callus of mature embryos of *Picea abies* (Norway spruce). *In Vitro*, 22:685-688.

11. Gupta, P.K., Dandekar, A.M. and Durzan, D.J., 1988, Somatic proembryo formation and transient expression of a luciferase gene in Douglas fir and loblolly pine protoplasts, Plant Sci., 58:85-92.

12. Hakman, I. and von Arnold, S., 1985, Plantlet regeneration through somatic embryogenesis in *Picea abies* (Norway spruce). J. Plant Physiol., 121:149-158.

13. Hakman, I., Fowke, L.C., von Arnold, S. and Ericksson, T., 1985, The development of somatic embryos of *Picea abies* (Norway spruce). Plant Sci. Letters, 38:53-59.

14. Klekowski, E.J. Jr., 1988, "Mutation, Developmental Selection, and Plant Evolution." Columbia Univ. Press, N.Y.

15. Lindgren, D., 1975, The relationship between self-fertilization, empty seeds and seeds originating from selfing as a consequence of polyembryony. Studia Forestalia Succica, 126:1-24.

16. Litvay, J.D., Johnson, M.A., Verma, D.C., Einspahr, D.W. and Weyrauch, K., 1981, Conifer suspension culture medium development using analytical data from developing seeds. Inst. Paper Chem. Tech. Pap. Ser., 115:1-17.

17. Neale, D.B. and Sederoff, R.R., 1988, Inheritance and evolution of conifer organelle genomes. In: "Genetic Manipulation of Woody Plants," J.W. Hanover and D.E. Keathley, eds. Plenum Press, N.Y.

18. Singh, H., 1978, "Embryology of Gymnosperms." Enc. Plant Physiol., Vol. 10, Part 2. Gebrüder, Borntraeger, Berlin.

19. Sorensen, F.C., 1982, The roles of polyembryony and embryo viability in the genetic systems of conifers. Evolution, 36:725-733.

NECKING IN EMBRYONAL TUBE CELLS AND ITS IMPLICATIONS FOR

MORPHOGENIC PROTOPLASTS AND CONIFER TREE IMPROVEMENT

Don J. Durzan

Department of Environmental Horticulture
University of California
Davis, CA 95616 USA

ABSTRACT

The breakup of viscous protoplasm into alternate smaller and larger protoplasts (necking) occurs naturally as a product of stress in elongated cells of an embryonal-suspensor mass. Necking can also be induced as an artifact of handling these actively streaming cells. Protoplasts, formed as a product of necking inside cells, can be released and recovered by cell-wall digesting enzymes. At least eight size classes of protoplasts can be recovered from cell suspension cultures of the embryonal-suspensor mass. This range of size classes is based on the different cell types (proembryonal, embryonal tube, embryonal suspensor, upper suspensor) and by limited spontaneous fragmentation and fusion of protoplasts among recovered size classes. Staining properties among recovered protoplast size classes reveal that: i) those derived from proembryonal cells may have morphogenic potential; ii) more than one protoplast inside cells of the embryonal tube and suspensor may contain a nucleus derived from a free nuclear stage; iii) the protoplast population represents a varied and fractional genetic potential based on organelles trapped in protoplasts during the necking and fusion processes; and iv) necking in conifer cells may explain illustrations in the literature showing multiple migrating nuclei just after fertilization or in cylindrical cells of the embryonal-suspensor mass. These nuclei produce a new cytoplasm or neocytoplasm associated with the establishment of the classical "basal plan" for proembryonal development. This neocytoplasm interacts in unknown ways to select for chloroplast genomes from pollen and mitochondria of the female parent.

INTRODUCTION

This study examines the origin, properties, and significance of a physico-chemical necking phenomena in streaming protoplasm of elongated cells in rescued embryonal-suspensor masses (ESMs) of three conifers.

The origins of necking are based in protoplasmic properties and most evident in the elongated cylindrical cells derived from the primary proembryonal cell (pE). These cells are predominantly the embryonal tube (Et) and embryonal suspensor (Es). While necking may occur in the highly elongated cells of the upper suspensor system, these cells are not normally morphogenic and are not selected for culture nor examined in this study. Notations and developmental origins of all cells are based on zygotic embryogenesis as described by Doyle,[16] Sinnott,[34] Singh,[33] Dogra[12,13,14] and more recently for somatic polyembryogenesis by Durzan[17] as defined for cell culture systems.

Protoplasmic systems that demonstrate necking are found in highly elongated cylindrical cells. In physico-chemical terms, most flowing liquid masses become unstable beyond a critical length.[1,6] When these physico-chemical and geometrical parameters are applied to protoplasmic streaming in embryonal tube or suspensor cells, the circumference of protoplasm becomes critical. When the protoplasmic mass becomes unstable, it tends to constrict at one end and to bulge out at the other end and to form a series of beads that trap organelles along a protoplasmic strand over the length of the cell. In physico-chemical models, a flowing cylindrical mass promptly collapses into characteristically smaller and larger spheres or beads. Taken together these phenomena represent necking. The physical aspects of necking are readily demonstrated by a cylinder of liquid, as in a stream emerging from a circular nozzle[1] and by spider silk coated with water.[40] Necking is also well-known as a product of vertical density currents in limnology[8] and has been recognized as a factor influencing the form and growth of plant cells.[38] This study recognizes necking as a property of cylindrical cells in the embryonal-suspensor mass. Necking becomes complicating factor in the application of protoplast biotechnologies to tree improvement. The impact of necking may have to be considered in evaluating genotypic and phenotypic variances among plants recovered from protoplasts.[3,22]

MATERIALS AND METHODS

Source Materials

Improved seeds (mixed half-sibs) of loblolly pine (*Pinus taeda* L.) were collected in June, 1985, and obtained as gifts from Weyerhaeuser's forest seed orchard (Lyons, GA). Seed cones were stored in darkness at 4°C for no longer than 3 weeks until used in this study. Cones of Douglas fir (*Pseudotsuga menziesii*), representing four genotypes from Weyerhaeuser (Tacoma, Washington), were treated in the same way. Seeds of Norway spruce (*Picea abies*, one genotype) were obtained from Peter Krogstrup in 1983 and originated from DDR Thuringerwald, Streufdorf (Lot No. 4-1347B).

Cell Suspension Cultures

ESMs, when induced on the surface of the zygotic embryo or when transplanted from an immature (pine, fir) or mature (spruce) seed and grown aseptically as a cell suspension culture, were suitable for the recovery of somatic embryos by polyembryony.[22,23,24]

Cell suspension cultures of Douglas fir, Norway spruce and loblolly pine were grown in darkness with 7 to 10 day subcultures. All media were adjusted initially to 200 to 250 mOsmolality and pH 5.7 at 25°C, as described by Durzan and Gupta[19] and Gupta and Durzan.[22,23] After 1 to 5 days of culture, samples of embryonal-suspensor masses (ca 10 mg fresh weight at pH 5.0 to pH 5.5 in culture media now having less than 200 mOsmoles) were removed from flasks (Erlenmeyer or nippled) and placed on glass slides with a cover slip, or double-stained sequentially, first with acetocarmine (ca 230 mOsm), and then with Evan's blue (ca 320 mOsm).[23]

ESM Heterogeneity

ESMs do not contain callus as judged by microscopic inspection, cytochemical staining and patterns of daughter-cell growth. ESMs in darkness exhibit two types of somatic polyembryony, *viz.* cleavage and budding.[17,18] Polyembryony in the zygotic and somatic situations derives by cleavage from the proembryonal (pE) cells or by budding from proembryonal daughter cells *viz.* the embryonal tube and embryonal suspensor (Fig. 1).

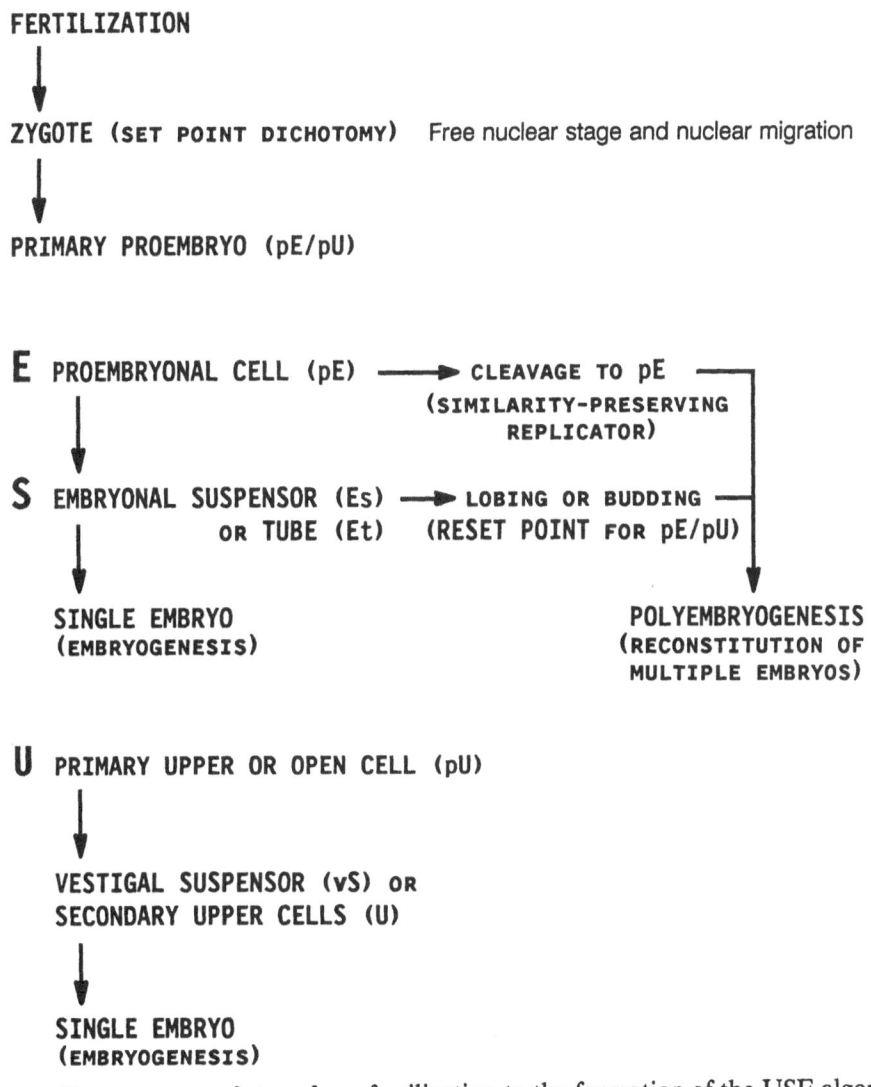

Fig. 1. The sequence of steps from fertilization to the formation of the USE algorithm for conifers.[14,33] The primary proembryo is composed of two cells (pE/pU) derived from free nuclei. Each cell in turn divides and differentiaties to contribute to the structure of the proembryo. Alternatively, pE cells and their daughter cells (Es, Et) can cleave or bud to produce multiple embryos by polyembryogenesis. Necking is observed in the cylindrical Es and Et cells.

The highly elongated suspensor cells of the upper tier (U), derived from the primary proembryo, and the somewhat elongated embryonal-tube and -suspensor cells (S) in this mass, when combined with the dense spherical proembryonal cells (pE), comprise the major "histogenic subroutines" in a complex developmental "USE" algorithm.[14,15,17] The cytoplasms and nuclei among these cells are distinguished not only by their size, protoplasmic contents and different developmental fates, but by their differential reactivity to acetocarmine and Evan's blue. The small, dense, proembryonal cells have large nuclei and react strongly with acetocarmine (see below). Cylindrical suspensor cells are highly permeable to Evan's blue (see below). Cells, derived from the proembryonal cell, react moderately with acetocarmine and are becoming progressively vacuolated.

Necking Characteristics

When cells are collected in a drop of culture medium on glass slides and stained cytochemically for microscopic viewing, the integrity of the protoplasm, associated with the cylindrical suspensor cells, changes. Viscous fingering of cell masses and protoplasts due to excessive pressure on cell masses by the cover glass should be avoided.[39] Necking can be observed directly in individual cells by bright field or phase contrast light microscopy. The exposure of cells in a drop of fresh culture medium (200 to 250 or greater mOsmolality at 25°C) on a glass slide to the drying effect of the air and to cytochemical stains creates the experimental condition where necking and/or collapse (plasmolysis) of protoplasm are often observed. Shearing and osmotic forces are associated with necking.[28] Cylindrical cells of varying lengths (pE, pU, Es, Et) tend to have varying degrees of collapse (plasmolysis) and permeability to Evan's blue. However, it is the elongated and cylindrical daughter-cell products of the dense, spherical pE cell that exhibit the necking phenomenon.

In flowing, liquid systems, the predicted products of necking are spheres of at least two alternate sizes.[1] In ESMs, the protoplasm of highly cylindrical Et and Es cells and the upper suspensor cells is characterized by large vacuoles and transvacuolar strands of protoplasm that contain subcellular organelles. Organelles are carried by a peristaltic protoplasmic streaming process, which complicates the physico-chemistry of the system. Organelles, such as nuclei and amyloplasts serve as markers in the recovery of necking products.

Microscopy and Staining

Organelle behavior and the permeability of the cytoplasm to cytochemical stains can be followed continually by a double-staining procedure with acetocarmine and Evan's blue (cf. ref. 32). Under polarized light, amyloplasts with starch grains are distinguished by their shape, size and birefringence in polarized light. The acetocarmine, being a chromatin stain, reveals nuclei. Acetocarmine, however, also detects yet unidentified products in the neocytoplasm of the conifer zygote that enable the study of how daughter cells become polyembryonic. The neocytoplasm is the "new cytoplasm" that arises around the nucleus in the zygote after fertilization.[9] The neocytoplasm gives an intense, red-staining or reaction-product. The reaction-product can also be detected in proembryonal cells and in embryonal-suspensor cells undergoing lobing and budding.[17] The viability of protoplasts was judged by the intensity of fluorescence with fluorescein diacetate.[41]

Protoplast Preparation

Protoplasts were prepared as described by Gupta and Durzan.[21,22] The behavior of protoplasts, cells, and their contents was monitored by light microscopy and photomicrography before and after staining. The diameters of recovered protoplasts

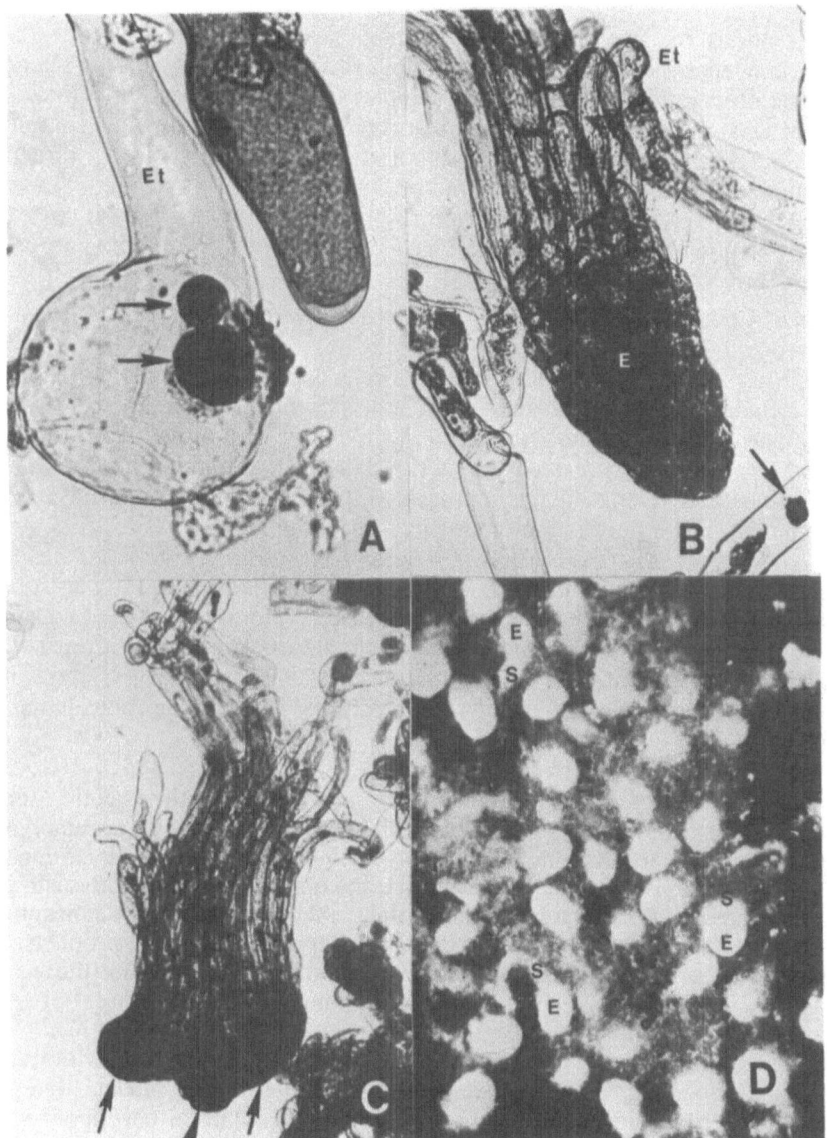

Fig. 2. Budding and cleavage polyembryogenesis in cell suspension cultures of Norway spruce (*cf.* ref. 17). **A.** Embryonal tube (Et) with two nuclei, each enclosed in a sphere of protoplasm (arrows), stained with acetocarmine and Evan's blue. The upper right is suspensor cell with a nucleus and a cytoplasm that is permeable to Evan's blue. x 56. **B.** Single somatic embryo (E), stained with acetocarmine and Evan's blue. Et, embryonal tube. Arrow points to a product of necking. × 44. **C.** Cleavage polyembryogenesis (arrows) occurs *in vitro*. Norway spruce is normally a noncleavage type species (*cf.* ref. 14). × 44. **D.** The development of multiple individual somatic embryos, formed earlier by cleavage and budding, is encouraged by abscisic acid in the medium.[7] × 25.

were referenced to the smallest visible protoplast and expressed arithmetically as multiples of the smallest diameter, as is done with particles in fluid dynamics.[28]

Experimental Design

Observations on the products of necking have been taken directly and from photographic records over a four-year period. Tracing of the origin of the phenomenon is based on observations with harvested samples from cell suspension cultures of each species, at least in triplicated experiments using weekly subcultures over a one-month period. Measurements of relative protoplast diameters were taken directly from photographs of fields of freshly prepared protoplasts.

OBSERVATIONS

Behavior of Cells in the ESM During Suspension Culture

In cell suspension cultures, the Et and Es cells derived from a pE cell elongate and become cylindrical. Moreover, they may "bud off," with the formation of free nuclei, to start a new wave of polyembryogenesis of the lobing or budding type (Figs. 2, 3, 4). The two types of somatic polyembryogenesis (cleavage and budding) in Norway spruce, loblolly pine and Douglas fir represent different configurations (tactical cell displacements) of the developing embryonal-suspensor mass.

Literature on zygotic embryonic development indicates that Norway spruce and Douglas fir are not considered as cleavage species.[2,14] However, rescue and subculture of the ESM show that the cleavage process occurs in cell suspension cultures.[7,19,23] The longitudinal, cleavage patterns of polyembryogenesis (Figs. 2C, 3E, 4D) are distinctly different from the typically multifractal patterns of budding polyembryogenesis (Fig. 3G).

Various degrees of cleavage and budding are known in the zygotic literature.[13] Budding, in this study, refers to the formation of a new proembryo or proembryonal cell through a free nuclear stage involving polar, nuclear migration. Observations on the fate of cells in suspension culture indicate that the observer is faced with distinguishing among cell types and their histogenic subroutines. Aberrations in these subroutines may lead to callus formation and to abnormal cleavage and budding products. These products can be selected against by careful selection of cells for subculture routines. In this study, all ESMs were visually and cytochemically free of callus.

During budding polyembryogenesis, the cylindrical Et and Es cells, with their active protoplasmic streaming, branch off and bear at least two free nuclei. These nuclei migrate to opposite poles to form a new proembryo (pE) and a new upper suspensor (pU). In this process, there is no need to involve a callus phase. Cleavage polyembryogenesis was especially dominant in the Douglas fir and Norway spruce genotypes used in this study. Budding polyembryony requires careful microscopic observation and a reasonable synchrony of cell suspension cultures to facilitate observations of the free nuclear stage. Partial synchrony can be established by routinely sizing cell masses for innocula during subsequent subcultures.

Necking

The budding and necking process was observed in all species (Figs. 2A, 3A,B, 4A,B). The cylindrical Et and Es cells that display necking (< 3%) are more prominent at the start of the ten-day subculture cycle or with freshly transplanted embryonal-tube or -suspensor masses. Their protoplasm may also collapse with the formation of ergastic material. Necking occurred during protoplast preparation, where high levels of *myo*-inositol and sugars are used to maintain the osmotic integrity of protoplasts.

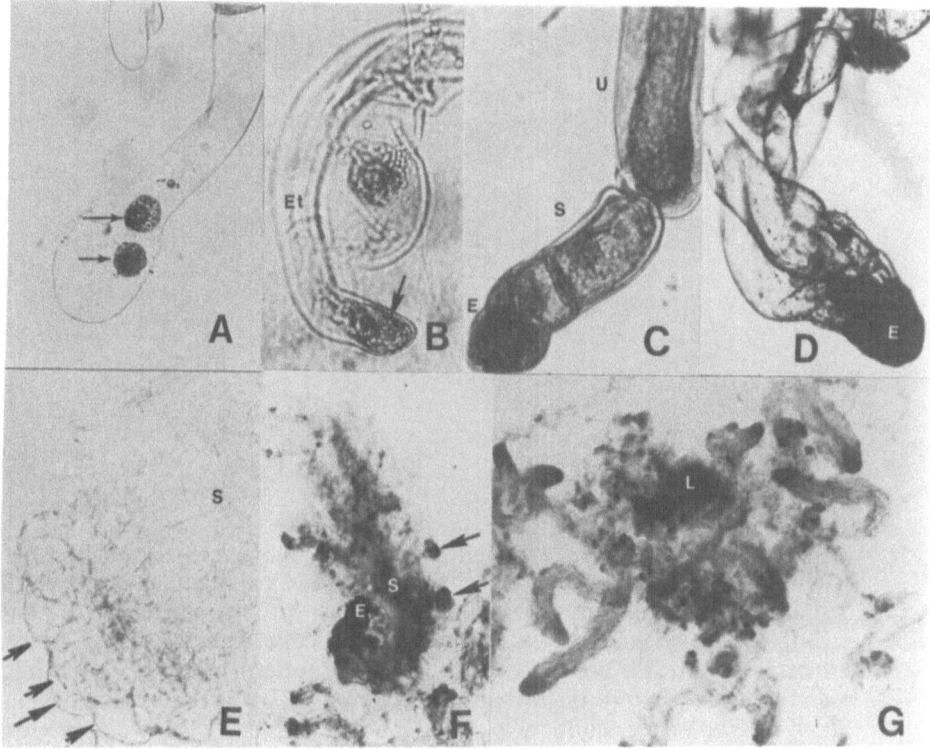

Fig. 3. Examples of budding and cleavage polyembryogenesis in cell suspension cultures of loblolly pine (after refs. 17 and 23). **A.** Freely suspended embryonal-tube cell with two free nuclei surrounded by cytoplasm (arrows), stained with acetocarmine. × 56. **B.** Elongated embryonal tube (Et) where one (arrow) of the two free nuclei has migrated to one end of the cell. **C.** A new embryo, stained with acetocarmine and Evan's blue, is reconstituted from an embryonal tube and the developmental pattern is of the USE type. × 56. **D.** An acetocarmine-stained single embryo (E), showing the elongated suspensor cells. × 44. **E.** An aggregate of multiple, cleaved, unstained and unseparated embryos with their suspensors (S). × 44. **F.** New embryos (arrows) arise along the suspensor system (S) of a single embryo (E) by budding polyembryogenesis in a double-stained preparation. × 25. **G.** The development of a polyembryonic mass, largely by budding as distinct from cleavage, as revealed by double-staining. Note pattern differences between E and G. × 25.

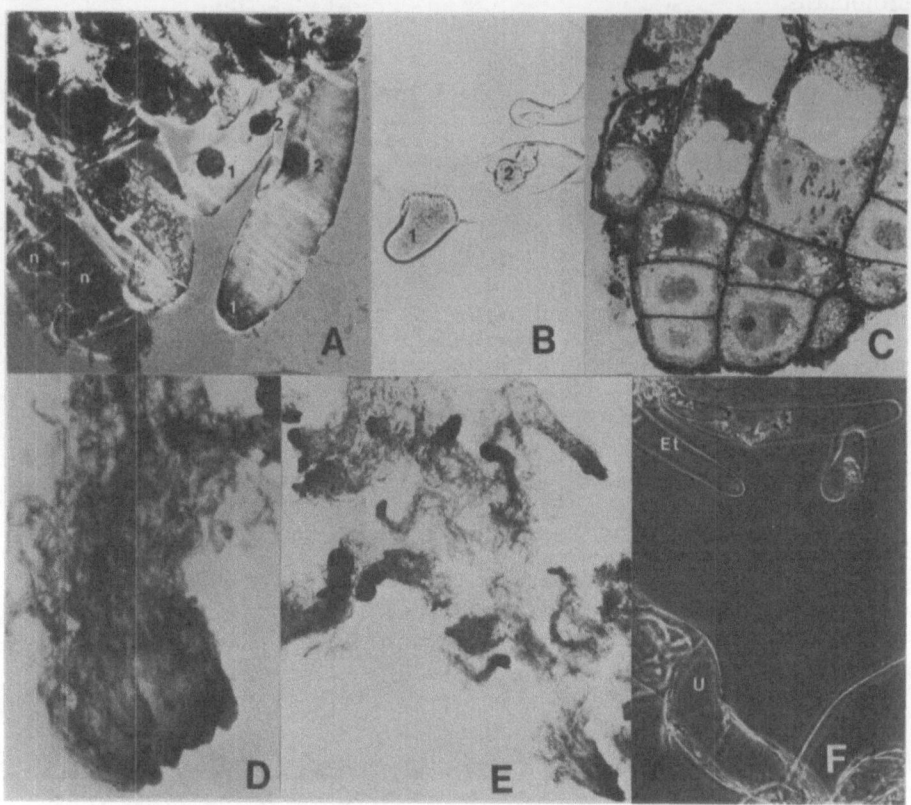

Fig. 4. Examples of budding and cleavage polyembryogenesis in cell suspension cultures of Douglas fir (after 17 and 19). **A.** An embryonal-suspensor mass stained with orcein reveals budding cells, each with two nuclei (pE [1] and pU [2]) and a cell undergoing the necking process, forming two protoplasmic masses (n), with a bridge between the two (bottom left). × 64. **B.** Budding embryonal tube with migrating nuclei [1] and [2], covered with neocytoplasm and reacted with calcofluor. × 64. **C.** Thin section of a zygotic embryo to show tiered development and longitudinal files of cells that contribute to the cleavage phenomenon *in vitro*. × 80. **D.** Cleavage polyembryogenesis in cell suspension culture. Proembryonal cells have cleaved longitudinally to form a mass of 12 or more embryos each with their suspensor systems. The differential elongation of cells leads to an asynchrony in the process, so that not all cleavage embryos are at the same stage of development. The preparation was stained with acetocarmine and Evan's blue. × 56. **E.** A double-stained and freely suspended mass of proembryos undergoing the cleavage process, with some budding. Contrast this pattern with loblolly pine in Figs. 3F,G. × 44. **F.** Cells of the upper tier (U) elongate and sometimes produce spurs (arrow) that would normally attach to the female gametophyte or other embryos to push the attached developing embryo towards nutrient reserves. Et, embryonal tube that contributes to the budding process. × 56.

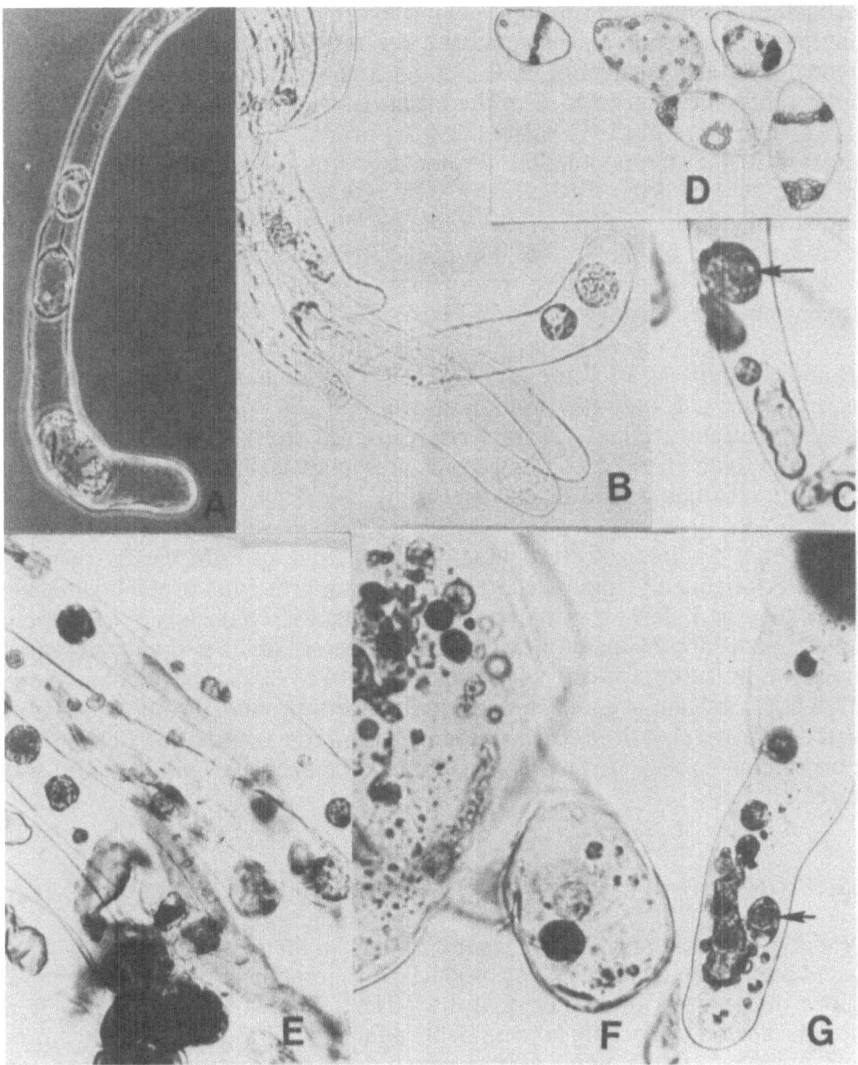

Fig. 5. Necking in embryonal tube cells of Douglas fir and loblolly pine embryonal-suspensor masses in suspension cultures. **A.** Start of instabilities leading to necking in an elongated cylindrical embryonal tube cell of Douglas fir. In this unstained preparation, spheres of protoplasm with organelles are still connected to one another by a thin thread (bridge) of protoplasm. × 56. **B.** The collapse of the bridge leads to spheres of protoplasm with and without a nucleus. Protoplasmic masses stain differently with acetocarmine and Evan's blue, as evidenced by the two differently shaded spheres of protoplasm. × 56. **C.** Collapse of the bridges leads to different-size spherical masses of protoplasm, and to irregularly shaped masses. Arrow indicates an acetocarmine-reactive protoplasmic mass bearing a nucleus. × 56. **D.** Nonembryogenic callus cells derived from the same isolated embryonal-suspensor mass show different patterns of subcellular disruption. × 56. **E.** Necking in embryonal-tubes of Douglas fir stained with acetocarmine and Evan's blue reveal two major sizes of protoplasmic masses that can be recovered by cell wall digestion. × 56. **F.** Necking and fragmentation products in an embryonal tube of loblolly pine, stained with acetocarmine and Evan's blue. × 64. **G.** Necking and fragmentation products of different sizes in loblolly pine. Acetocarmine brings out the protoplasmic masses bearing a nucleus (arrow). × 56.

One predictable and characteristic result of necking is that cells display large and small globules of cytoplasm (protoplasts) inside cells (Fig. 5). After double-staining of loblolly pine embryonal-suspensor masses, the distribution of sizes of protoplasts in embryonal-tube and -suspensor cells was as follows: protoplasts with the smallest diameter represented up to 65% of the products, whereas protoplasts with diameters 3 to 5X greater were in the minority (33%). For Douglas fir, the smallest protoplasts represented 20% of the population. Protoplasts with diameters of 1.5 and 3X greater contributed 21 and 55%, respectively. Results varied from batch to batch with variations not exceeding 80% of the reported mean. For Norway spruce, necking products are not as prominent because of the shorter embryonal-tube and -suspensor cells (Fig. 2B). Representative protoplasts created by necking and bearing an assortment of organelles are shown in Fig. 5.

Protoplast preparations represented all cell types found in the embryonal-suspensor mass. However, it is only the protoplasm from pE, Es and Et protoplasts that react strongly with acetocarmine. Protoplasts recovered from all cell types were highly permeable to Evan's blue, even though such protoplasts exhibit protoplasmic streaming and a reaction to fluorescein diacetate (Fig. 6).

The overall diameter of protoplasts obtained from an ESM varied, depending on their cellular origin and from batch to batch. Protoplasts from proembryonal cells are small (ca 20 μ dia.), densely cytoplasmic, and contain large nuclei. Proembryonal cell protoplasts are approximately twice the diameter of the dominant miniplasts. Protoplasts from suspensor cells of the upper tier (6 to 8x dia.) are among the largest. They are highly vacuolate and less reactive with acetocarmine. Protoplasts (miniplasts) recovered after necking or protoplast fragmentation represented the smallest diameter class considered in this study (Fig. 6). Ergastic materials, while smaller, were not measured.

DISCUSSION

Products of necking originated as predicted, from the actively streaming and highly elongated cells of the ESM. Necking products were trapped as protoplasts with variable inclusions inside walls of cylindrical cells. The redistribution of organelles among protoplasts created by necking is presumably a function of the position and entrapment of each organelle at the point of fragmentation of the protoplasmic mass and the alternate size distribution of protoplasmic beads. Deterioration of the smaller protoplasts and clusters of subcellular organelles or secondary products could account for ergastic materials sometimes shown in figures of embryonal tubes at the free nuclear stage and as illustrated for *Ephedra* by Bold, et al.[5]

An important physico-chemical attribute of protoplasmic interfaces is that properties (viscosity, density, interfacial tension) may vary along their boundary layers and perpendicular to the cell wall; i.e., a multiple directionality of forces is experienced by the streaming protoplasm (*cf.* ref. 4). Necking begins as shown in Fig. 5A. The constrictions indicate that the interfacial tension of the protoplasmic mass initially becomes metastable. As the protoplasm breaks up into protoplasts inside the cell, it is assumed that the surface free energy for each new mass increases. The term "surface free energy" implies only that work is required to form more surface, i.e. to bring molecules from the interior of the phase to the surface region.[1]

Assuming that the smallest necking products are equivalent to the smallest protoplasts recovered, it can be postulated that necking contributes to the 1 to 3X range of protoplast diameters that can be recovered (Figs. 6, 7). Within a population of equal-sized protoplasts, some stain strongly with acetocarmine (i.e. pE-derived protoplasts) and others do not. The latter represent necking or fragmentation products from pU cells or pE-derived cells (Es, Et) (Figs. 5, 6A).

Some budding embryonal tubes contain at least two nuclei that migrate with an attached neocytoplasm (Figs. 2A, 3A, 4A,B). The necking products that are seen inside these cells or recovered as protoplasts may stain blue or red, depending on their origin and contents. In zygotic and somatic embryogenesis, the multiple nuclei that arise during the free-nuclear stage migrate to establish the proembryonal "basal plan" (Fig. 1).

Since the neocytoplasm in embryonal-tube and -suspensor cells is highly reactive to acetocarmine, as are nuclei, it is easy to believe that protoplasts created by necking are

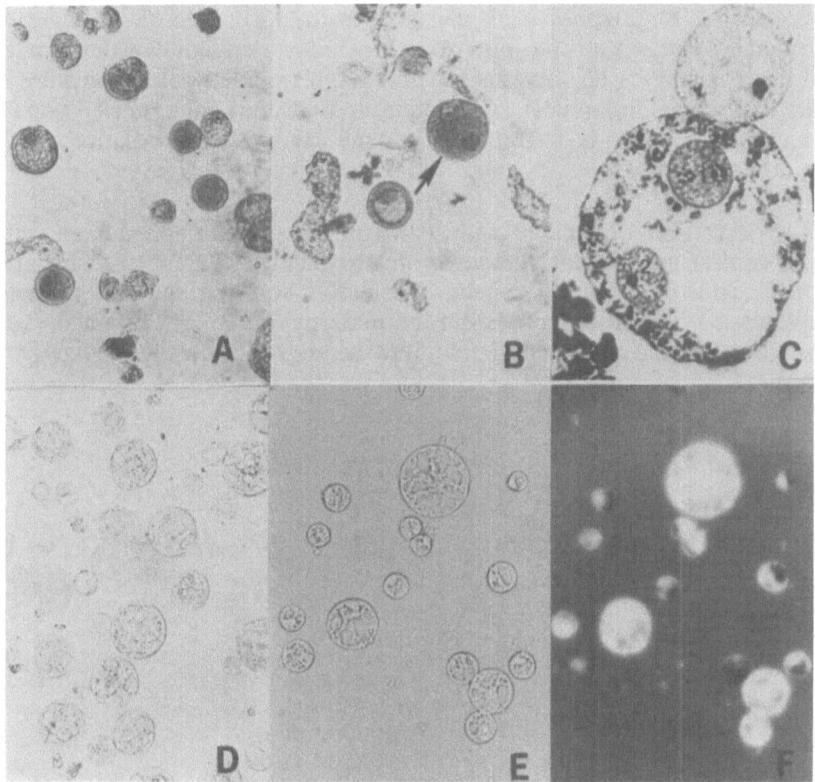

Fig. 6. Recovery of protoplasts, miniplasts and cytoplasts from conifer embryonal-suspensor masses in loblolly pine and Douglas fir. **A.** Protoplasts from small proembryonal cells of loblolly pine can be distinguished from other protoplasts by their strong acetocarmine reactivity. × 44. **B.** Protoplasts may contain two nuclei, and are possible results of spontaneous fusion of two protoplasts or, less likely, the early free nuclear stage associated with embryonal tubes. × 44. **C.** Thin section of a Douglas fir protoplast, to show two nuclei (1 and 2) and an attached cytoplast. × 64. A, B, and C are fresh, unpurified preparations. **D.** Protoplast preparation of Norway spruce showing multiple classes of large and small protoplasts that arise from embryonal-suspensor cells. × 44. **E.** Purified loblolly pine protoplast preparation, to show different size classes of protoplasts, before staining with fluorescein diacetate. × 44. **F.** The viability of protoplast populations in Fig. E can be distinguished by bright yellow fluorescence with fluorescein diacetate. × 44.

indeed free nuclei. These protoplasts may even be mistaken for relict nuclei (*cf.* ref. 24,33). In reality, we have a mixture of protoplasmic globules, some with nuclei and others without nuclei (Fig. 5). The occurrence of nuclei is supported by thin section of a stained protoplast of loblolly pine (Fig. 6C). Protoplasts and miniplasts, not having nuclei, are probably derived by necking or fragmentation during the preparation of protoplasts.

Further understanding of the distribution of sizes becomes complicated when protoplasts of various sizes fuse spontaneously to give intermediate diameters. It is significant for the study of conifer proembryology that the protoplasts, containing two large nuclei (e.g. Fig. 6BC) and a thin layer of highly acetocarmine-reactive neocytoplasm, may indeed represent the proembryonal sheath described by Dogra.[13]

In zygotic proembryogenesis, the proembryonal sheath contains free nuclei and is a precursor of the proembryo. It can lead to the formation of "cells" inside cells, as illustrated by Bold et al.[5] (*cf.* Figure 3A). Dogra[15] has pointed out that the ventral canal nucleus and the egg nucleus can also fragment. The role of necking in trapping nuclei or fragmented nuclei or miniplasts that look like nuclei and stain with acetocarmine needs to be considered in some cases as an alternative explanation for these observations.

The occurrence of necking, with its "organelle trapping" phenomena also involves the chloroplast genome that derives from the paternal parent at fertilization.[31] The paternal chloroplasts are somehow selected for during the formation of a neocytoplasm.[9,33] Mitochondria from the maternal parent are retained. Chloroplasts (amyloplasts or leucoplasts) of the mother tree are somehow digested away. In zygotic

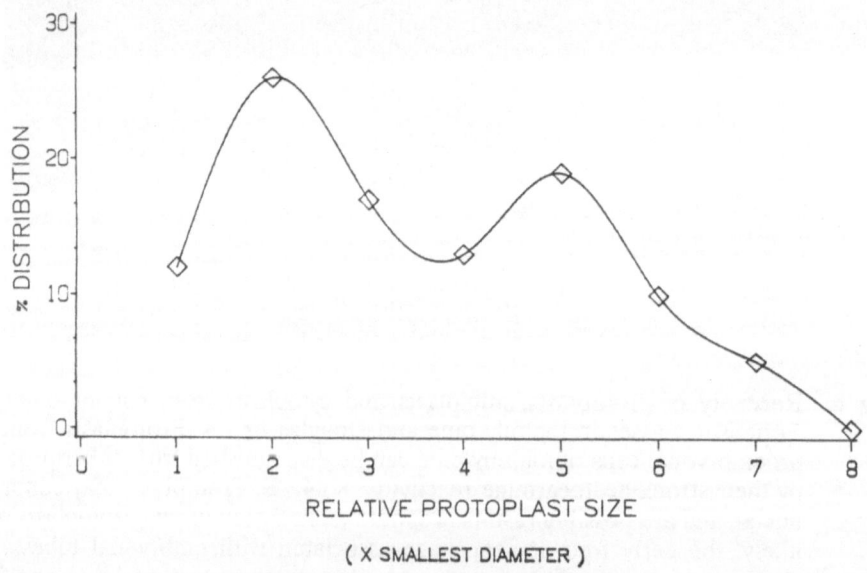

Fig. 7. Percent distribution of protoplasts of varying diameters prepared from cell suspension cultures of embryonal-suspensor masses of Norway spruce (*cf.* Fig. 6D). The variation around each point usually does not exceed 80% of the mean. Variances can be substantially greater, although the same general pattern holds, depending on the number, diameter, and length of the cylindrical cells that contribute to protoplasts in each preparation.

embryony, the reallocation of cytoplasmic organelles by necking to migrating nuclei, in advance of the first division, can add further genotypic variation to embryos. The variation arises from organelle redistribution in polyembryony.

Zygotic polyembryogenesis is more frequent in stressful environments.[35] Environmental stress, in turn, could lead to conditions that enhance the necking phenomenon in nature. The migration of egg mitochondria and their clustering around the egg nucleus favors their inclusion in the neocytoplasm of the proembryo. In clonal propagation programs employing rescued ESMs, these phenomena in their extreme could become a source of epigenetic defects[26] and in this way explain some of the opportunism and lethality associated with conifer seed reproduction.[11,20,29,35] Through recurrent selection, lethals could be rouged from the population, especially if ESMs were derived from haploid cell lines where genes are more fully expressed.

Protoplast sorting techniques should enable the isolation of discrete ranges of size classes bearing varying degrees of morphogenic and/or biosynthetic ability. Sorting out will facilitate the evaluation of products by display of the phenotypic variances (e.g. biosynthetic potentials) associated with this technology.

Necking may in some cases complicate the recovery of morphogenic protoplasts as carriers of new genetic information. Ideally, protoplasts representing the elite genotype should be recovered mainly from pE cells. Poorly controlled protoplast fusion technology, involving uncontrolled necking and fusion of protoplasts, should be avoided. Indiscriminate protoplast fusion could be enhanced during electroporation to introduce foreign genes (e.g. luc gene in Douglas fir and loblolly pine.[21] Taken together, these observations will force a reevaluation of the variances in genotype and phenotype among embryos regenerated from conifer protoplasts.[3]

One related physical factor, not investigated in this study, is the collapse of protoplasm in an electric field. Apparently, an infinitesimal change in electrical potential across a polyelectrolyte gel produces a discrete, but reversible, volume change.[36] Collapses of this type may contribute to the reduction of the viability of morphogenic protoplasts and to the appearance of ergastic bodies in protoplasts that are subjected to electroporation.[21] More study is needed to understand the importance of the rich variety of interactions and interaction potentials among protoplasts and their fluids in ways that depend on the properties of surfaces and intervening liquids (cf. refs. 27,39).

In summary, unawareness of necking may lead to misinterpretations about the genotypes of embryos recovered from protoplasts or polyembryony. The misinterpretations may arise from a) the occurrence of a free-nuclear stage, b) relict nuclei, i.e. nuclei that have entered the cell from an adjacent cell, as postulated by several investigators (cf. ref. 10), and c) the origin of different-sized protoplasts from fresh preparations of the embryonal-suspensor mass or their suspension cultures. Necking helps to explain the origin of different-sized protoplasts from small proembryonal and large suspensor cells. Necking also may occur naturally during the migration of nuclei during the free nuclear stage in embryonal tubes.[5,14] The free nuclear stage establishes the developmental "set point" for the basal plan and subsequent USE histogenic algorithms (Fig. 1). These considerations have been included in attempts at establishing bioreactor-based process controls.[18]

Necking contributes to the production of enucleated protoplasts (cytoplasts) and miniprotoplasts[30] from morphogenic embryonal-suspensor masses. Control of this phenomenon should enable the study of developmental variances. Through the recovery of different morphogenic types, it may be possible to establish genotypic deletions that could facilitate molecular analyses at the genetic and epigenetic levels.

Coupled with the use of restriction fragment length polymorphism,[37] this technology should enable a better understanding of foreign gene expression during somatic proembryogenesis. In the long run, control over these phenomena could be of value to tree improvement programs.[18,25]

ACKNOWLEDGEMENTS

Special thanks go to Drs. P.K. Gupta and M.H. Boulay for use of illustrative stages of development for sugar pine, loblolly pine, Douglas fir and Norway spruce. The author benefited by assistance from Mssrs. F. Ventimiglia and Lin Hong and from discussions with Dr. P. Dogra.

REFERENCES

1. Adamson, A.W, 1982, "Physical Chemistry of Surfaces," 4th Edition. J. Wiley and Sons, N.Y.
2. Allen, G.S. and Owens, J.N., 1972, "The Life History of Douglas Fir." Environment Canada, Forestry Service, Ottawa.
3. Attree, S.M., Dunstan, D.I. and Fowke, L.C., 1989, Plantlet regeneration from embryogenic protoplasts of white spruce (*Picea glauca*). Bio/Tech. 7:1060-1062.
4. Batchelor, G.K., 1967, "An Introduction to Fluid Dynamics." Cambridge University Press, Cambridge.
5. Bold, H.C., Alexopoulos, C.J., and Delevoryas, T., 1980, "Morphology of Plants and Fungi." Harper and Row, New York.
6. Boucher, E.A., 1980, Capillary phenomena: properties of systems with fluid/fluid interfaces, Rep. Prog. Phys. 43:497-546.
7. Boulay, M.P., Gupta, P.K., Krogstrup, P. and Durzan, D.J., 1988, Development of somatic embryos from cell suspension culture of Norway spruce (*Picea abies* Karst), Plant Cell Reports 7:134-137.
8. Bradley, W.H., 1965, Vertical density currents. Science 150:1423-1428.
9. Camefort, H., 1969, Fécondation et proembryogénèse chez les Abietaceés (notion de neocytoplasme), Rev. Cytol. Biol. Vég. 32:253-2711.
10. Chamberlain, C.J., 1935, "Gymnosperms. Structure and Evolution." University of Chicago Press, Chicago.
11. Dawkins, R., 1981, "The Extended Phenotype," W.H. Freeman and Co., San Francisco.
12. Dogra, P.D., 1966, Observations on *Abies pindrow* with a discussion on the question of occurence of apomixis in gymnosperms. Silvae Genetica 15:1-32.
13. Dogra, P.D., 1980, Embryogeny of gymnosperms and taxonomy - an assessment. In: "Glimpses in Plant Research," P.K.K. Nair, ed., Vikas House Ltd., New Delhi, 5:114-128.
14. Dogra, P.D., 1978, Morphology, development and nomenclature of conifer embryo, Phytomorphology 28:307-322.
15. Dogra, P.D., 1967, Seed sterility and disturbances in embryogeny in conifers with particular reference to seed testing and tree breeding in *Pinaceae*. Stud. Forestal. Suecica 45:1-97.
16. Doyle, J., 1963, Proembryogeny in *Pinus* in relation to that in other conifers -- a survey. Proc. Roy. Irish Acad. 62B:181-216.
17. Durzan, D.J., 1988, Process control in somatic polyembryogenesis. In: "Molecular Genetics of Forest Trees," Franz Kempe Symp., June 14-16, 1988, Swedish Univ. Agric. Sci, Umea, Rept. No. 8, 147-186.

18. Durzan, D.J., 1988, Somatic polyembryogenesis for the multiplication of tree crops, Biotech Genetic Eng. Revs. 6:339-376.

19. Durzan, D.J. and Gupta, P.K., 1987, Somatic embryogenesis and polyembryogenesis in Douglas fir cell suspension cultures, Plant Sci. 52:229-235.

20. Eberhard, W.G., 1980, Evolutionary consequences of intracellular organelle competition, Quart. Rev. Biol. 55:231-249.

21. Gupta, P.K., Dandekar, A.M. and Durzan, D.J., 1988, Somatic proembryo formation and transient expression of luciferase gene in Douglas fir and loblolly pine protoplasts. Plant Sci. 58:85-92.

22. Gupta, P.K. and Durzan, D.J., 1987a, Somatic embryos from protoplasts of loblolly pine proembryonal cells. Bio/Tech. 5:710-712.

23. Gupta, P.K. and Durzan, D.J., 1987b, Biotechnology of somatic polyembryogenesis and plantlet regeneration in loblolly pine. Bio/Tech. 5:147-151.

24. Gupta, P.K. and Durzan, D.J., 1987c, Plantlet regeneration via somatic embryogenesis from subculture callus of mature embryos of *Picea abies* (Norway spruce). *In Vitro* Cell. and Devel. Biol. 22:685-688.

25. Haissig, B.E., Nelson, N.D. and Kidd, G.H., 1987, Trends in the use of tissue culture in forest improvement. Bio/Tech. 5:52-57.

26. Holliday, R., 1987, The inheritance of epigenetic defects. Science 238:163-170.

27. Israelachvil, J.N. and McGuiggan, P.M., 1988, Forces between surfaces in liquids. Science 241:795-800.

28. Kay, J.M. and Nedderman, R.M., 1985, "Fluid Mechanics and Transfer Processes." Cambridge Univ. Press, Cambridge.

28. Klekowski, E.J. Jr., 1988, "Mutation, Developmental Selection, and Plant Evolution." Columbia Univ. Press, New York.

30. Lörz, H., 1984, Enucleation of protoplasts: Preparation of cytoplasts and miniprotoplasts. In: "Cell Culture and Somatic Cell Genetics of Plants," Vol. 1, "Laboratory Procedures and Their Applications," I.K. Vasil, ed., Academic Press, New York, p. 448-453.

31. Neale, D.B. and Sederoff, R., 1988, Inheritance and evolution of conifer organelle genomes. In: "Genetic Manipulation of Woody Plants," J.W. Hanover and D.E. Keathley, eds., Plenum Press, pp. 251-264.

32. Powledge, T.M., 1984, Biotechnology touches the forest. Bio/Tech. 2:763-772.

33. Singh, H., 1978, "Embryology of Gymnosperms." Enc. Plant Physiol., Vol. 10, Part 2. Gebrüder, Borntraeger, Berlin.

34. Sinnott, E.W, 1960, "Plant Morphogenesis." McGraw-Hill Inc., New York.

35. Sziklai, O., 1986, Polyembryony of *Pinus contorta* Doug. in central Yukon. In: "Provenances and Forest Tree Breeding for High Latitudes," D. Lindgren, ed., Proc. F. Kempe Symp., Swedish Univ. Agric. Sci., Umea, Rept. 6., p. 251.

36. Tanaka, T., Nishio, I., Sun, S.-T. and Ueno-Nishio, S., 1982, Collapse of gels in an electric field. Science 218:467-469.

37. Tanksley, S.D., Young, N.D., Paterson, A.H. and Bonierbale, M.W., 1989, RFLP mapping in plant breeding. New tools for an old science. Bio/Tech. 7:257-264.

38. Thompson, D.W., 1963, "On Growth and Form," Vol. 1. Cambridge Univ. Press, Cambridge.

39. Vicsek, T., 1989, "Fractal Growth Phenomena," World Scientific, Singapore.

40. Vollrath, F. and Edmonds, D.T., 1989, Modulation of the mechanical properties of spider silk by coating with water. Nature 340:305-307.

41. Widolm, J.M., 1972, The use of fluorescein diacetate and phenosafranine for determining viability of cultured plant cells. Stain Tech. 47:189-194.

COMPARISON OF LARCH EMBRYOGENY IN VIVO AND IN VITRO

Patrick von Aderkas[1]

Jan Bonga[2]

Krystyna Klimaszewska[3]

John Owens[1]

[1] Department of Biology
University of Victoria
Victoria, B.C. Canada
V8N 1X9

[2] Maritime Regional Forestry Centre
Forestry Canada
P.O. Box 6000
Fredericton, N.B. Canada
E3B 5P7

[3] Petawawa National Forestry Centre
Forestry Canada
Chalk River, Ont. Canada
K0J 1J0

ABSTRACT

Larch species have been induced to form embryoids both from both haploid and diploid explants. The developmental steps in embryogenesis of diploid explants of Larix leptolepis, L. decidua, L. occidentalis and L. x eurolepis are outlined and compared with the embryogeny of zygotic embryos. The various terms commonly found in the classical embryological descriptions are discussed in terms of their usefulness in describing events in vivo. A number of tissue culture terms, such as proembryo(id), embryonal suspensor mass, and callus are discussed as well. This comparative embryological study is extended to haploid embryoid development, which is initially different from both somatic and zygotic embryogenesis.

Woody Plant Biotechnology, Edited by M.R. Ahuja
Plenum Press, New York, 1991

INTRODUCTION

In the relatively short time since european larch
(12) and Norway spruce (10) were induced to form somatic
embryos in culture, similarly induced embryogenesis has
been obtained in other species and genera (1). Larch
species presently in culture and producing somatic embryos
include Larix decidua, (3,19) L. leptolepis, (19) L. x
eurolepis (11), and L. occidentalis (von Aderkas,
unpubl.). Larch species presently in culture and
producing gynogenic haploid embryoids include L. decidua
(12), L. leptolepis and L. x eurolepis (19).

This paper has two goals: 1) to describe the
developmental pathways associated with embryogenesis in
vitro and in vivo, and 2) to assess which terminology
presently being used in the conifer tissue culture
literature is appropriate for larch in vitro development.
It is not meant to be an exhaustive review on Larix
embryogeny. The curious reader should refer to Schopf
(1943) for detailed information.

In describing the developmental pathways, haploid and
diploid embryogenesis will be treated separately as they
differ significantly from one another in their early
stages (19). The development of somatic embryos will be
compared with published zygotic embryogeny, and then
gynogenic embryoid development will be compared with both
somatic and zygotic embryogenesis.

In his monograph on embryogenesis in Larix decidua,
Schopf (14) listed a number of developmental stages for
zygotic embryos. His stages, pro-, meta-, ana- and telo-
never gained general acceptance. The generally acceptable
embryogenic sequence for conifer zygotic embryos which
will be used in this account consists of three stages: a)
proembryogeny - all stages before the elongation of the
suspensor, b) early embryogeny - all stages after the
elongation of the suspensor and before the establishment
of a root generative meristem and c) late embryogeny -
root and shoot meristem formation and subsequent events
leading to a mature embryo wth cotyledons (15). Part of
this sequence has much merit since embryo development
makes sense in the context of the surrounding tissue as
well. Proembryogeny progresses within the confines of the
archegonium. When the suspensor expands, the embryo is
thrust into the corrosion cavity, which is a
morphologically and physiologically very different
environment.

The second stated goal of addressing terminological
difficulties is meant to tackle a currently tendentious
problem. Terminology is often considered quite
secondarily when descriptions of in vitro cultures are
published. In consequence, developmental stages may be
assessed incorrectly, occasionally leading to erroneous
conclusions. Alternately, lack of awareness of the
subtleties of the embryogenic process may mean that many
problems well-suited to study in vitro may go
unrecognized. As larch is the only system to date in
which both haploid and diploid embryogenesis have been

induced in vitro, examples from larch development will be
used to outline the appropriateness of certain types of
terminology. This does not mean the suggested terminology
is universal to gymnosperms in vitro. A short glance
through gymnosperm embryology literature should dispel any
such notion.

DIPLOID DEVELOPMENT

Proembryogeny

    Proembryogeny in vivo. Most gymnosperm orders and
genera show great regularity in early division sequences,
so much so that embryogeny formulae have been devised.
Larix, for example, follows a pattern described by the
following formulation

primary proembryo stage
             4pU   4pE
after internal division
             4U   4S   E4 + 4
early embryogeny
             4U   4dS   4E's1,4E's2...e't...e'

This list briefly summarizes the following events (Fig.
1). Following fertilization the zygote nucleus divides
twice. The four nuclei then descend to the base of the
proembryo where they form a layer. A further vertical
division of each of the nuclei results in 8 nuclei in two
layers, none of which are separated by walls. The two
tiers are known as the primary upper tier (pU) and the
primary embyronal tier (pE). Secondary spindles form and
cell walls develop between the nuclei. The upper tier
remains open, the embryonal tier becomes completely
enclosed. Each of these layers divide once more to form a
total of four tiers: the two embryonal (E), suspensor (S)
and an upper (U) tiers. The latter two tiers degenerate.
The suspensor does not elongate and is considered a
disfunctional suspensor (dS). In most Pinaceae the
suspensor cells which develop are not derived from
suspensor tier, but from the embryonal layer (E).   The
first sets of suspensor cells (E's1, E's2) are known as
embryonal suspensors. The expansion of the embryonal
suspensor cells (which make up the primary suspensor)
heralds the end of proembryogeny and the beginning of
early embryogeny (Fig. 2A). Similar types of formulae
have been derived for almost all the gymnosperms, and
though there are documented variations in the Pinaceae on
such formulae (4), proembryogeny tends to be conservative
and the formulae are quite a reliable guide.

    Primary suspensor cells have been known to give rise
to meristematic clusters of cells, the so-called rosette
embryos (15), which have never been reported to result in
embryos (2, 4, 5). Such clusters are also known to occur
in disfunctional suspensor cells of other conifers in

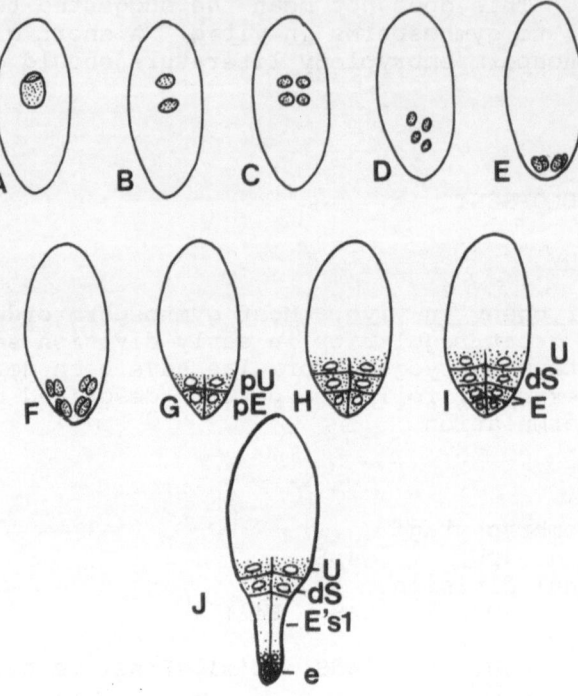

Fig. 1.   Proembryogeny of larch in vivo. Following syngamy
(A), the zygote nucleus divides twice (B, C)
resulting in a tetranucleate coenocyte. The nuclei
migrate together to the chalazal end of the
archegonium (D, E). The four nuclei divide again to
form 8 nuclei. At this point secondary spindles are
formed and cell walls are formed between the various
nuclei (F). The two tiers are known as the pU -
primary upper tier, and the pE - primary embryo
tier. The pU tier divides again (G) to form an upper
(U) tier and a suspensor (S) tier. This suspensor
tier, which in some other gymnosperms produces the
primary suspensor, is dysfunctional (dS) in larch
and does not divide or expand any further (H). The
embryonal tier (E) divides once again (I), and
becomes two-tiered (e). It is the inner of the two
tiers which behaves as a primary suspensor (E's1),
pushing the embryonal tier into the megagametophyte
(J).

the family Pinaceae, but not Larix (14).   The
disfunctional suspensor cells of Larix are
characteristically lacking in any meristematic qualities.

     Development from explant: proembryogeny or early
embryogeny? Explants of zygotic embryos have been
cultured in either one of three ways (Fig. 2B-D). The
first method involves co-culturing the immature zygotic

142

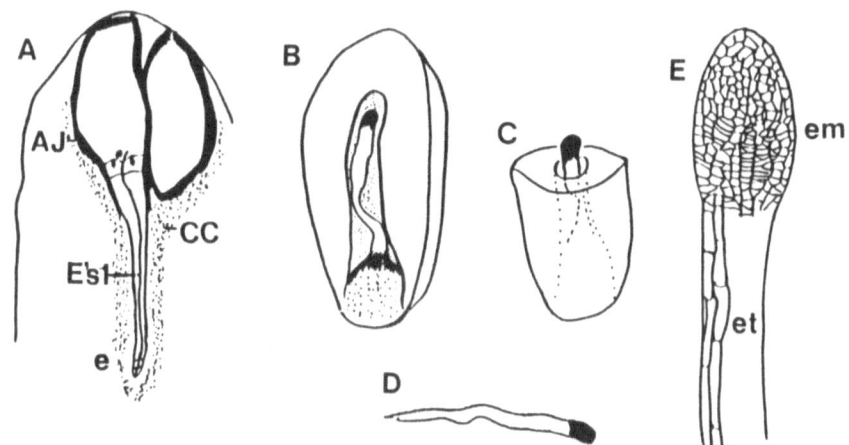

Fig. 2A. The primary suspensor (E's1) of the zygotic embryo
pushes the embryonal tier (E) through the
archegonial jacket (AJ) and into the corrosion
cavity (CC). This marks the end of proembryogeny and
the beginning of early embryogeny. B-D. Explant
preparation of _Larix_ zygotic embryos. Coculture is
either with the the megagametophyte cut lengthwise
(B) or crosswise (C). Culture methods sometimes use
the embryo on its own (D). E. Drawing of
longitudinal section of zygotic embryo at time of
explanting. The embryonal mass (em) is in an early
stage of late embryogeny. The secondary suspensor is
composed of embryonal tubes (et).

embryo with half of the megagametophyte. The chalazal
portion is cut away and the embryo remains nested in the
micropylar half of the megagametophyte (11,19).
Alternatively, the megagametophyte is cut longitudinally
and the embryo left in one half (3). Yet another method is
to remove the immature embryo from the surrounding
megagametophyte and to culture it on its own on the semi-
solid medium (3). To date mature embryo explants have not
been induced to form embryoids (11). The method of explant
isolation has no influence on induction and early embryoid
development. The embryo explants are all pre-cotyledonary
and represent stages in early embryogenesis (not
proembryogenesis), or late embryogenesis (Fig. 2E).
Embryoids develop from cells of the suspensor of the
zygotic embryo. Small centres of cytoplasmically dense
cells develop among the suspensor cells and give rise to
organized meristematic centres, themselves producing
suspensors (Fig. 3). Cleavage of these centres leads to
further numerical increase in embryoids. There are,
therefore, three processes at work: 1) cells of the
secondary suspensor dedifferentiate and divide unequally
resulting in small cells which in turn produce small

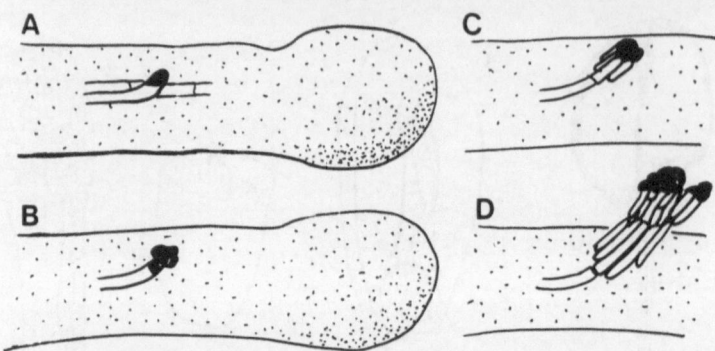

Fig. 3. Drawing of development of embryonal mass from a
zygotic embryonal tube cell. The embryonal tube
cells at the lateral margins of the suspensor
often become dissassociated with their immediate
neighbours. A polar division can occur (A), and
the small cell which results gives rise to the
embyronal mass (B, C), which produces embryonal
tube cells (D). The embyronal mass is capable of
cleavage (E).

cells, until the mass of small cells results in an embryonal
mass capable of producing a suspensor composed of embryonal
tube cells (Fig. 2E); 2) cleavage of any embryonal mass by
differentially expanding meristematic centres, each producing
a suspensor which looks like a secondary suspensor of an in
vivo embryo in that the cells are not all of an even length.
We have observed these in all the somatic embryoids we have
produced over the years(11,19) and this has been confirmed by
Cornu and Geoffrion's (3) work on somatic embryognesis of
european larch. A third possibility is the release of
repressed embryonal apices formed during delayed
cleavagepolyembryony of the zygotic embryo. No evidence has
been put forward for the latter, but it is a theoretical
possibility (Fig. 4E).

We can compare this with the previously described
conservative embryogenic scheme recorded from zygotes
(section A1). There does not appear to be any similarity
between the proembryogeny or basal plan (5) of zygotic
embryos and the early development of somatic embryoids.
Therefore use of the terms somatic proembryo and
proembryoid to describe development in tissue culture is
inappropriate.

As a smaller, more trivial matter, there is also a
widespread practice of calling somatic embryoids, embryos.
The term somatic embryoid, rather than embryo, should be
used to reflect the distinctly different morphological
development observed in vitro. The study of somatic

embryogeny, remains a sub-discipline of embryology to which it must conform.

The cells of the suspensor of the immature zygotic embryo explant are not E'sl type cells, or put differently, they do not belong to the primary suspensor. They belong to the secondary suspensor, which is composed of embryonal tube cells. On the ability of embryonal tube cells to produce cell clusters the conifer embryological literature is silent. It is, however, common in conifers that lagging embryonal masses develop embryonal tubes. Somatic embryogenesis from small cell clusters developed from embryonal tubes of explants were very evident for three species of larch (<u>L</u>. <u>decidua</u>, <u>L</u>. <u>leptolepis</u>, <u>L</u>. x <u>eurolepis</u>)(19).

Early embryogeny

<u>Zygotic early embryogenesis</u>. At the end of proembryogeny the primary suspensor elongates and pushes the embryonal tier through the archegonial jacket and into the corrosion cavity. The E layer continues to divide and produce more cells of the same type (e'). In <u>Larix</u> one of these cells usually outgrows the others (Fig. 4A). An apical cell develops which has two cutting faces (4B). As the mass of cells increases, a third cutting face arises and the apical cell becomes tetrahedral in shape (4C). Eventually the apical cell is subsumed within the mass

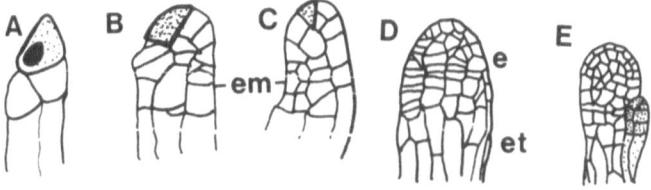

Fig. 4. Early embryogeny. One of the embryonal cells outgrows the others and gives rise to an apical cell, which is here pictured following an oblique division (B). As the embryonal mass (EM) increases the apical cell develops a second (C) and then a third (D) cutting face (after Schopf, 14). D. Early embryogeny. The embryonal mass continues to increase and the tetrahedral apical cells are subsumed in the cells (e') of the tip. The secondary suspensor is composed of embryonal tube cells (et) produced by the embyronal mass (after Schopf, 14). E. <u>In</u> vivo zygotic embryogenesis shows a phenomenon called delayed cleavage polyembryony, or incipient polyembyrony, in which one part of the original embryo tier in the proembryo outgrows the rest. This can lead to two centres of growth occurring as illustrated here (after Schopf, 14). The smaller stippled group of cells is dominated by an apical cell with two cutting faces.

of actively dividing cells. These small apically situated cells are known as the embryonal mass (Fig. 4B,D). As it increases so too does the production of secondary suspensor cells (embryonal tube cells - e't).

<u>Somatic early embryogenesis</u>. By comparison, earliest developmental stages <u>in vitro</u> properly fall under early embyrogeny. In the previous section on development from the explant, embryogeny was described as occurring in one of three ways. All of these represent forms of polyembryony. Development of meristematic clusters from suspensor cells is a form of adventitious polyembryony (15,20). Differential growth of embryonal mass on medium supplemented with 2,4-D represents somatic polyembryony (7). Differential growth of regions of the zygotic embryonal mass represent delayed cleavage polyembryony (14).

The cluster of apically situated cytoplasmically dense cells corresponds to an embryonal mass of zygotic origin. The somatic embryonal mass produces both suspensor and embryo. In zygotic embryogeny the suspensor may be one of two kinds: 1) an embryonal suspensor, a primary suspensor made up of expanded cells derived from the E tier at the end of proembryogeny, which are all equal in length, and only represent a brief stage in suspensor development; 2) embryonal tubes which are unequally elongating secondary suspensor cells derived from the embryonal mass and which make up the bulk of the suspensor. As the embryonal mass enlarges, so too does the suspensor. The suspensor which develops from an enlarged embryonal mass is a secondary suspensor and is composed entirely of embryonal tube cells. Embryos at this point are composed of two different regions, one in which most of the divisions are transverse (proximal end) and another in which the divisions are in all planes (distal end). The distal end is organized, since it has a cone or dome shape (Fig. 2E, 6B). The only meristematic cells at this point are not the cells on the upper surface, but are cells found just above the suspensor (the row initials). This tissue is located between the suspensor and the central mass of cells. Initially, most of the cells formed proximally from the row initials develop into embryonal tubes, but as the entire embryo increases, there are many cells which do not contribute directly to the suspensor, but to intercalary growth.

Embryogenic cultures are generally full of embryonal masses and their suspensors. The cultures, therefore, show quite a bit of organization and are most uncallus-like. Some authors have chosen to use the term embryonal suspensor masses (ESM's), which is descriptively appealing, but unnecessary. We do not think this is applicable to larch cultures, for one major reason: these organized structures already represent embryonal masses. Each embryonal mass has, automatically, a suspensor. The newly coined term ESM - embryonal suspensor mass - implies that there is a mass of embryonal suspensors, but embryonal suspensors, are technically the primary suspensors (15) which are only found in zygotic embryos. The suspensor cell type seen in our cultures is a secondary suspensor cell type, the embryonal tube cell.

It has been established that conifer somatic embryogenic cultures are highly organized. As is often the case, convenience dictates terminological use, and tissue culturists so used to using the term callus have frequently chosen to call their cultures callus cultures. This certainly has the effect of confusing graduate students and people coming into the field.

During this stage in development in vitro, which is essentially the first stage that somatic embryogenic lines of Larix exhibit, it is reportedly possible to test for embryogenicity (8). This test is based on differential staining. The two step method involves: 1) acetocarmine, a general nuclear stain and 2) Evan's Blue, a vital stain which only penetrates cells which have lost their membrane integrity. Tissues are stained in acetocarmine, then they are heated for a short duration, and stained with Evan's Blue. Evan's Blue passes into the embryonal cells, but is excluded from the smaller cells of the embryonal mass. The result is a very clear visual separation of the red embryonal mass from the blue embryonal tube cells. As interpreted by Gupta and Durzan (9), this test shows the difference between embryogenic and non-embryogenic portions of the embryoid. It is not an exclusive measure of embryogenicity. If embryoids are first stained with Evan's Blue, the stain only enters the dead portion of the suspensor. The younger living embryonal tube cells and embryonal mass remain unstained. Following this staining step with acetocarmine leads to a colour distribution of red embryonal mass, red nuclei of living embryonal tube cells and a blue region in the dead portion of the suspensor. The suspensor is therefore not all non-embryogenic. Our results bear this out (Fig. 5). Durzan and Gupta's interpretation of the cytoplasm which stains positively in acetocarmine has been that this represents neocytoplasm. This is open to question since it is difficult to compare with the previous concept of neocytoplasm which, by definition, is a mixture of 1) perinuclear cytoplasm of the egg with 2) some of the cytoplasm of the pollen tube or male gamete.

There are terminological problems in relation to types of cleavage. Direct numerical increase of embryonal masses is due to cleavage polyembryony. Larix zygotic embryos in vivo exhibit an unusual form of cleavage which has been termed incipient cleavage (6), or delayed cleavage (14). Both terms mean the same thing - favored growth of some cells of the original four cells in the primary embryonal group over the others. These cells divide more frequently, produce more cells, and overtop their original neighbours. Figure 4E shows a repressed initial. It has an apical initial having two cutting faces. The cells derived from this unit are still in association with the embryo. Repression of such localized centres of growth may occur because of competitive inhibition within the corrosion cavity. In tissue

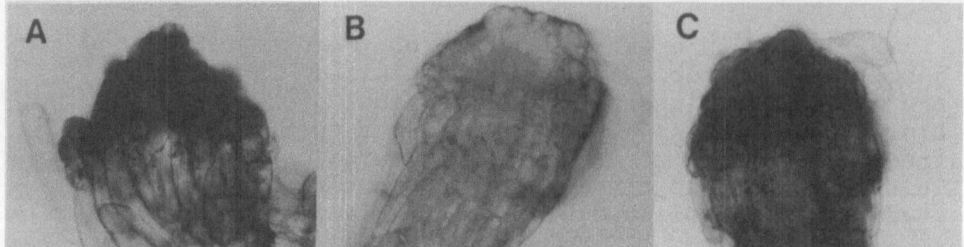

Fig. 5.   Early embryogeny in vitro. A. Embryonal mass
          stained with Evan's Blue and acetocarmine according
          to the schedule devised by Gupta and Durzan (8).
          This technique stains the embryonal tube cells blue
          and the embryonal mass cells red. B. Evan's Blue
          stained embryonal mass. The upper portion of the
          suspensor and the embryonal mass do not take up the
          stain. C. Acetocarmine stained embryonal mass.
          The stain is concentrated in the cytoplasmically
          dense cells, but nuclei of the upper portion of the
          suspensor take up the stain as well. D. Embryonal
          mass stained by Evan's Blue, then with acetocarmine,
          resulting in a three part distribution of colour.
          The embryonal mass stains red with the acetocarmine,
          the upper portion of the secondary suspensor takes
          up little stain (weak staining of the nuclei) and
          the lower, dead portion of the suspensor takes up
          the Evan's Blue.

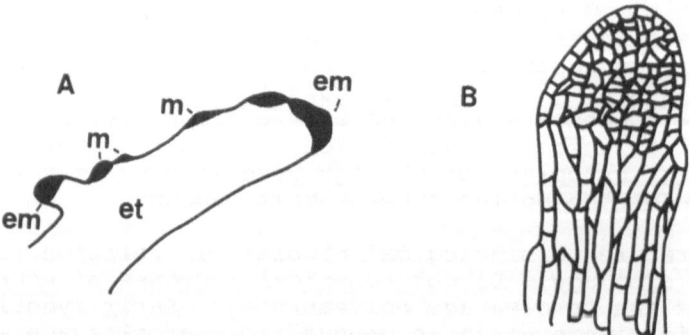

Fig. 6A. Polyembryogenesis in vitro. L. x eurolepis. A number
          of meristematic centres have developed by unequal
          expansion of cells in the embryonal mass. These
          centres develop into embryonal masses themselves
          (after von Aderkas et al., 19).   B.   Early
          embryogeny.   Cross-section of L. decidua zygotic
          embryo at the end of early embryogeny. The cells of
          the distal portion divide in one direction, whereas
          the cells of the proximal portion divide in all
          directions (after Schopf, 14).

cultured embryoids localized differential growth is often seen. The slower areas are not eliminated but prosper on their own. Most of the polyembryony we have documented in vitro is due to such lobing or budding of the embryonal mass (Fig. 6B). Schopf (14) documents a suspected case of cleavage in vivo where two units of the embryonal tier have formed their own embryos, but he noted that this occurred rarely. Dogra (4) considers in vivo phenomenon to be false cleavage in zygotic embryos, since it leads to degeneration. Readily apparent is the fact that events leading to degeneration within the confines of a corrosion cavity do not apply to explants placed in tissue culture, which releases explants from local constraints and competition.

Late embryogeny

Zygotic late embryogenesis. Late embryogeny is characterized by the development of distinct meristems. The first to appear is the root meristem. The intercalary portion eventually develops into the root cap, the area distal to it produces a generative meristem, which in turn becomes the root meristem. This proximal portion of the embryo matures before the shoot meristem. The shoot meristem has a very convoluted origin. The cells of the rounded portion of the embryonal mass divide in all directions. The surface cells contribute to the suspensor at the edges, and later to the root cap, but the cells in the middle eventually become the focal zone (Fig. 7A), which will give rise to the embryonic shoot apex. Cytohistological differentiation becomes evident with formation of a procambium and a stele promeristem at the site of the future node. During this stage the embryo develops a system of secretory elements which are not to be confused with resin ducts (Fig. 7B). Their function is not known. Cotyledons develop, an epidermal layer becomes evident, and a cone of cells develops between the cotyledons (Fig. 7C). This will become the shoot apical meristem sometime during germination, but at this point it is a cluster of largely inactive cells (14,15).

Somatic late embryogenesis. Irrespective of the in vivo embryogenic conifer system, the differentiation of the embryonal mass has not been well-studied. In somatic embryogenesis, it has not been studied at all. Cell segmentation patterns preceding meristem formation remain undescribed. As Singh (15) writes 'critical studies on this aspect of gymnosperms should be rewarding and can hardly be overemphasized.'

Once the embryonal mass shows some organization, there has been a tendency of tissue culturists to apply angiosperm embryological terminology, such as globular and torpedo to conifer embryo development. These terms simply do not exist in the corpus of gymnosperm embryology and should be dropped immediately, to avoid further confusion. The shape of these embryoids may well be globular, but no conifer embryologist has yet to call it the globular stage, and attempts to do so have met with little acceptance.

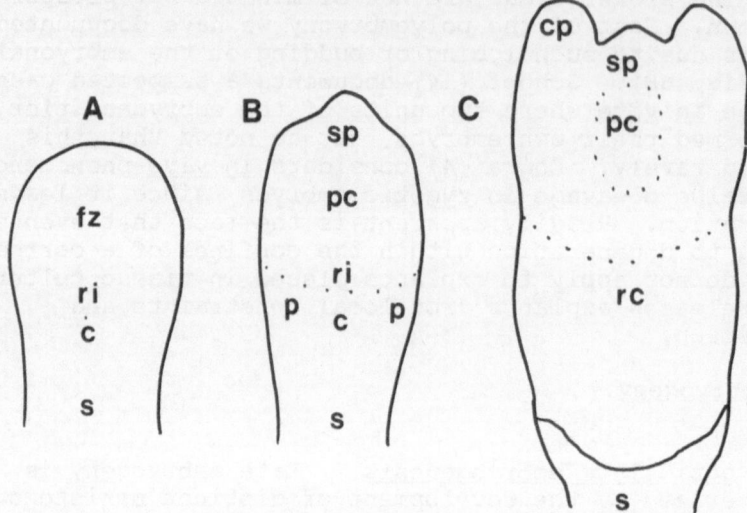

Fig. 7A. Late embryogeny. Cross-section of L. decidua in which the root generative meristem has been initiated. The root initials and calyptoperiblem have begun to develop. Embryonal tube cells are still produced at the bottom. The focal zone of central cells has begun to develop (after Schopf, 14). B. Late embryogeny. Cross-section of L. decidua in which the root cap and stele promeristem have begun. In the root cap there are clearly defined column and pericolumn regions. The stele promeristem is an area above the elongate cells of the procambium (after Schopf, 14). C. Late embryogeny. Cross-section of L. decidua in which the cotyledonary primordia have been initiated. The epicotyl develops after germination (after Schopf, 14).

The mature embryoid is similar to zygotic embryos in that it has two polar meristems, epidermis, cortex, procambium, pith and a root cap. Embryo development in vitro lead to plants which could be outplanted and raised as small trees (11).

HAPLOID DEVELOPMENT

Since there is no evidence of in vivo haploid plant development in the genus Larix, this description is confined to in vitro events, which will be compared with somatic embryogenesis as well as zygotic embryogenesis.

Development from explant

In larch, gynogenic (megagametophyte-derived) embryoids were first induced by Nagmani and Bonga (12). Further investigation has shown the importance of the

phenocritical state of the explant (16,18). The
description of the system has been given in detail
(13,17). 2,4-D induces elongation of the cells of the
megagametophyte (Fig. 8A). These cells must undergo a
polar division if any further development is to occur
(Fig. 8B-D). The result is a small, cytoplasmically dense
cell capable of further division. It seems fairly simple
to elicit long cell development, but we cannot control the
occurrence of the polar division very well. As a result
few of the thousands of explants we have cultured have
actually produced mature embryoids. If small cells
proliferate, they continue to produce similar cells, but
also produce elongate cells. As a result the culture is
composed of a network of long and short cells.
Frequently, cells separate completely from their neighbors
and then continue to divide, resulting in many clusters of
cells. The absence of organization allows one to call
this a callus or a microcallus, since the clusters are so
small. It is of a particularly loose type and composed of
two different cell types. The long cells are also capable
of division. Following this microcallus stage, the
developmental pathway forks. The usual path is formation
of an embryonal mass by multiplication of small cells and
subsequent production of long cells in one general
direction (17), until elongate cell production becomes
quite prolific, at which point it could be considered a
suspensor composed of embryonal tubes (13). We are wary of
considering the first appearance of elongate suspensor-
like cells as proof of embryogenesis as there is often
quite a bit of latitude (Fig. 8C) in the directional
orientation of these elongate cells. For instance, at
this point it is common to find clusters of small cells
from which elongate cells radiate in two or more
directions, quite unlike a cluster of cells from which
elongate cells are produced in one direction only and
which would be considered to be an early embyronic stage.
Another route to embryonal mass formation is the formation
of free nuclei in large ovate cells, which migrate to one
end, where cell walls form (Fig. 8E-G). There is
increasing evidence that this is not an imitation of
proembryogeny, but a rather plastic phenomenon. In a
variation on this theme of subdivision of a larger cell,
the nucleus may migrate to one end divide repeatedly, but
karyokinesis is accompanied by cytokinesis. Whatever the
derivation of the cell cluster at the pole of the original
cell may be, the cells continue to proliferate until they
burst the confines of the original cell and form a mass of
cells which eventually forms elongated cells (16). Either
way, an embryonal mass is formed which produces embryonal
tube cells.

Early embryogeny

With the advent of an embryonal mass development of
haploid gynogenic embryoids proceeds in a fashion
identical to the diploid somatic embryoid, or for that
matter the zygotic embryo. All of them seem to share
developmental pathways from this point onwards. It can be
appreciated that the embryological terminology from
classical conifer embryology is really only applicable
after this point. In vitro, cleavage polyembryony is

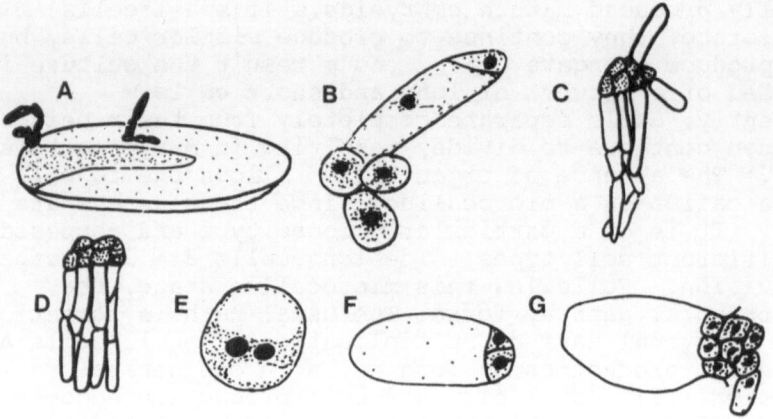

Fig. 8. Haploid embyrogeny. A. Megagametophytes in culture
are induced to produce long cells. These cells
arise from cut surfaces, and may occur near the
micropylar end (M), along the corrosion cavity (CC),
or from any cut surface. B. Long cells produce
small cytoplasmically dense cells by unequal
divisions. These polar divisions continue and the
result is a network of cells composed of long and
short cells. C. A cluster of small cytoplasmically
dense cells may produce long cells in a number of
directions. D. A cluster of small cytoplasmically
dense cells may produce long cells in one direction,
in which case the small cells behave more in keeping
with an embyronal mass, and the long cells are
similar to embryonal tubes of a secondary suspensor.
Note that there were no stages equivalent to
proembryogeny in haploid embryogeny. E. Coenocyte
development. Larger, rounded cells are frequently
seen to divide freely, resulting in multinucleate
conditions. F. The nuclei come to rest at one end
of the cell and form up walls around themselves.
G. The small cytoplasmically dense cells which
arise, divide and burst the confines of the original
cell, producing a mass of cells capable of forming
long cells.

important for regeneration, and there appears to be no difference between haploid and diploid cultures in this regard. Regeneration also occurs in all species from embryonal tube cells dividing to form small clusters of cells which give rise to an embryonal mass (13).

Late embryogeny

Late embryogeny _in vitro_ appears to be very similar to the development of _in vivo_ embryos. This remains, however, an unstudied aspect. Attention has usually focussed on whether an embryo completed its development. It is our observation that the range of sizes of mature embryoids is quite variable, depending on conditions imposed during maturation.

Embryo development _in vitro_ lead to plants which could be outplanted and raised as small trees (Bonga and von Aderkas unpubl).

CONCLUSION

Haploid gynogenic and diploid somatic embryogenic larch cultures are developmentally comparable to zygotic embryos at the stages which are known as early and late embryogeny in the classical conifer embryological literature. However, _in vitro_ there is no evidence of proembryogeny similar to the sequence found _in vivo_. This stage is simply missing in culture, and therefore the term proembryoid should not be used. In all cases, early embryogeny is similar. When embryonal mass is discussed in any of the three systems it means the same thing. By contrast embryonal suspensor mass (ESM) implies that the early embryogenic stage of these cultured tissues is somehow radically different. Durzan and Gupta (7) have stated that somatic embryogenesis is really somatic polyembryogenesis, and this is correct. Somatic polyembryogenesis in larch includes three types of polyembryony. Early embryogenic structures all have the ability to regenerate numerous clusters of cells from the embryonal mass, whether by cleavage polyembryony, adventitious polyembryony or by delayed cleavage. In vivo variations on cleavage polyembryony are very few, but in tissue culture, the physical and biochemical constraints imposed by the corrosion cavity have been removed and polyembryony is quite extensive. In zygotic embryos regeneration is unknown from embryonal tube cells, but is widespread in both haploid and diploid embryoids. In haploid cultures, regeneration from embryonal tube cells may lead to microcallus formation, the only occurrence of a true callus in any of these stages.

The importance of using terminology correctly is that it avoids attributing characteristics of _in vivo_ development to _in vitro_ cultured material. There are evidently many ways an embryo may develop. In those cases which have been studied in larch, we have attempted to draw together the _in vivo_ and _in vitro_ terminology where similarity in development warrants unity. The goals of this paper have been to show how inappropriate use of terms obscures some of these developmental differences.

REFERENCES

1. Attree, S.M. and Fowke, L.C., in press, Somatic embryogenesis in conifers, in, "Biotechnology in agriculture and forestry," vol. 17, Y. P. S. Bajaj, ed., Springer Verlag, Berlin and New York.
2. Berlyn, G.P., 1972, Seed germination and morphogenesis, in, "Seed biology," vol. I, pp. 223-312, T. T. Kozlowski, ed., Academic Press, New York and London.
3. Cornu, D., and Geofrrion, C., in press, Aspects de l'embryogénèse somatique chez le mélèze.
4. Dogra, P.D., 1967, Seed sterility and distrubances in embryogeny in conifers with particular reference to seed testing and tree breeding in Pinaceae, St. For. Suecica, 45: 5-97.
5. Doyle, J., 1963, Proembryogeny in Pinus in relation to that in other conifers - a survey - Proc. Roy. Irish Acad., 62A: 181-216.
6. Doyle, J., and Brennan, M., 1971, Cleavage polyembryony in conifers and taxads - a survey. II. Cupressaceae, Pinaceae and conclusions, Sci. Proc. Roy. Dublin Soc., 4A: 137-158.
7. Durzan, D.J., and Gupta, P.K. 1987, Somatic embryogenesis and polyembryogenesis in Douglas-fir cell suspension cultures, Plant Sci., 52: 229-235.
8. Gupta, P.K.. and Durzan, D.J., 1987a, Biotechnology of somatic polyembryogenesis and plantlet regeneration in loblolly pine, Bio/technology, 5:147-151.
9. Gupta, P.K., and Durzan, D.J., 1987b, Somatic embryos from protoplasts of loblolly pine eproembryonal cells, Bio/technology 5:710-712.
10. Hakman, I., Fowke, L.C., von Arnold, S., and Eriksson, T., 1985, The development of somatic embryos in tissue cultures initiated from immature embryos of Picea abies (Norway spruce), Plant Sci., 38:53-59.
11. Klimaszewska, K., 1989, Plantlet development from immature zygotic embryos of hybrid larch through somatic embryogenesis, Plant Sci., 63:95-103.
12. Nagmani, R., and Bonga, J.M., 1985, Embryogenesis in subcultured callus of Larix decidua, Can. J. For. Res., 15:1088-1091.
13. Rohr, R., von Aderkas, P., and Bonga, J.M., 1989, Ultrastructural changes in haploid embryoids of Larix decidua during early embryogenesis, Am. J. Bot., 76:1460-1467.
14. Schopf, J.M., 1943, The embryology of Larix, Illinois Biological Monographs, 19:1-97.
15. Singh, H., 1978, Embryology of gymnosperms, Gebrüder Borntraeger, Berlin and Stuttgart.
16. von Aderkas, P., and Bonga, J.M., 1988a, Formation of haploid embryoids of Larix decidua: early embryogenesis, Am. J. Bot., 75:690-700.
17. von Aderkas, P., and Bonga, J.M., 1988b, Morphological definition of phenocritical period for initiation of haploid embryogenic tissue, in, "Somatic cell genetics of woody plants", pp. 29-38, M. R. Ahuja, ed., Kluwer Academic, Dordrecht.

18. von Aderkas, P., Bonga, J.M., Nagmani, R., 1987, Promotion of embryogenesis in cultured megagametophytes of <u>Larix</u> <u>decidua</u>, Can. J. For. Res., 17: 1293-1296.

19. von Aderkas, P. Klimaszewska, K., and Bonga, J.M., 1990, Diploid and haploid embryogenesis in <u>Larix</u> <u>leptolepis</u>, <u>L.</u> <u>decidua</u>, and their reciprocal hybrids, Can. J. For. Res. 20: 9-14.

20. Williams, E.G., and Mahashwaran, G., 1986, Somatic embryogenesis: factors influencing coordinated behaviour of cells as an embryogenic group, Ann. Bot., 57: 443-462.

Application of Somatic Embryogenesis to Clonal Propagation of Interior Spruce

D.R. Roberts, F.B. Webster, B.S. Flinn,
W.R. Lazaroff, S.M. McInnis and B.C.S. Sutton

Forest Biotechnology Centre, British Columbia
Research Corporation, Vancouver, British Columbia
Canada V6S 2L2

ABSTRACT

Seventy one lines (genotypes) of embryogenic cultures from six open pollinated families were obtained by culturing immature embryos of interior spruce. Induction frequencies varied among families and was also affected by cone collection date. Comparison of the SDS-PAGE protein profiles of embryo explants from different collection dates revealed that embryos were most competent for embryogenesis during the period of development that directly preceded significant accumulation of storage protein. In order to optimize the maturation protocol, cultures from each genotype were screened through a range of abscisic acid (ABA) concentrations. The ABA-dependent developmental profile (the proportion of shooty embryos, precociously germinating embryos and mature embryos) differed among genotypes but, in general, production of somatic embryos was highest at 40 and 60 μM ABA. At these concentrations of ABA, embryos from most genotypes did not germinate precociously and entered a period of quiescence. Germination of eight genotypes tested was markedly enhanced after partial drying of mature embryos at high relative humidity. After one week, root emergence averaged 67% for somatic embryos treated with partial drying compared with 0% for non-treated controls. Emblings derived from this method were vigorous and acclimatized well to soil. Survival of emblings following transfer to soil, acclimatization and the first season's growth in the nursery was 80% or greater for most genotypes and averaged 83% overall. Growth of emblings during the first season averaged 8.5 cm and was equivalent to that of seedlings.

INTRODUCTION

There have been considerable advances in conifer tissue culture over the past four years since the first report of somatic embryogenesis (16). Embryogenic cultures have been induced from a wide range of conifer species and maturation protocols for embryo development have been improved such that embryo production and quality have been enhanced (3,9,13,14,17,27). However, utilization of somatic embryogenesis for clonal propagation has been limited by the low frequency of root emergence and poor vigour of the resulting plantlets (3,8,12). The term "embling" has been recommended to describe plants produced from somatic embryogenesis and will be used throughout this report (Libby, Univ. of Calif. Berkeley, personnel communication).

Embryo drying is a natural feature of seed development and it has been proposed that desiccation has a role in the developmental transition between maturation and germination (19). Desiccation has been reported to stimulate germination of alfalfa and soybean somatic embryos (24; Seneratna, T. Univ. of Guelph, Guelph, Ontario). We have found that partial drying of spruce somatic embryos promotes root emergence and enhances embling vigour. This advance has facilitated the production of a large number of emblings for evaluation of acclimatization and nursery performance. In this report we have evaluated the application of somatic embryogenesis to propagation of interior spruce from the induction of embryogenic cultures, through maturation, germination, acclimatization and the first season's nursery growth.

MATERIALS AND METHODS

Interior spruce represents a mixture of two closely related species <u>Picea glauca</u> [Moench] Voss and <u>Picea engelmannii</u> Parry, from the interior of British Columbia where they hybridize with one another (23). Cones were collected from six open pollinated families (source trees 10,81,103,118,133,171) on three collection dates (13th and 21st of July and August 24, 1987) by British Columbia Ministry of Forests, Red Rock Nursery, Prince George, British Columbia. Embryogenic cultures were induced from immature embryos of interior spruce according to established protocols (16,28). Cultures have been maintained in the dark on (VE) von Arnold and Erikkson (2) basal medium containing 2.2 mg/l 2,4-dichlorophenoxyacetic acid, 1.13 mg/l benzyladenine, 1% sucrose, 0.54% agar noble, pH 5.5 at 27°C. Cultures were subcultured every two weeks.

Culture weight was determined prior to differentiation for later comparisons. In order to induce differentiation, cultures were transferred from maintenance media to VE basal media containing 1% charcoal and 3.4% sucrose for one week in the light (16 h photoperiod at 25-35 μmol m^{-2} sec^{-1}). In order to optimize the maturation protocol for each genotype the cultures were then screened through a range of abscisic acid (ABA) concentrations (0, 20, 40, 60 μM). Each ABA treatment also included 1 μM indole-butyric acid (IBA). Cultures were subcultured every two weeks and embryos were allowed to mature for five weeks at which time the number of rooty embryos, shooty embryos, precociously germinating embryos and mature embryos were determined. Rooty and shooty embryos were aberrant structures characterized by a root protruding from a callus mass and by stunted cotyledon/hypocotyl development with a basal callus, respectively. Bipolar embryos were characterized as precociously germinated if they exhibited greening or elongation of the cotyledon/hypocotyl and mature if they remained opaque with well developed cotyledons.

Mature embryos (45/genotype) were removed from the culture and placed in 6 wells of a 12 well petri plate in which the other six wells contained 2.5 mls of sterile distilled water. The plate was wrapped with Parafilm and incubated in the dark at 27°C for 16 days. Embryos were removed and placed in shell vials on slants of 1/2 VE basal medium containing 2% sucrose and 0.54% agar. Root emergence was recorded after one and five weeks on this germination medium. After five weeks the emblings with roots were removed from the vials and shoot and root length were recorded. Emblings were then transferred to styroblocks containing a mixture of peat:sawdust which had been pre-wetted with tap water and treated with 5 ml of 0.2 Hoaglands solution. Plastic wrap was stretched over the styroblocks (forming a tent-like cover held open along the sides by about 1 inch with wooden toothpicks). The styroblocks were placed under low light (10 μmol m^{-2} sec^{-1}) for the first 3 days and then transferred to high light (70 μmol m^{-2} sec^{-1}) for 3 weeks (plastic wrap was removed after 1 week at high light). Styroblocks were then transferred to an operational forest nursery (Peltons Reforestation, Maple Ridge, British Columbia) and placed in an open sided greenhouse where they received mist irrigation and fertilization alongside seedlings of interior spruce. Interior spruce seedlings were germinated at the nursery during the same period as emblings were germinated in vitro and grown alongside the emblings in the greenhouse. Height measurements were made at the end of the growing season.

RESULTS AND DISCUSSION

The percentage of zygotic embryo explants that formed somatic embryos differed among source trees and collection date (Fig. 1). Zygotic embryos from the first collection date produced somatic embryos at the highest frequency, with those from 118 showing the best response followed by 171, 10,103 and 81. The competence of the embryos to form embryogenic cultures declined markedly in collections two and three. These results are similar to those reported previously where it was found that the embryogenic response of immature spruce embryos was related to collection date and a specific stage of embryo development (5,16). Only a two-three week period exists during the cotyledonary stage of embryo development where a high frequency of embryogenesis can be achieved (5). In our study, embryos from each collection date had well developed cotyledons and it was difficult to distinguish explants with higher competence for embryogenesis based on morphological traits. The major seed storage proteins of interior spruce were identified from their presence in purified protein bodies and used as biochemical markers for embryo maturation (Fig. 2). SDS-PAGE analysis of embryo explants revealed that cotyledonary embryos from the first collection date, with a higher capacity to undergo embryogenesis had not accumulated significant levels of storage proteins and could be distinguished from embryos of the later collection dates, with decreased competence, based on the storage protein levels. This provides a biochemical criterion for explant selection and may give some insight into the types of biochemical changes that occur during embryo maturation which are associated with explant competence. Recent reports

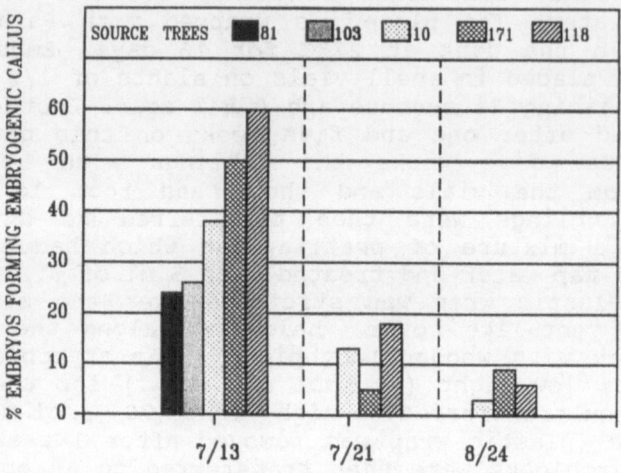

Fig. 1. Formation of embryogenic callus by embryo explants from three collection dates.

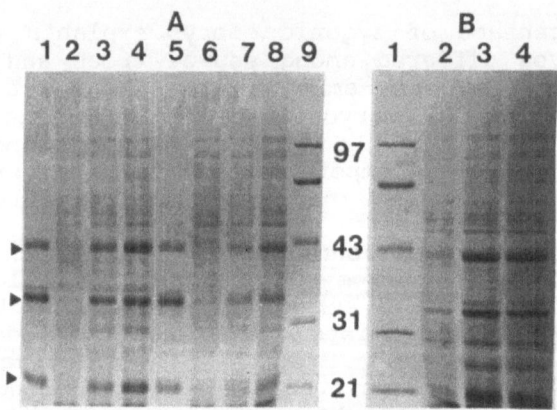

Fig. 2. Protein profiles of embryo explants from the three collection dates. A. Lane 1, purified protein bodies; Lanes 2,3 and 4, protein profiles from source tree 10 for collection dates 7/13, 7/27 and 8/24, respectively; Lane 5, purified protein bodies; Lanes 6,7 and 8, protein profiles from source tree 171 for collection dates 7/13, 7/27, and 8/24, respectively; Lane 9 Mw marker proteins. B. Lane 1, Mw protein markers; Lanes 2,3,4, protein profiles from 118 for collection dates 7/13, 7/27, and 8/24, respectively (arrows denote storage proteins).

have demonstrated embryogenesis from mature and germinated embryo explants of spruce (4,21). However, the frequency of embryogenesis is low compared to that obtained from the immature embryo and in order to apply this system to forestry it will be important to be able to capture a diversity of genetic backgrounds. This will require a embryogenic response by the maximum number of explants and therefore requires identification and utilization of the most competent explant available.

Significant improvements have been made in the protocols for maturation of the proembryos found in the proliferating callus (3,13,17). These improvements have enabled a much greater yield of embryos to be obtained from the culture mass. We have found that by increasing ABA beyond the concentrations found optimal by others (20-60 μM), precocious germination was inhibited, embryos accumulated storage proteins and then entered a period of quiescence. Embryo types found on spruce calli were classified as shooty, precociously germinating and mature embryos (Fig 3). At low levels of ABA rooty and shooty embryos (SE) are produced. Increasing ABA (10-20 μM) promoted normal bipolar embryo development but these germinated precociously (PE) on the maturation medium. Levels of ABA between 20 and 60 μM inhibited precocious germination such that embryos accumulated storage proteins and entered a period of quiescence (ME). A similar relationship between ABA concentrations and embryo maturation was found in cultures of caraway (1). Protein bodies have been identified in spruce somatic embryos developed in the presence of ABA and recently ABA has been found to promote the accumulation of storage lipid in spruce embryogenic callus (15,17). Accumulation of sufficient levels of storage reserves by somatic embryos is most likely a prerequisite for vigorous germination (26). The poor germination and root elongation that has characterized spruce somatic embryos may, at least in part, be attributed to precocious germination which can have an adverse effect on subsequent development (7,25). The mature, quiescent somatic embryo more closely resembles the zygotic embryo during latter stages of maturation compared with those that germinate precociously (Flinn et al., Forest Biotechnology Centre, British Columbia, unpublished data).

A developmental profile was established for each of the 71 genotypes by recording the types of embryos produced by the cultures at different concentrations of ABA. The response to the maturation protocol differed among genotypes (Fig. 4). The bar graph depicts the number of different embryo types found on the callus of genotypes 66 and 70 at 0, 20, 40 and 60 μM ABA. Genotype 66 exhibited a much greater propensity for shooty and precociously germinating embryos at lower levels of ABA whereas genotype 70 mainly produced mature embryos at each of the ABA concentrations tested. The optimum concentration for production of mature embryos was 40 μM for genotype 70 while genotype 66 produced few mature embryos below 60 μM. This data demonstrates that in order to optimize the production of mature embryos for each genotype, each must be tested for its individual maturation profile. The number of mature somatic embryos produced by different genotypes varied between 1 and 440 per

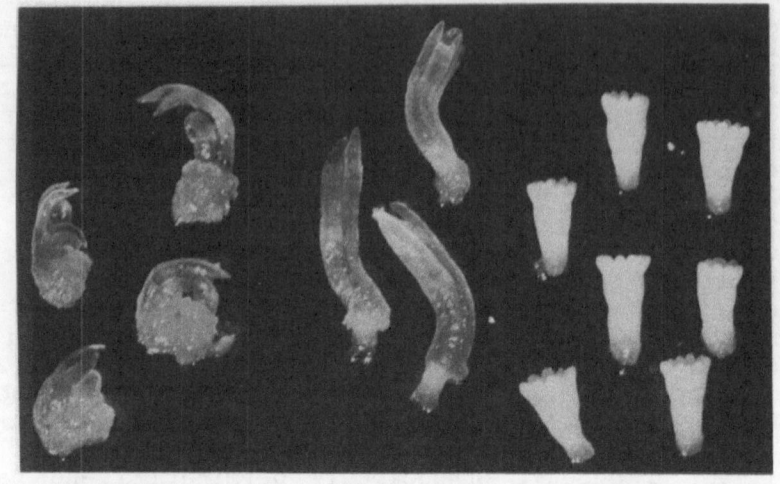

SE   PE   ME

Fig. 3. Effects of abscisic acid on the morphology of embryo types. Shooty embryos (SE) that predominated at low levels of ABA (1-10 μM), precociously germinating embryos (PE) formed on 10-20 μM ABA and mature somatic embryos (ME) produced on concentrations of ABA above 30 μM. This profile is from genotype 11.

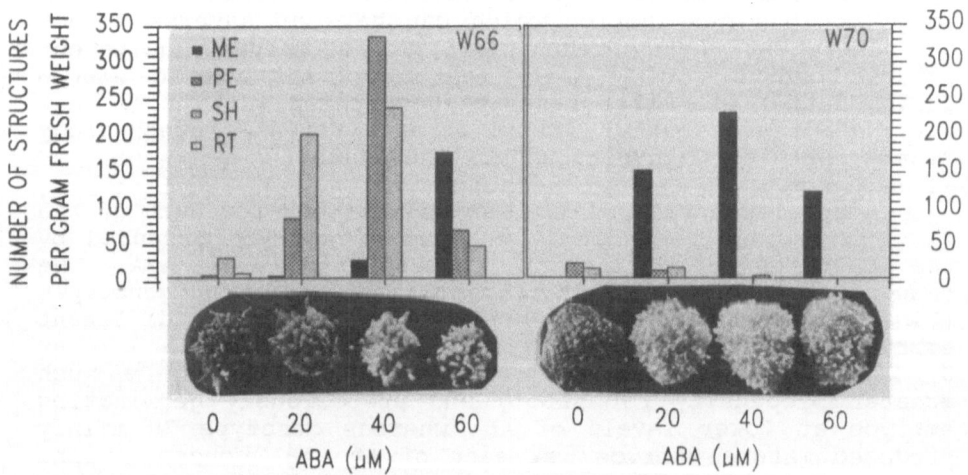

Fig. 4. Profile of embryo types produced on calli from two genotypes at different concentrations of abscisic acid.

gm fresh wt of culture and no mature embryos were produced by 3 genotypes (data not shown). In a similar study, it was found that differentiation varied among genotypes of Norway spruce (6). However, at the time of that study maturation protocols were less developed and the callus potential was not as fully realized. In figure 5 the response of all 71 genotypes to the ABA screening is summarized and shows that abnormal development (rooty and shooty embryos) predominated at low concentrations of ABA, embryos germinated precociously at 20 μM ABA and that the optimum level of ABA for production of mature embryos for most genotypes was 40 and 60 μM ABA.

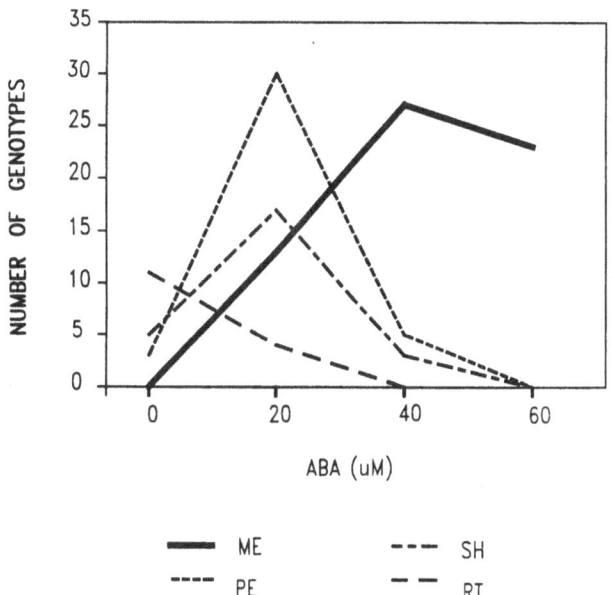

Fig. 5. Distribution of optimal abscisic acid concentration for the production of specific embryo types.

The use of somatic embryogenesis for clonal propagation of conifers has been limited by low germination rates and poor plantlet performance (3,8,12). This is especially evident for somatic embryos of interior spruce by their low frequency of root emergence (Fig. 6, Table 1). Root emergence of somatic embryos placed directly on germination medium reached only 24% and occurred over a period of two to three weeks. We have found that a partial drying of the somatic embryos can greatly increase the frequency of root emergence and root vigour. Furthermore, there is a marked decrease in the time period over which root emergence occurs such that instead of 3 weeks it occurs within 1 week of placing the embryo on germination medium. The partial

Table 1. The effects of partial drying on germination of mature somatic embryos.

	EMBRYOS WITH ROOT EMERGENCE (%)			
	1 WEEK		5 WEEKS	
CLONE	NON-DESICCATED	DESICCATED	NON-DESICCATED	DESICCATED
74	0	87	31	91
70	0	82	24	100
68	0	73	4	93
29	0	65	0	53
11	0	60	29	98
51	NA[a]	46	16	52
58	0	40	7	21
76	0	24	4	49

[a]   Data not available.

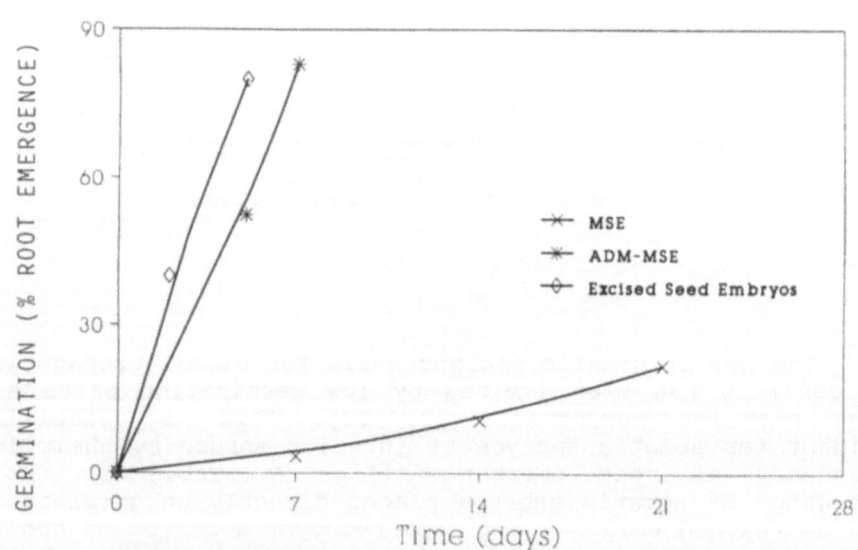

Fig. 6. Effects of partial drying on germination of spruce somatic embryos. Hydrated embryos (MSE), partially dried embryo (ADM-MSE) and excised seed embryo.

drying treatment promoted synchronized germination such that root elongation coincided with elongation of the hypocotyl/cotyledons. This process is similar to germination of excised spruce zygotic embryos (Fig. 6). Partial drying of castor bean seed embryos can stimulate germination in lieu of desiccation (20). Following germination, the vigour of root elongation is also enhanced in spruce somatic embryos treated with partial drying. The effects of partial drying on germination of spruce embryos is similar to those obtained through more severe desiccation of alfalfa and soybean somatic embryos (24; Seneratna, T. University of Guelph, Ontario, personal communication). However, spruce embryos were not as tolerant to drying as these other species and did not survive drying in low (81% and less) humidities (Roberts et al., unpublished data).

The effects of partial drying on the germination of spruce somatic embryos was compared to non-treated controls for 8 different genotypes (Table 1) and for each genotype germination frequency was improved. However, in some cases although germination frequency was improved it remained low and we believe this reflects a technical barrier in using desiccated or partially dried embryos. The poor germination appears to result from rapid uptake of water from the germination medium by the naked embryo. This phenomena termed "imbibition damage" is well documented for germination of excised seed embryos (29). For some reason this occurred more often in some genotypes than others and most likely reflects variation in the effects of partial drying or germination treatments.

Sucrose was a critical component of the germination medium and at low levels (less than 1%), germination was very slow and the frequency was diminished. Carbohydrates are also required for the germination of excised seed embryos (10) and this most likely reflects the absence of the megagametopyte which zygotic embryos utilize as a nutrient source during germination.

Acclimatization of tissue culture propagules has proven to be problematic and the transition between the test tube and the ambient environmental conditions can be the most difficult stage of micropropagation (11). Emblings which resulted from the germination of partially dried embryos were transferred after 5 weeks of in vitro growth to styroblocks containing a standard peat:sawdust soil mix used by local forest nurseries. The acclimatization of the emblings to soil was quite successful and demonstrates the vigour of emblings following the partial drying treatment. Initially, it was necessary to provide the emblings with protection from the lower ex vitro humidity by covering the styroblock with a "plastic wrap tent". This combined with a short period at low light intensities resulted in survival frequencies of greater than 80% for most of the genotypes tested and an overall average of 83% (Table 2). This survival value also includes losses that occurred during the first season's growth in the nursery due to natural phenomena such as damping off, insect damage, etc.

Table 2. Survival of emblings after acclimatization and one growing season in the nursery.

Survival Category (%)	No. of Genotypes
> 90	26
> 80	20
> 70	10
> 60	7
< 50	4

Observations were made on the emblings throughout the growing season and there was a high degree of phenotypic uniformity within the clones, which may indicate limited somaclonal variation. However, it is possible that an entire culture line could be derived from a somaclonal variant, which would not be apparent from this experiment. Interior spruce seed was germinated during the same period as somatic embryos and their growth compared at the end of the first growing season. The average height of the emblings at the end of the first growing season was 8.5 cm, which equalled the average height of the control seedlings (Table 3). In addition, the average height of some genotypes was greater than seedlings after one growing season. These growth rates can be contrasted with those reported for conifer plantlets derived through organogenesis, where in general, early growth of plantlets is much slower when compared to seedlings (18,22). At the end of the first growing season it was difficult to distinguish between the average embling and seedlings based on visual observations of shoot, branch and root development (Fig. 7). During the fall emblings developed buds characteristic of seedlings and changes in foliage colour of emblings and seedlings indicated that they had cold hardened in a similar manner.

Table 3. Height of emblings and seedlings after one growing season in the nursery.

Average Height Category (cm)	No. of Genotypes
10 - 14	20
7 - 9.9*	29
5 - 6.9	12
<5	5

* Average height of seedling controls was 8.5 cm.

1   2   3   4   5   6

Fig. 7. Photograph of emblings and
seedlings after one growing season.
(From left to right, emblings 1,2,3
from genotypes 51,29 and 11,
respectively; control seedlings 4,5,6).

SUMMARY

Somatic embryogenesis has been applied to the
propagation of interior spruce. Maturation of the somatic
embryos was optimized by screening each genotype through a
range of ABA concentrations. The optimum ABA concentration
for production of mature embryos varied among genotypes.
Germination and embling vigour was markedly enhanced by
partial drying of the embryos at high relative humidity.
Emblings acclimatized to soil and _ex vitro_ conditions at a
high frequency and the first season's nursery growth of the
emblings was comparable to that of seedlings.

ACKNOWLEDGEMENTS

The authors greatly appreciate the assistance of Dr.
D.T. Webb, Mr. S. Pelton, Mr. F. Rey and Mr. R. Folk in
carrying out this research. The authors would like to
acknowledge the assistance of Mr. G. Kiss and the British
Columbia Ministry of Forests and Lands for cone material and
funding provided by Forestry Canada and the British Columbia
Ministry of Forests through the Forest Development
Agreement. A special thanks to Ms. Kim Gowe for her help in
preparation of this manuscript.

REFERENCES

1. Ammirato, P.V. 1974. The effects of abscisic acid on the development of somatic embryos from cells of caraway (Carum carvi L.). Bot. Gaz. 135:328-337.

2. von Arnold, S. and Erikkson, T. 1977. A revised medium for growth of pea mesophyll protoplasts. Physiol. Plant 39:257-260.

3. von Arnold, S., and Hakman, I. 1988. Regulation of somatic embryo development in Picea abies by abscisic acid (ABA). J. Plant Physiol. 132:164-169.

4. von Arnold, S., and Woodward, S. 1988. Organogenesis and embryogenesis in mature zygotic embryos of Picea sitchensis. Tree Physiol. 4:291-300.

5. Becwar, M.R., Wann, S.R., Johnson, M.A., Verhagen, S.A., Feirer, R.P. and Nagmani, R. 1988. Development and characterization of in vitro embryogenic systems in conifers. In: M.R. Ahuja [Ed.], Somatic cell genetics of woody plants, Kluwer Academic Publishers, Dordrecht, The Netherlands, pp. 1-18.

6. Becwar, M.R., Noland, T.L. and Wann, S.R. 1987. A method for quantification of the level of somatic embryogenesis among Norway spruce callus lines. Plant Cell Rep. 6:35-38.

7. Bewley, J.D. and Black, M. 1985. Maturation Drying-The effects of water loss on developmental events. In: Seeds: Physiology of Development and Germination. Plenum Publishing Corp. New York, NY. pp 70-73.

8. Boulay, M.P., Gupta, P.K., Krogstrup, P. and Durzan, D.J. 1988. Development of somatic embryos from cell suspension cultures of Norway spruce (Picea abies Karst.). Plant Cell Rep. 7:134-137.

9. Bourgkard, F., and Favre, J.M. 1988. Somatic embryos from callus of Sequoia sempervirens. Plant Cell Reports 7:445-448.

10. Brown, C.L., Gifford, E.M. 1958. The relation of the cotyledons to root development of pine embryos grown in vitro. Plant Physiol. 33:57-64.

11. Driver, J.A. and Suttle, G.R.L. 1987. Nursery Handling of Propagules. In: Cell and Tissue Culture in Forestry. Vol. 2. Specific Principles and Methods: Growth and Developments. Edts. J.M. Bonga and D.J. Durzan. Martinus Nijhoff Publishers, Dordrecht. The Netherlands.

12. Dunstan, D.I. 1988. Prospects and progress in conifer biotechnology. Can. J. For. Res. 18:1497-1506.

13. Dunstan, D.I., Bekkaoui, F., Pilon, M., Fowke, L.C. and Abrams, S.R. 1988. Effects of abscisic acid and analogues on the maturation of white spruce (Picea glauca) somatic embryos. Plant Science 58:77-84.

14. Durzan, D.J. and Gupta, P.K. 1987. Somatic embryogenesis and polyembryogenesis in Douglas-fir cell suspension cultures. Plant Science 52:229-235.

15. Feirer, R.P., Conkey, J.H. and Verhagenm S.A. 1989. Triglycerides in embryogenic conifer calli: a comparison with zygotic embryos. Plant Cell Reports 8:207-209.

16. Hakman, I. and von Arnold, S. 1985. Plantlet regeneration through somatic embryogenesis in *Picea abies* (Norway spruce). J. Plant Physiol. 121:149-158.

17. Hakman, I. and von Arnold, S. 1988. Somatic embryogenesis and plant regeneration from suspension cultures of *Picea glauca* (White spruce). Physiol. Plant. 72:579-587.

18. Horgan, K. 1987. *Pinus Radiata*. In: Cell and Tissue Culture in Forestry. Vol. 3. Case Histories: Gymnosperms, Angiosperms and Palms. Edts. J.M. Bonga and D.J. Durzan. Martinus Nijhoff, The Netherlands. pp 128-145.

19. Kermode, A.R. and Bewley, J.D. 1985. The role of maturation drying in the transition from seed development to germination. J.Exp. Bot. 36:1906-1915.

20. Kermode, A.R. and Bewley, J.D. 1989. Developing seed of *Ricinus communis L.*, detached and maintained in an atmosphere of high relative humidity, switch to germinative mode without the requirement for complete desiccation. Plant Physiol. 90:702-707.

21. Krogstrup, P. 1986. Embryo-like structures from cotyledons and ripe embryos of Norway spruce (*Picea abies*). Can. J. For. Res. 16:664-668.

22. McKeand, S.E. and Allen, H.L. 1984. Nutritional and root development factors affecting growth of tissue culture plantlets of loblolly pine. Physiol. Plant. 61:523-528.

23. Owens, J.N. and Molder, M. 1984. The reproductive cycle of interior spruce. British Columbia Ministry of Forests, Information Services Branch; Victoria, British Columbia.

24. Parrott, W.A., Dryden, G., Vogt, S., Hilderbrand, D.F. Collins, G.B. Williams, E.G. 1988. Optimization of somatic embryogenesis and embryo germination in soybean. In Vitro Cell. and Dev. Biol. 24:817-820.

25. Raghaven, V. 1986. Experimental embryogenesis. In: Embryogenesis in Angiosperms. A Developmental and Experimental Study. Cambridge Univ. Press. pp. 103-110.

26. Redenbaugh, K., Paasch, B.D., Nichol, J.W., Kossler, M.E., Viss, P.R. & Walker, K.A. 1986. Somatic Seeds: Encapsulation of asexual plant embryos. Bio/tech. 4:797-781.

27. Roberts, D.R., Flinn, B.S., Webb, D.T., Webster, F.B. and Sutton, B.C.S. 1990. Abscisic acid and indole-3-butyric acid regulation of maturation and accumulation of storage proteins in somatic embryos of interior spruce. Physiol. Plant. 78:355-360.

28. Webb, D.T., Webster, F., Flinn, B.S., Roberts, D.R. and Ellis, D.E. 1989. Factors influencing the induction of embryogenic and caulogenic callus form embryos of *Picea glauca* and *P. engelmannii*. Can. J. For. Res. 19 (10):1303-1308.

29. Woodstock, L.W. 1988. Seed Imbibition: A critical period for successful germination. J. Seed Technol. 12:1-15.

EFFECT OF EXPLANT AND MEDIA ON INITIATION, MAINTENANCE, AND MATURATION OF SOMATIC EMBRYOS IN PSEUDOTSUGA MENZIESII (MIRB.) FRANCO (DOUGLAS-FIR)

R. Nagmani, M. A. Johnson, and R. J. Dinus

The Institute of Paper Science and Technology
Atlanta, GA 30318

ABSTRACT

Embryogenic calli were initiated from immature zygotic embryos of Pseudotsuga menziesii (Mirb.) Franco (Douglas-fir). Only precotyledonary embryos gave rise to embryogenic cell lines that could be maintained and manipulated to yield cotyledonary somatic embryos. Neither intact ovules nor cotyledonary explants were responsive to the four test media. Embryogenic cell lines were initiated on modified MS medium supplemented with L-glutamine (500 mg/l), casein hydrolysate (1000 mg/l), and 2,4-D (5 (mg/l), either alone or in combination with BA (2.5 mg/l). Three of the four embryogenic cell lines originated from one source tree and were initiated in the light, and the other was derived from a second source tree and initiated in darkness. Responsive explants were collected between July 14 and 25, 1988, in western Washington. All four lines have been maintained for 10 or more months. Limited numbers of cotyledonary somatic embryos were obtained following transfer to a transition medium with no growth regulators and casein hydrolysate but with L-glutamine and activated charcoal, and subsequent culture on maturation medium lacking casein hydrolysate and supplemented with abscissic acid and L-glutamine.

Abbreviations: 2,4-D = 2,4-dichlorophenoxy acetic acid, BA = benzyladenine, MS = Basal medium of Murashige and Skoog (1962).

INTRODUCTION

Initiation of embryogenic callus and development of somatic embryos has been reported for a number of conifer species since somatic embryogenesis was first described for Norway spruce [Picea abies (L.) Karst] (9) and European larch (Larix decidua L.) (12). Workable protocols have since been defined for Norway spruce (9, 1), white spruce (Picea glauca), (10), white pine (Pinus strobus L.), (6) and sugar and loblolly pines (Pinus lambertiana L. and P. taeda L.) (7, 8).

In other species such as Douglas-fir, however, initiation, maintenance, and development are not nearly so straightforward. Small amounts of embryogenic callus have been observed on occasion in our laboratory, but all such cultures quickly turned brown and eventually died. Durzan and Gupta (5) reported initiation from immature embryos, but their report omitted some details of the experimental protocol; e.g., explant type and developmental stage, number of explants used per treatment, and relative importance of media and growth regulator combinations. In addition, the proportion of cultures that could be maintained over time was not specified. The present investigation was undertaken to further examine these and related issues and to explore protocols for development and maturation.

MATERIALS AND METHODS

Plant Materials

Developing female cones of Douglas-fir at pre- and post-fertilization stages of ovule development were supplied by Weyerhaeuser Corp., Federal Way, WA, in June, July, and August, 1988. Open-pollinated cones were collected at weekly intervals from each of five unrelated mother trees (WTC-566-69 and 571) in a commercial seed orchard, and shipped to our laboratory in insulated containers via airfreight. Seeds were removed from cones within 24-48 hours of arrival, surface-sterilized with commercial bleach (Hilex, 25% v/v) containing 6% sodium hypochlorite, for 15 min., and then rinsed three times with sterile distilled water. Seed coats were aseptically removed, and explants excised for culture. Explant types were: intact pre- and post-fertilization ovules collected between June 3 and July 15, isolated precotyledonary embryos collected between July 14 and 28, and isolated cotyledonary embryos collected between July 29 and August 11. Relative size and developmental status of isolated embryos were noted prior to culture.

MEDIA AND CULTURE CONDITIONS

Initiation

Test media were variations of MS basal medium (11). Inorganic, vitamin, and amino acid components as well as growth regulator concentrations and combinations are given in Table 1. Salt composition of MSCG medium was altered per Brown and Lawrence (4). DCR and DG were used by Gupta and Durzan to obtain initiation for sugar pine (7) and Douglas-fir (5). Composition of MS medium is shown for comparison.

Media for initiation were gelled with 0.8% agar and their pH adjusted to 5.5 prior to autoclaving at 121 psi for 15 min. L-glutamine was filter-sterilized and added to warm (50°C) medium. Eight to 10 ml of medium were dispensed into 50 mm diameter Falcon Petri dishes.

Five explants were cultured in each dish, with the number of dishes per treatment ranging from 2 to 10. Over 80% of individual treatment combinations were represented by at least 8 dishes and 40 explants. A total of 7055 intact ovules were cultured on MSCG 5/0, MSCG 5/2.5, and DG media. The 5693 isolated precotyledonary and cotyledonary embryos were cultured on all test media.

Cultures were incubated at 24-26°C, half in the light and half in the darkness. Light during the 16 hr photoperiod was provided by incandescent and cool-white fluorescent lights, 15-50 μE m^{-2} s^{-1} at culture level.

Cultures were examined at weekly intervals, and explants were transferred to new positions in the same Petri dish every two or four weeks. Scoring for embryogenic callus was begun shortly after cultures were started. Once putatively responsive cultures were noted, entire explants and/or associated calli were transferred to fresh media every week.

To determine if cultures were embryogenic, explants and associated calli were placed on glass slides, stained with 1% acetocarmine or 0.5%

toluidine blue in glycerol, and pressed gently with a cover glass. The entire preparation was then observed under a Zeiss photomicroscope fitted with phase contrast optics. Similar procedures were used to ascertain sites of initiation and monitor embryo development.

Table 1. Basal media, supplements, and growth regulator combinations used for initiation and maintenance. Composition of standard MS medium is shown for comparison.

Components, mg $\cdot$ 1^{-1}	MS	MSCG 5/0	MSCG 5/2.5	DCR 3/0.5	DG
NH_4NO_3	1650	—	—	400	275
KNO_3	1900	100	100	340	2337
$MgSO_4 \cdot 7H_2O$	370	370	370	370	185
KH_2PO_4	170	170	170	170	85
$CaCl_2 \cdot 2H_2O$	440	440	440	85	220
$Ca(NO_3)_2 \cdot 4H_2O$	—	—	—	556	—
KCl	—	745	745	—	—
KI	0.83	0.83	0.83	0.83	0.41
H_3BO_3	6.2	6.2	6.2	6.2	3.1
$MnSO_4 \cdot H_2O$	16.9	16.9	16.9	22.3	8.45
$ZnSO_4 \cdot 7H_2O$	8.6	8.6	8.6	8.6	4.3
$Na_2MoO_4 \cdot 2H_2O$	0.25	0.25	0.25	0.25	0.125
$CuSO_4 \cdot 5H_2O$	0.025	0.025	0.025	0.025	0.0125
$CoCl_2 \cdot 6H_2O$	0.025	0.025	0.025	0.025	0.00125
$NiCl_2 \cdot 6H_2O$	—	—	—	0.025	—
$FeSO_4 \cdot 7H_2O$	27.8	27.8	27.8	27.8	13.9
Na_2EDTA	37.3	37.3	37.3	37.3	18.65
Inositol	100	100	100	200	1000
Glycine	—	—	—	2	—
Nicotinic acid	0.5	0.5	0.5	0.5	0.5
Pyridoxine	0.1	0.1	0.1	0.5	0.1
Thiamine HCl	0.1	0.1	0.1	0.1	1.0
Sucrose	30,000	30,000	30,000	30,000	30,000
Glutamine (G)	—	500	500	250	450
Casein Hydrolyzate (C)	—	1000	1000	500	500
Agar	0.8%	0.8%	0.8%	0.8%	0.8%
Growth Regulators					
2, 4-D	—	5	5	3	11
BA	—	—	2.5	0.5	4
Kinetin	—	—	—	—	4.3

MAINTENANCE

Media composition and culture conditions were the same as those used for initiation. In addition, maintenance media suggested by Durzan and Gupta (5) were tested on a small scale. Embryogenic cultures were shifted to and tested on each such media as sufficient amounts became available. Attempts were made to rescue and preserve browning or otherwise declining cultures by transfer to MSCG 5/0 and MSCG 5/2.5 media. Early results suggested that these were best for maintenance.

DEVELOPMENT AND MATURATION

Development and maturation trials involved two media and steps.

The first or transition medium was MSCG without casein hydrolysate (hereafter referred to as MSG medium) and supplemented with 1% activated charcoal. The second, or maturation medium, was MSG medium supplemented with L-glutamine and abscissic acid (ABA). Levels of L-glutamine ranging from 1-30 mM and ABA, 0-10 mg/l were evaluated in factorial combination. Both media were gelled with 0.3% gelrite. Embryogenic calli from one cell line were transferred, as available, from maintenance media to MSG or transition medium, incubated for 7-10 days, and then cultured on maturation medium until embryos developed to the cotyledonary stage or cultures declined. All cultures were grown in the same light and temperature environment as used for initiation and maintenance.

RESULTS AND DISCUSSION

Initiation

None of the 7055 intact ovule explants produced embryogenic callus. Female gametophyte portions of roughly 1000 of these explants, however, did produce white callus that was nonembryogenic and later turned yellow or green.

Four of the 2813 isolated precotyledonary embryos produced white, mucilaginous callus, both from embryonal head and suspensor regions after 10-15 days of culture (Figure 1A, B). Responsive explants had either bullet-shaped heads and long massive suspensors or clearly visible shoot apices. All had been collected between July 14 and 28, indicating that mid-July is an appropriate time for collecting responsive explants in western Washington. This finding enlarges upon earlier data showing that optimal collection time is four to five weeks after fertilization (5).

Time of fertilization is a useful marker, but the event is difficult to document and more convenient markers are needed. The "window of opportunity" can be expected to vary among years with temperature and other environmental factors, but our data should facilitate future testing of media and other factors.

When entire explants and associated calli were transferred to fresh medium, proliferation continued from the embryonal head portions whereas cultures derived from suspensors turned brown. Isolated embryos at similar stages of development also proved most responsive in loblolly pine, where 7 of 10 embryogenic callus lines were derived from precotyledonary embryos (2,3).

Three of the four responsive explants were from source tree WTC-571 and were initiated in the dark on MSCG medium containing 2,4-D (5 mg/l) and BA (2.5 mg/l). Microscopic evaluation of acetocarmine squashes showed fair numbers of early stage somatic embryos (Fig. 1C), and overall nature and composition of a typical cell mass is shown in Fig. 1D. The other embryogenic callus line was initiated from an explant of source tree WTC-569 cultured in the light on MSCG with 2,4-D only. Somatic embryos in this culture generally were visible to the naked eye within eight weeks. (Fig. 1E).

The distinct phenotypic differences between cultures initiated on MSCG medium containing 2,4-D alone and 2,4-D plus BA underscore the importance of growth regulators in the various steps of embryogenesis. Both 2,4-D alone and in combination with BA supported initiation, but somatic embryos clearly reached larger and more advanced stages in the presence of only 2,4-D. These findings are supported by similar observations made during later steps (described below), and BA therefore does not appear necessary for maintenance or maturation.

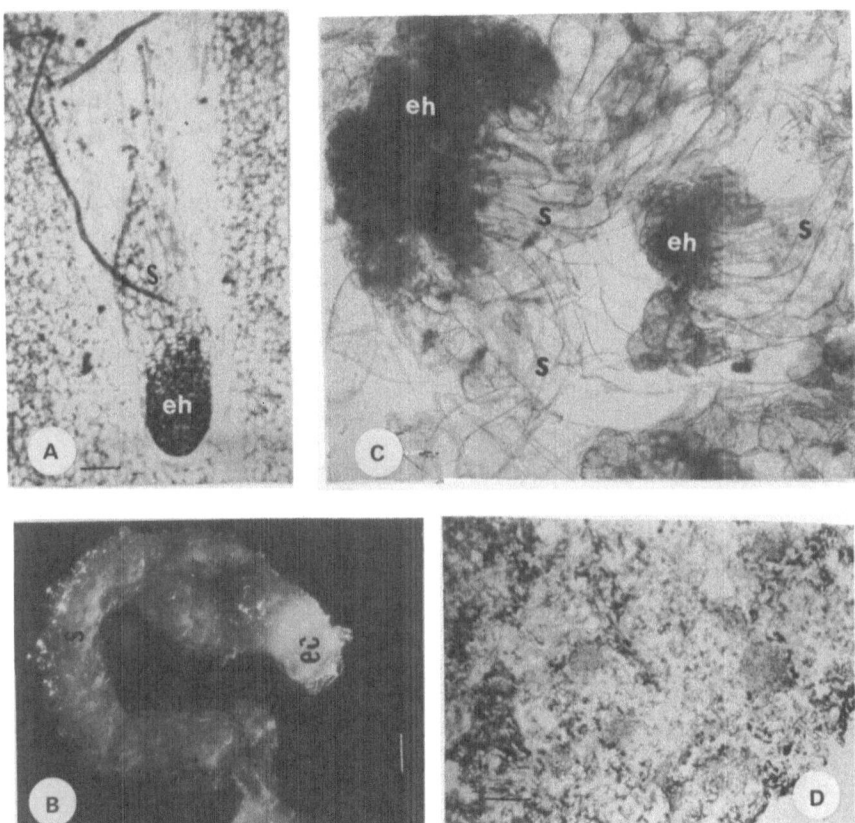

Figure 1. A.Precotyledonary zygotic embryo explant with smooth bullet-
shaped embryonal head (eh) and long massive suspensor (s).
Bar = 1 mm.
B. Typical precotyledonary embryo explant after two weeks of
culture on MSCG media containing either 2,4-D alone or 2,4-D
and BA. Note callus formation from embryonal head and
suspensor. Bar = 1 mm.
C. Acetocarmine squash of embryogenic callus grown on MSCG
5/2.5 medium containing somatic embryos at early developmental
stages. Bar = 100μ.
D. Section of embryogenic callus initiated on MSCG 5/2.5
medium. Bar = 100μ.

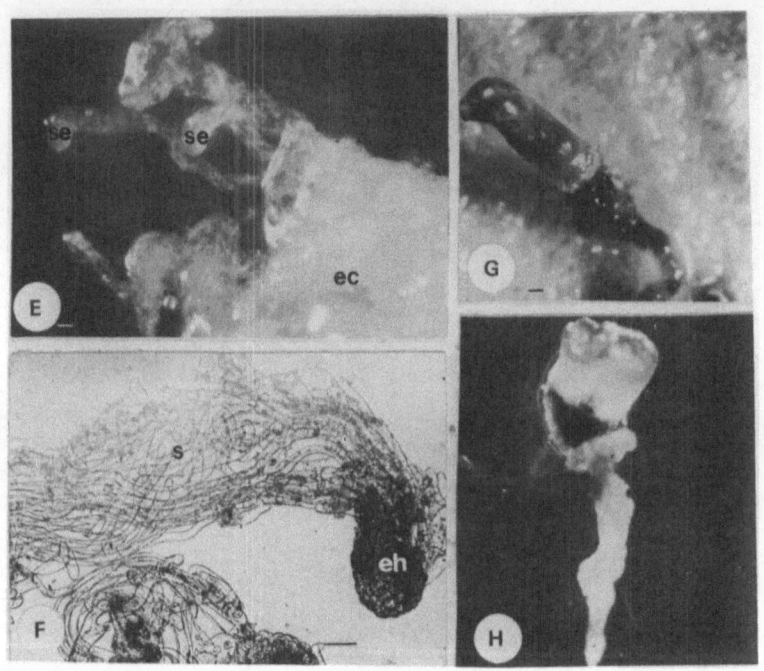

Figure 1. E.   Embryogenic callus (ec) with visible somatic embryos (se)
on MSCG 5/0 medium after eight weeks of culture.   Bar = 1 mm.
F.   Somatic embryo with bullet-shaped head and massive
suspensor (s) on MSCG 5/0 medium.   Bar = 100μ.
G.   Cotyledonary somatic embryo.   Bar = 1 mm.
H. Fused cotyledonary somatic embryos.   Bar = 1 mm.

Both light and dark conditions supported initiation; three of the four embryogenic cell lines were initiated in the dark. This difference, however, could be more genotypic than environmental in that it was also associated with different source trees. Thus, firm conclusions cannot be drawn, and such factors should be explored in greater depth. Successful initiation from a larger variety of source trees seems particularly important.

Suspensors of another 1000 precotyledonary explants formed a glossy, translucent callus within 3-4 weeks of culture. The phenotype initially resembled embryogenic callus, but quickly turned brown and could not be maintained regardless of media or frequency of transfer. Another 250 explants formed white or yellow callus that was clearly nonembryogenic and later turned green. The remaining 1559 explants turned brown and died soon after cultures were started.

Response of the 2880 cotyledonary explants was initially encouraging. Approximately 50% produced white, glossy callus from the hypocotyl. The phenotype appeared embryogenic when first observed, but soon turned yellow or green. Microscopic examinations confirmed absence of somatic embryos, regardless of source tree, explant developmental stage, media and growth regulator treatment, or time after culture. That neither cotyledonary embryos or intact ovules of Douglas-fir produced somatic embryos contrasts with reports for spruce and pine species. Embryogenic cultures have been obtained from cotyledonary stage embryos in Norway and white spruce (9,10,13), and intact ovule explants of <u>Pinus radiata</u> (14) and loblolly pine (2,3).

Observations regarding origin of the four embryogenic cell lines indicated that all were derived from embryonal heads. Origin from suspensors in future experiments, however, may be possible as most callus derived from suspensors in the present study browned and died soon after culture and was therefore unavailable for further assessment.

MAINTENANCE

Over 20% of explants initially produced callus that appeared embryogenic, but most such cultures could not be maintained. Browning and death were both frequent and rapid. Thus, the initiation frequency reported here, 4 of 5693 explants, refers to only those cell lines that could be maintained for 10 or more months. Successful maintenance required subculturing to MSCG 5/0 or MSCG 5/2.5 media every 7-10 days, and periodic removal of brown areas from otherwise white and growing portions. Differences in proliferation were noted among cell lines, with the best producing over 500 cell masses each weighing 150-200 mg every three or four months. Culture on other maintenance media, including DCR 3/0.5, DG, and another suggested by Durzan and Gupta (5), resulted in browning and eventual death.

DEVELOPMENT AND MATURATION

Embryogenic calli maintained on MSCG 5/0, regardless of initiation media, showed somatic embryos emerging from callus surfaces (Fig. 1E). These closely resembled their zygotic counterparts at developmental stages having either bullet-shaped heads and long massive suspensors or visibly distinct shoot apices (Fig. 1F). One such cell line was tested extensively on transition and maturation media, and somatic embryos in it developed to the cotyledonary stage in a relatively short time (Fig. 1G). Fusion of embryos at the hypocotyl (Fig. 1H) was observed with considerable frequency. Subjective evaluation indicated that best development occurred on media supplemented with 20 mM L-glutamine and 0.1-2.6 mg/1 ABA.

ACKNOWLEDGMENTS

We thank Shirley Verhagen, Debbie Hanson, Judy Wyckoff, Lynnea Armstrong, Thomas Merchant, and Sonja Ozturk for their kind and thorough technical assistance.

REFERENCES

1. Becwar, M. R., Noland, T. L., and Wann, S. R., 1987, A method for quantification of the level of somatic embryogenesis among Norway spruce callus lines. Plant Cell Reports **6**:35-38.
2. Becwar, M. R., Wann, S. R., Johnson, M. A., Verhagen, S. A., Feirer, R. P., and Nagmani, R., 1988, Development and characterization of in vitro embryogenic systems in conifers. in: "Somatic cell genetics woody plants," pp 1-18, M. R. Ahuja, ed., Kluwer Academic Publishers, Dordrecht, The Netherlands.
3. Becwar, M. R., Nagmani, R., and Wann, S. R., 1990, Initiation of embryogenic cultures and somatic embryo development in loblolly pine (Pinus taeda L.). Can. J. For. Res. **20**:810-817.
4. Brown, C. L., and Lawrence, R. H., 1968, The culture of pine callus on defined medium. For. Sci. **14**:62-64.
5. Durzan, D. J., and Gupta, P. K., 1987, Somatic polyembryogenesis and Polyembryogenesis in Douglas-fir cell suspension cultures. Plant Sci. **52**:229-235.
6. Finer, J. J., Kriebel, H. B., and Becwar, M. R., 1989, Initiation of embryogenic callus and suspension cultures of eastern white pine (Pinus strobus L.). Plant Cell Rep. **8**:203-206.
7. Gupta, P. K., and Durzan, D. J., 1986, Somatic polyembryogenesis from callus of mature sugar pine embryos. Bio/Tech. **4**:643-645.
8. Gupta, P. K., and Durzan, D. J., 1987, Biotechnology of somatic polyembryogenesis and plantlet regeneration in loblolly pine. Bio/Tech. **5**:147-151.
9. Hakman, I., Fowke, L. C., von Arnold, S., and Eriksson, T., 1985, The development of somatic embryos in tissue cultures initiated from immature embryos of Picea abies (Norway spruce). Plant Sci. **38**:53-59.
10. Hakman, I., and Fowke, L. C., 1987, Somatic embryogenesis in Picea glauca (white spruce) and P. mariana (black spruce). Can. J. Bot. **65**:656-659.
11. Murashige, T., and Skoog, F., 1962, A revised medium for rapid growth and bioassay with tobacco tissue cultures. Physiol. Pl. **15**:473-497.
12. Nagmani, R., and Bonga, J. M., 1985, Embryogenesis in subcultured callus of Larix decidua L. Can. J. For. Res. **15**: 1088-1091.
13. Nagmani, R., Becwar, M. R., and Wann, S. R., 1987, Single-cell origin and development of somatic embryos in Picea abies (Norway spruce) and P. glauca (white spruce). Plant Cell Reports **6**:157-159.
14. Smith, D. R., Singh, A. P., and Wilton, L., 1985, Zygotic embryos of Pinus radiata in vivo and in vitro. in: Int. Conifer Tissue Culture work group. Abstracts. D. R. Smith, ed., Rotorua, New Zealand.

MATURATION OF YELLOW-POPLAR SOMATIC EMBRYOS

Scott A. Merkle

School of Forest Resources
University of Georgia
Athens, GA 30602

ABSTRACT

Since somatic embryogenesis in yellow-poplar (Liriodendron tulipifera)
was first reported, a major problem with the system has been the very low
conversion frequencies of somatic embryos. Although large numbers of
embryos could be produced in liquid medium, conversion percentages were even
lower than for those differentiated on solid medium, where conversion
averaged less than 1 percent. Proembryogenic masses (PEMs) removed from
auxin-supplemented medium and placed in hormone-free medium generally
produced malformed and clustered somatic embryos that were asynchronous in
their development and that failed to mature properly, retaining fused or
malformed cotyledons, even though radicles would elongate. To promote
production of somatic embryos capable of conversion to plantlets, we tested
physical and chemical treatments on embryogenic suspension cultures.
Production of synchronous, mature, well-formed embryos was promoted by
fractionating embryogenic suspensions on sieves and culturing in medium
supplemented with abscisic acid (ABA). High levels of sucrose in the medium
could substitute for ABA to a certain extent in aiding embryo maturation.
However, conversion of the embryos was inhibited once they were transferred
to solid medium. A modification of this technique, in which PEMs were
plated on physical supports on solid medium immediately following
fractionation, raised average conversion frequencies for mature embryos to
over 70 percent.

INTRODUCTION

Yellow-poplar (Liriodendron tulipifera), a member of the Magnoliaceae,
attains the greatest height of any hardwood tree in eastern North America
and may also reach the largest diameter (2). The species is characterized
by rapid growth, straight form and wood of exceptional working quality (5).
We first reported somatic embryogenesis in yellow-poplar in 1986 (3). We
found that a small percentage of immature embryos explanted onto a Blaydes
(6) conditioning medium containing 2 mg/l 2,4-dichlorophenoxyacetic acid
(2,4-D), 0.25 mg/l 6-benzyladenine (6BA) and 1 g/l casein hydrolysate (CH)
would produce a fast-growing, nodular callus, which was later found to be
composed of proembryogenic masses (PEMs). As long as the PEMs were
maintained on conditioning medium, they did not further differentiate.
However, when PEMs were transferred to induction medium (same as
conditioning medium minus growth regulators), somatic embryos would appear

Woody Plant Biotechnology, Edited by M.R. Ahuja
Plenum Press, New York, 1991

within 5-6 weeks and undergo the classic stages of embryogeny (globular, heart, torpedo).

Prior to the research described here, our standard procedure to produce plantlets from the somatic embryos was to pick individual converting embryos from the cultures and transfer them to a Risser and White's (4) plantlet development medium. Plantlets would produce 3-4 leaves and were ready for transfer from in vitro conditions within 8 weeks. Following potting in soil mix, plantlets were acclimatized in a humidifying chamber for 6-8 weeks, grown for 2-4 months in the greenhouse and transplanted to raised nursery beds. Over the past 2 years, we have produced hundreds of somatic embryo-derived plantlets in this manner. A few hundred of these trees were planted in the field in 1989. However, the number of plantlets obtained to date represents only a tiny percentage of the millions of somatic embryos produced by our cultures. Although a single plate could produce thousands of embryos, a typical plantlet yield per plate was only 10-20. Thus our conversion percentage was substantially less than 1% and the potential of our system for mass propagation of yellow-poplar trees was far from being realized.

The low numbers of yellow-poplar plantlets produced from somatic embryos were the result of a major problem with our embryogenic system--low conversion percentage. Most embryos allowed to develop from PEMs on induction medium were malformed, with fused cotyldons. A high percentage of malformed embryos would germinate when transferred to plantlet development medium, but shoots failed to develop. This was especially a problem with embryos allowed to develop in liquid induction medium, where radicles elongated without formation of distinct cotyledons. Embryos also tended to form secondary embryos by repetitive embryogenesis, resulting in clusters of embryos. Since the embryos germinated without forming separated cotyledons, we believed the problem was a failure of the embryos to mature properly, and that once we were able to obtain mature embryos, they would readily convert to plantlets. Thus the objective of our research was to discover methods for the high-frequency production of mature, well-formed somatic embryos with a high conversion percentage.

MATERIALS AND METHODS

Our experiments were divided into 3 sets. In the first set, we tested the hypothesis that embryos selected for maturity from asynchronous cultures would convert at a high frequency. In the second set of experiments, we tested chemical and physical techniques for the production of synchronous, mature embryos in suspension culture. The third set of experiments tested a compromise protocol between the maturity-selection technique and the suspension culture technique for efficiency of production of mature embryos and plantlets. All three sets of experiments were conducted with 4 embryogenic yellow-poplar lines, initiated as described earlier, and established as suspension cultures by transferring PEMs into liquid conditioning medium. Suspension cultures were grown on a gyratory shaker at 90 rpm in the light and subcultured every 3 weeks.

Early Selection of Mature Embryos

As described in the introduction, prior to the research reported here, our standard procedure was to pick converting embryos following their differention from PEMs on solid medium and transfer them to plantlet development medium. This method resulted in a conversion rate of less than 1%. Our first experiment was to test the hypothesis that embryos selected for maturity before germination began and moved to a secondary medium away from the influence of other embryos and PEMs would convert at a significantly higher rate than 1%.

Using a dissecting scope, we picked 1-1.5 mm-long embryos (the size of mature zygotic embryos) that appeared to be mature, with separated cotyledons, from clusters of asynchronously-developing somatic embryos on solid induction medium. Embryos were transferred to plates of fresh medium (here designated "secondary medium") at a density of approximately 20 embryos per plate. We tested 4 different secondary media for their impact on conversion. These were regular induction medium (containing 1 g/l CH, 4% sucrose and 0.8% phytagar), induction medium without CH, induction medium without CH and only 1% sucrose, and induction medium without CH and solidified with 0.8% agarose instead of phytagar. Following 1 week on secondary medium, germinants were transferred to plantlet development medium in test tubes and scored for conversion to plantlets after 1 month.

Suspension-Cultured Embryo Maturation

We were especially interested in inducing liquid culture-grown somatic embryos to mature and convert at high frequencies, since suspension cultures gave us the potential to synchronize embryo development by fractionation on sieves. Synchronized cultures had two important potential advantages: (1) Any scale up for mass-production of somatic embryos (e.g. in a bioreactor) would only be useful if large numbers of mature embryos were ready for use at the same time (e.g. for encapsulation to produce artificial seeds). (2) By working with synchronized embryos, large numbers of embryos at a given developmental stage would be available for maturation and conversion experiments. Thus we would be able to apply our treatments to a uniform population of embryos, making it easier to discern if our treatments were effective.

Our fractionation/synchronization protocol was based on a method reported for synchronization of carrot somatic embryos (1), and is outlined in Figure 1. Briefly, for each synchronization, approximately 1 g of PEMs were transferred from liquid conditioning medium to liquid induction medium. Then, the PEMs were sieved on a 140 µm pore size stainless steel screen and the fraction that passed was saved. The saved fraction was sieved on a

SOMATIC EMBRYO SYNCHRONIZATION

Place 1 g of PEMs into induction medium

↓

Fractionate PEMs on sieves and save fraction
that passes 140µm but not 38µm

↓

Culture PEMs for one week in induction medium

↓

Fractionate developing embryos on sieves and save
fraction that passes 230µm but not 140µm

↓

Culture embryos for 7-10 days in induction medium to
obtain synchronized heart-torpedo stage embryos

Figure 1. Protocol for synchronization of yellow-poplar somatic embryos grown in suspension culture.

38 μm screen and the fraction that failed to pass was cultured in induction medium. After 1 week, the developing embryos were fractionated on a 230 μm screen, and the fraction passing through was sieved on a 140 μm screen. The fraction caught on the last screen was cultured in induction medium.

Fractionated PEMs and embryos were exposed to a number of treatments, either during synchronization or following synchronization. Those treatments tested during synchronization were primarily aimed at obtaining mature, well-formed embryos, and included addition of abscisic acid (ABA) to

SYNCHRONIZATION, MATURATION, GERMINATION AND CONVERSION
Compromise Method

Place 1 g of PEMs into induction medium
↓
Fractionate PEMs on seives and save fraction
that passes 140µm but not 38µm
↓
Backwash saved fraction from screen onto filter paper;
Allow medium to drain; Place filter paper with embryos
on solid induction medium
↓
Incubate embryos on filter paper 12-14 days
at 30° C to obtain synchronous, mature embryos
↓
Harvest mature embryos and distribute on secondary medium;
Incubate at 30° C in light for 1 week to germinate
↓
Transfer germinating embryos to plantlet development
medium in tubes or Magenta boxes

Figure 2. Compromise protocol for synchronization, germination and maturation of yellow-poplar somatic embryos by plating fractionated PEMs.

the induction medium in concentrations ranging from 10^{-7} M to 10^{-5} M, and addition of osmotica to the medium, including mannitol, sorbitol and sucrose. Treatments tested following synchronization were aimed at inducin mature embryos to convert, and included addition of ABA, gibberellic acid (GA) or 6BA to the medium, desiccation, cold stratification and leaching in water or solutions of GA prior to transfer to solid medium.

Compromise Protocol

The last set of experiments tested a protocol which was a compromise between the two methods described above, combining the fractionation of suspension cultured PEMs and early selection of mature embryos on solid medium. The "Compromise Protocol" is outlined in Figure 2. As with the earlier synchronization method, approximately 1 gram of PEMs from liquid conditioning medium was transferred to liquid induction medium and fractionated on stainless steel sieves. The fraction passing the 140 μm sieve but not the 38 μm sieve was saved and backwashed off of the 38 μm sieve and onto a 60 mm disk of filter paper, using liquid induction medium. After the excess liquid drained through the filter paper, the filter with the PEMs was placed directly on solid induction medium in a 60 mm petri dish. Thus the PEMs were not in contact with the medium, but were "buffered" by the layer of filter paper. PEMs were cultured on the filter paper for 12-14 days, after which total embryos and mature embryos were counted. Then a subsample of at least 20 embryos each was transferred to the same 4 secondary media as were tested with the early-selected embryos described above. Following 1 week on secondary media, embryos were transferred to plantlet development medium in test tubes or Magenta GA-7 vessels. The number of embryos converting to plantlets was scored after one month. In addition to the test of secondary media, we also varied the medium upon which the filter paper was placed immediately following fractionation (primary medium). The primary media tested were the same 4 media as were tested as secondary media.

RESULTS AND DISCUSSION

Early Selection of Mature Embryos

By selecting mature embryos and transferring them to a scondary medium, we were able to attain an average conversion percentage of 37% over all secondary media and over the 4 tested clones. It should be noted that this

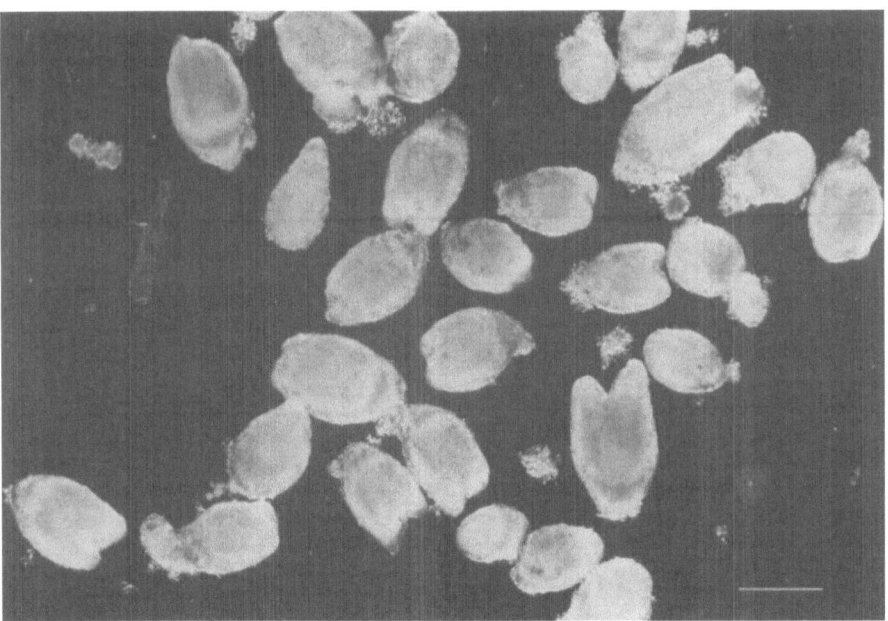

Figure 3. Heart-stage yellow-poplar somatic embryos derived from fractionated PEMs, 7 days following second sieving, in liquid induction medium with 5 x 10^{-7} M ABA. Bar represents 500μm.

figure does not constitute an overall conversion rate for all embryos, but only the conversion rate for mature embryos. An overall rate could not be computed for this experiment, because the lack of synchrony of the somatic embryos precluded an accurate count of all embryos produced on each plate. The secondary medium resulting in the highest conversion percentage was induction medium minus CH (63.2%), while the lowest conversion resulted from using regular induction medium as the secondary medium (16.2%). Induction medium minus CH and solidified with agarose and induction medium minus CH, but with only 1% sucrose also produced significantly lower conversion rates than induction medium minus CH (38.6% and 28.7% respectively).

Suspension Cultured Embryo Maturation

We found that fractionation of PEMs on sieves worked quite well in gross synchronization of the developing embryos. Furthermore, fractionation apparently speeded up the process of embryony, since globular stage embryos could be obtained in less than 2 weeks (versus 5-6 weeks with asynchronous cultures). Although sieving itself appeared to promote the production of a single embryo per PEM, there were still some multiple embryo clusters when synchronization was carried out using regular induction medium. Furthermore, most embryos failed to develop distinct cotyledons before elongating while still in the liquid medium. Thus fractionation did roughly synchronize embryo development but did little to promote maturation.

Of the treatments applied to modify embryo development and promote maturation, ABA appeared to have the greatest impact. At 10^{-7} M, ABA slightly slowed development of the fractionated embryos and delayed elongation, but did not promote maturation. ABA at 10^{-6} M (later lowered to 5×10^{-7} M) was optimal for preventing formation of embryo clusters and for promoting maturation, giving a high percentage of outwardly mature embryos with separated cotyledons (Fig. 3). However, these embryos appeared to produce vacuolate cells on their outer surfaces, giving them a fuzzy appearance. This effect was even more apparent when fractionated PEMs were cultured in medium containing 10^{-5} M ABA. In most cases PEMs were almost completely inhibited from developing into mature embryos, remaining at the globular stage and producing a thick surface layer of vacuolate cells. When mannitol or sorbitol were substituted for ABA, they had little effect of delaying elongation, except for the highest level tested (8%), which inhibited elongation, but failed to promote the formation of separated cotyledons. The highest level of sucrose tested (10%) also delayed elongation and resulted in a number of embryos with separated cotyledons. However the impact of the high sucrose concentration was not as pervasive as that the ABA, and embryos did not remain synchronous (Fig. 4).

We expected that embryos that appeared to mature properly in suspension culture would behave similarly to mature embryos produced on solid medium and convert to plantlets at a high frequency, once placed on solid secondary medium (i.e. induction medium minus CH). However, although embryos removed from liquid medium containing 5×10^{-7} M ABA enlarged and cotyledons greened when placed on secondary medium (and subsequently plantlet development medium), apical development did not follow. Furthermore, most ABA-treated embryos failed to even germinate. Embryos matured in induction medium supplemented with high levels of mannitol, sorbitol or sucrose displayed similar inhibition. Since we thought it possible that residual ABA or some other factor had induced a dormancy or quiescence in the embryos, we tested a number of treatments designed to break the presumed dormancy, which were listed earlier as treatments applied following synchronization. None of these treatments (GA, 6BA, cold stratification, desiccation, leaching) had a significant impact on embryo conversion. From all the fractionated embryos differentiated in liquid medium under all treatments, less than 100 total plantlets were obtained. Thus even though embryos differentiated in liquid medium could be induced to take on a mature appearance, there appeared to be some factor linked to the long period the embryos spent in liquid medium, which ultimately prevented conversion. Therefore, we adopted the compromise protocol, in which the fractionated PEMs would be removed from the liquid medium as soon as possible.

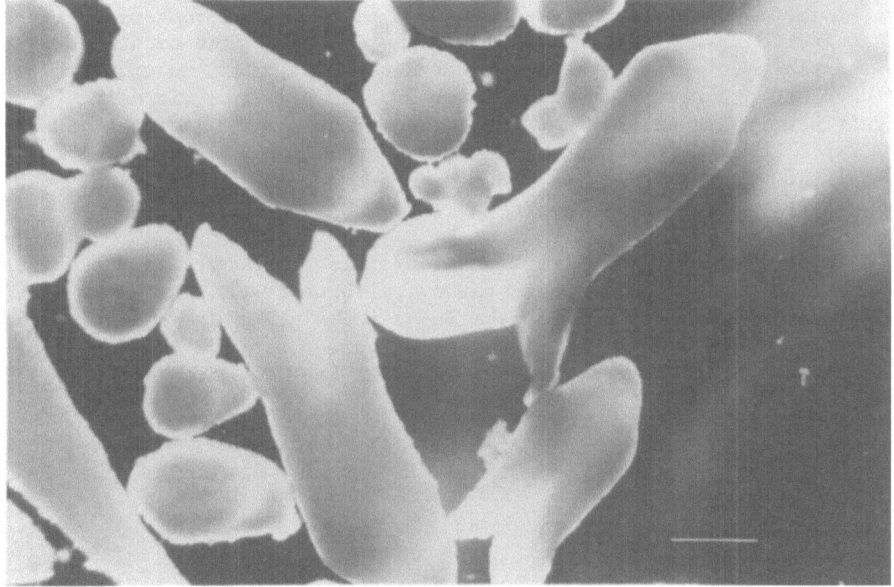

Figure 4. Torpedo-stage yellow-poplar somatic embryos derived from fractionated PEMs, 7 days following second sieving, in liquid induction medium with 10% sucrose. Bar represents 500µm.

Compromise Protocol

As was the case with fractionated PEMs cultured in liquid medium, fractionated PEMs plated immediately on filter paper produced globular stage embryos very rapidly (approximately 8 days) compared to asynchronous PEMs cultured on solid induction medium (5-6 weeks). These embryos continued to develop in a roughly synchronous manner resulting in the production of hundreds of torpedo or cotyledon stage embryos per 60 mm plate within 2 weeks following plating (Fig. 5). Best embryo development appeared to occur when PEMs were in a single layer on the filter paper. Again, fractionation appeared to promote the production of one embryo per PEM, resulting in a low number of multiple embryo clusters.

For the 4 embryogenic lines tested, primary medium type appeared to have no significant effect on either the percentage of embryos that appeared mature 2 weeks following plating or on the percentages of mature or total embryos that converted to plantlets. However, there are indications of a primary medium x clone interaction that has yet to be tested.

Upon transfer to secondary medium, mature embryos germinated rapidly, and radicle elongation was accompanied by cotyledon expansion and greening (Fig. 6). Within 1 week following transfer to secondary medium, germinants were large enough to transfer to plantlet development medium, where they were allowed to grow for one month before being scored for conversion.

Similar to the results with early-selected embryos, secondary medium had a significant impact on the conversion percentage of mature embryos. Induction medium minus CH again resulted in the highest conversion percentage (71.9%), which was significantly higher than the rate for regular induction medium (53.3%) or induction medium minus CH with 1% sucrose (44.6%). Replacing agar with agarose did not significantly change the conversion frequency of mature embryos (64.2%). When conversion frequency was computed as the percentage of all embryos (not just mature ones) counted on the plates of primary medium after 2 weeks, again induction medium minus CH gave the greatest conversion rate (32.5%), but this was only significantly higher than for induction medium minus CH with 1% sucrose (20.7%). Regular induction medium and induction medium minus CH and solidified with agarose gave conversion frequencies of 25.3% and 30.0%, respectively.

We believe that this fractionation/filter paper plating technique can be modified to further increase the overall conversion frequency of yellow-poplar somatic embryos and to make large scale plantlet production possible. Currently, we intend to apply this technique to produce 1000 trees of each of 7 clones for field testing. We estimate that the protocol will allow us

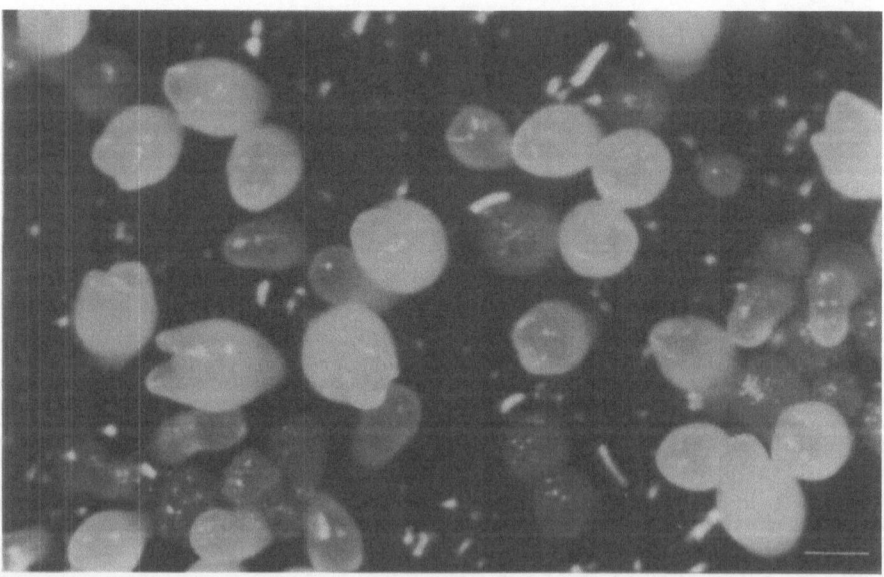

Figure 5. Roughly synchronous population of yellow-poplar somatic embryos, 10 days following plating of fractionated PEMs on filter paper on top of solid induction medium. Bar represents 500μm.

to produce 7000 plantlets ready to be transferred to soil mix in less than 10 weeks. The procedure may also prove to be a useful system for improving conversion of somatic embryos of other species that behave similarly to yellow-poplar in culture.

Figure 6. Germinating yellow-poplar somatic embryo 5 days following transfer to secondary medium. Bar represents 500μm.

LITERATURE

1. Giuliano, G., Rosellini, D., and Terzi, M., 1983, A new method for purification of the different stages of carrot embryoids. Plant Cell Rep. 2: 216-218.
2. Harlow, W.M., Harrar, E.S., and White, F.M. 1979, Textbook of Dendrology, Sixth Edition. McGraw-Hill, New York. 510 p.
3. Merkle, S.A., and Sommer, H.E., 1986, Somatic embryogenesis in tissue cultures of Liriodendron tulipifera. Can. J. For. Res. 16: 420-422.
4. Risser, P.G., and White, P.R., 1964, Nutritional requirements of spruce tumor cells in vitro. Physiol. Plant. 15: 620-635.
5. Wilcox, J.R., and Taft, K.A., 1969, Genetics of yellow-poplar. USDA Forest Service Research Paper WO-6. 12 p.
6. Witham, F.H., Blaydes, D.F., and Devlin, R.M., 1971, Experiments in Plant Physiology. Van Nostrand-Reinhold Co., New York. 245 p.

SOMATIC EMBRYOGENESIS IN TISSUE CULTURES

OF <u>EUTERPE</u> <u>EDULIS</u> MART. (PALMAE)

Miguel P. Guerra and Walter Handro

Department of Botany
Institute of Biosciences, University of São Paulo
C.P. 11461, 05499 São Paulo, BRAZIL

ABSTRACT

Somatic embryogenesis and further development of embryos to
plantlets was studied using: young and mature embryos excised from seeds
and inflorescences in 10 different stages until emergence. The influence
of several factors such as mineral salts, nitrogen source, growth
regulators, sucrose etc. were studied. The morphogenetical events were
also analysed in their histological aspects. The main results can be
summarized as follows: a) the embryogenic process has characteristic
steps: firstly, the development of an embryogenic tissue, producing
globular embryos, isolated or forming clusters. Depending on the hormonal
balance, either a secondary embryogenic process can occur, or the
development of the embryos to further stages until plantlets; b) very
young inflorescences and also those ones near the emergence were unable
to develop embryos; only certain intermediate stages can do this; c)
immature embryos showed higher embryogenetic potential than mature ones.
Requirements to control this process were: a) activated charcoal during
the first two weeks of culture, and dark in the case of inflorescences;
b) 2,4-D (50 mg.l^{-1}) is the best condition to induce embryogenesis; c)
further development of embryos, as well secondary embryogenesis occurred
after lowering 2,4-D to 10 mg.l^{-1} and adding 2-iP (5 mg.l^{-1}); d) further
transfer to medium lacking growth regulators leads to the development of the
embryos into plantlets.

INTRODUCTION

Economical and ecological aspects

Euterpe edulis Mart. is a single stemmed neotropical palm that occurs
in a narrow range of rain forest on the southern and southeastern Brazilian
coast. In the north of Brazil another Euterpe is found, Euterpe oleraceae
which produces suckers.

The importance of these palms is related to the production of "palmito"
(heart of palm), i.e., the growing apical bud surrounded by the young leaves.
In Brazil several palms produce edible heart of palm, but only the genus Euterpe is commercially explored in large scale. E. edulis produces the white
"palmito", a product of better quality destined for the international market.
Brazil is the world's major producer and exporter of "palmito". In view of

non-selective exploitation and increasing demand, the natural populations are rapidly disappearing.

Along with the importance of its final product, this species has favorable characteristics for management with sustained yield. As natural sources have been drastically reduced, a program for conservation, propagation and improvement of the remaining germplasm has become necessary.

The use of tissue culture techniques

Propagation of E. edulis occurs only through seeds, and the conventional methods of vegetative propagation cannot be applied to palms, in view of the absence of vascular cambium (22); the natural occurrence of vegetative propagation is limited to few species. The vegetative reversion of inflorescences or flowers in Elaeis guineensis (16) is rare and it seems to be difficult to establish a routine method for vegetative propagation from this occurrence (14). Thus the possibility of preserving and propagating elite plants depends on the establishment of non conventional methods of propagation through tissue cultures.

The morphogenetic studies in palms have received less attention than in other plant groups. This seems to be a consequence of the tropical origin of these plants, and of their peculiar growth patterns as compared with other perenial plants. Regarding the use of in vitro techniques, two basic procedures have been used for palm tissue culture: a) the obtention of plantlets through somatic embryogenesis or indirect organogenesis via calluses; b) attempts for inducing reversion of young flower meristems to a vegetative state. In a review of the research on palm tissue culture, Tisserat (21) reported that the induction of calluses from explants of several origins was achieved for 40 species of 29 genera. The initiation of embryo-like structures occurred in 11 species, but the regeneration of plantlets was unequivocally demonstrated only for Phoenix dactylifera, Elaeis guineensis and Cocos nucifera. Recently, we have demonstrated the regeneration of plantlets of Euterpe edulis through direct somatic embryogenesis from zygotic embryos (5). A report on somatic embryogenesis of Bactris gasipaes was done by Valverde et al (23).

Apparently the success of palm tissue culture is associated with the use of explants in physiological conditions able to follow a morphogenetic program. According to Blake (2) the basis of getting in vitro morphogenesis in palms is an adequate formulation of the culture medium leading to a rapid callus initiation and determination of embryogenic cells in such a way that embryos could develop when transferred to a secondary medium. It seems clear that auxins are among the critical components in the medium, specially 2,4-D.

This work deals with the basic procedures to establish a protocol for plant regeneration of E. edulis through somatic embryogenesis.

MATERIAL AND METHODS

Embryos

Zygotic embryos were employed in two different stages of development: they were excised either from completely mature seeds, or from immature ones. Before embryo isolation, fruits were submitted to a strong process of sterilization for 12 h under agitation in a solution with 4% sodium hypochloride, 0.15% streptomycin and 0.01% merthiolate. In aseptic conditions this procedure was repeated for 30 m, followed by 70% ethanol washing (5 m), and then washing in sterile water (3 times). Embryos were then rapidly excised and inoculated to prevent oxidation.

Inflorescences

Young inflorescences still recovered by spathes were used. In an adult plant of E. edulis ca. 10 young inflorescences in different stages of development can be found. The inflorescences are inserted in the leaf sheath base, which recovers them. In general inflorescences are free of contaminants, direct sterilization being unnecessary. To collect this kind of explant successive leaf sheaths are removed, after previous cleaning with a cotton padding embedded in 70% ethanol. Inflorescences about 5-10 cm long are then removed.

Culture medium

Mineral salts of Murashige and Skoog (12), and Eeuwens (4) proved adequate for allowing the initiation and progression of somatic embryogenesis. Vitamins (11) were also added to the culture media. Antioxidants proved to be necessary for instalation and further progression of cultures: activated charcoal was used at $1.5-3.0$ g.1^{-1}. Other compounds were not effective for preventing oxidation. Selected auxins and cytokinins (see results) were added to the basal medium according to different stages of culture.

Maintenance of cultures

All cultures were initially kept in the dark, to prevent oxidation. For zygotic embryos this period was 48 h, for inflorescences 15-20 d. Then the cultures were transferred to light (5 W.m^2, "Gro-Lux" Sylvania fluorescent lamps), 16 h/day, at 26 ± 1^0C.

RESULTS

Morphogenesis in zygotic embryos

Viable culture instalation and progression was possible only in semi-solid media with activated charcoal. The growth regulator type and concentration were determinants for morphogenetic activation of these explants.

Of the auxins tried (NAA, picloram, 2,4-D) only 2,4-D was able to induce embryogenesis. As shown in Table 1, only at 100 mg.1^{-1} inhibition of the embryo germination was observed and granulate structures arose on the cotyledonar petiole. These structures developed giving rise to globular and translucid somatic embryos (Fig. 1). While these embryos reached the pearly stage, the matrix tissue retained it potential for producing new globular embryos (Fig. 2). The transfer of embryo bunches to a secondary medium with 5.0 mg.1^{-1} 2iP plus 0.1 mg.1^{-1} NAA allow the development to a bipolar stage and the occurrence of an embryogenic process which characterizes a continuous, long-term, non-synchronous model. In these circumstances the embryogenic potential was maintained for periods longer than 12 months. It must be emphasized that this embryogenic model was direct, without callus stages, as shown in Fig. 3, where a globular embryo arises directly on the epidermis of the cotyledonar petiole. The development to plantlets (Fig. 4) was observed when bipolar somatic embryos were transferred to a third medium (liquid or semi-solid) in which sucrose and mineral salts were reduced by 50%. The germination of these somatic embryos were similar to the zygotic ones. When the plantlets reached a 10 cm height with a secondary root 5 cm long, they were transferred to pots, and kept for 1-2 weeks in a humid chamber before being transferred to the green-house (Fig. 5).

Differences were observed regarding the morphogenetic potential of embryos in relation to their development stage. Mature zygotic embryos produced a low frequency model of somatic embryogenesis, the primary events occurring 90 days after inoculation. Immature embryos on the contrary, produced a high frequency model in which primary events occurred 30 days after inoculation. An adequate sequence of transfers allows plantlet obtention 180 days after the inoculation of zygotic embryos in primary medium.

Table 1. Effect of 2,4-D on zygotic embryos of Euterpe edulis cultured in vitro (MS basal medium + 30 g.l^{-1} sucrose 1.5 g.l^{-1} activated charcoal, Morel and Wetmore vitamins, 3 g.l^{-1} 2iP, 5 g.l^{-1} agar).

2,4-D (mg.l^{-1})	Response after	
	30 days	60 days
0.0		
10.0		
50.0		
100.0		

cn - cotyledonary node
cs - cotyledonary sheath (haustorium)

Morphogenesis in inflorescences

Viable culture of inflorescences was possible only when these explants were inoculated in liquid medium with activated charcoal, and kept for 15 d in the dark. Other procedures produced intense oxidation of the cultures. After this initial treatment, rachyl segments 0.5-1.0 cm long were transferred to semi-solid media constituted of MS or Y3 salts (4), plus 40 g.l^{-1} sucrose, 1.5 g.l^{-1} activated charcoal, added of growth regulators. As verified for zygotic embryos, only 2,4-D (50 ml.l^{-1}) plus 2iP (3 mg.l^{-1}) promoted morphogenesis in this kind of explant. The characteristic sequence of a

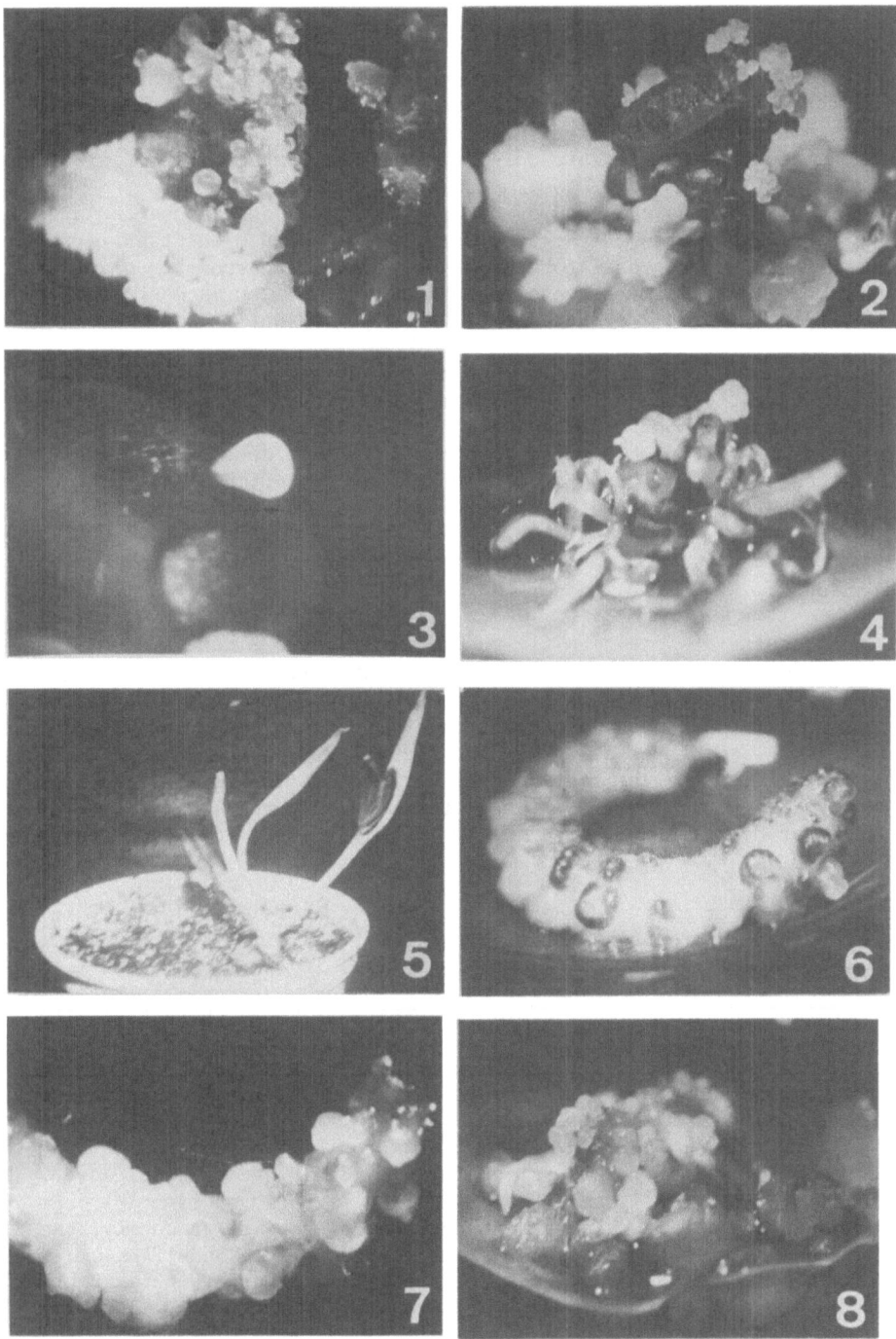

Fig. 1-8. Stages of somatic embryogenesis and plant regeneration in
zygotic embryos (Fig. 1-5) and inflorescences (Fig. 6-8) of
Euterpe edulis.

direct somatic embryogenic model was observed in a preferential way. Firstly, rachyls elongated and thickenned specially at the flower primordia level (Fig. 6). The differential cell proliferation in these regions resulted in cell aggregates or pro-embryogenic tissues (Fig. 7). In some favorable treatments, some translucid structures were formed 60 d after inoculation. Such structures arose directly on the pro-embryogenic tissues, corresponding to the formation of globular somatic embryos (Fig. 8). The isolation and transfer of the embryogenic regions of these cultures to medium with 5.0 mg.l^{-1} 2iP plus 0.1 mg.l^{-1} NAA showed the progression of a continuous, asynchronous, and high frequency embryogenic model, in which all stages were represented.

In many cultures the embryogenic potential was continuous for a period of one year. When parts of these cultures were isolated and transferred to medium without growth regulators, embryos germinated producing plantlets. Transference and acclimatation of plantlets was similar as described for those originated from zygotic embryos.

As in the zygotic embryos, the development stage of inflorescences determined activation and expression of the embryogenic route. In general, rachyl segments originated from highly differentiated inflorescences showed intense tissue oxidation, and in some cases, flower development. Very young inflorescences showed little oxidation and the formation of pro-embryogenic structures 60 d after inoculation, and plantlets in 200 d.

DISCUSSION

The results of E. edulis tissue cultures showed a direct somatic embryogenesis model, asynchronous, continuous, of high frequency. The induction and expression of this route were dependent strictly on the kind and stage of the explant, and on the type and concentration of auxin in the primary medium. According to Tisserat et al. (20), the type of the explant and its physiological qualities are the most determinant aspects for initiation os somatic embryogenesis. Christianson (3) suggested that only competent cells for somatic embryogenesis are receptive to stimuli related to this morphogenetic way.

Protocols for plant regeneration through somatic embryogenesis in palms were described for Phoenix dactylifera (1, 17, 19), Elaeis guineensis (7, 8, 13, 15, 17, 18, 19), and Bactris gasipaes (23). In all these cases the regeneration through the embryogenic way was preceded by callus stages, characteristic of an indirect route. On the other hand, the embryogenic process observed in E. edulis was direct, without intermediary callus stage. In this case, competent cells of receptive explants would be stimulated by 2,4-D to follow an embryogenetic program. Such model is of relatively rare occurrence; a similar model was described for Trifolium repens by Maheshwaran and William (9, 10), where the embryogenetic initiation was mediated by characteristic pro-embryogenetic complexes as described by Haccius (6). The relevance of these aspects must be emphasized: firstly, this model seems to be very adequate for studying experimental embryogenesis; secondly, the regeneration of plants through somatic embryogenesis in E. edulis seems to involve only organized meristematic tissues. This favors genetic stability, and the production of true-to-type plants.

From an applied point of view, the establishment of a protocol for plant regeneration of E. edulis through tissue culture makes possible the mass cloning of elite progenies and their re-introduction in the region of natural occurrence, allowing the conservation and improvement of the still remaining germplasm.

REFERENCES

1. Ammar, S. and Benbadis, A., 1977, Multiplication végétative du palmier-dattier (Phoenix dactylifera L.) par la culture de tissus de jeunes plantes issues de semis. C.R. Acad. Sci. Paris, 284: 1789 .

2. Blake, J., 1983, Tissue culture propagation of coconut, date and oil palm, in "Tissue Culture of Trees", J. J. Dodds, ed., AVI Publ., Westport.

3. Christianson, M. L., 1987, Casual events in morphogenesis, in "Plant Tissue and Cell Culture", C. F. Green, D. A. Sommers, W. P. Hackett and D. D. Biesbaer, eds, Allan Liss, New York.

4. Eeuwens, C. J., 1976, Mineral requirements for growth and callus initiation of tissue explants excised from mature coconut palms (Cocos nucifera) and cultured in vitro, Physiol. Plant. 36: 23.

5. Guerra, M. P. and Handro, W., 1988, Somatic embryogenesis and plant regeneration in embryo cultures of Euterpe edulis Mart. (Palmae), Plant Cell Rep. 7: 550.

6. Haccius, B., 1978, Question of unicellular origin of non-zygotic embryos in callus cultures, Phytomorphology 28: 74.

7. Jones, L. H., 1974, Propagation of clonal oil by tissue culture, Oil Palm News 17: 1.

8. Lioret, C., 1981, Vegetative propagation of oil palm by somatic embryogenesis, in "The Oil Palm in Agriculture in the Eighties", Inc. Soc. of Planters, Kuala Lumpur.

9. Maheswaran, G. and Williams, E. G., 1985, Origin and development of somatic embryos formed directly on immature embryos of Trifolium repens in vitro, Ann. Bot. 56: 619.

10. Maheswaran, G. and Williams, E. E., 1986, Direct secondary somatic embryogenesis from immature sexual embryos of Trifolium repens cultured in vitro, Ann. Bot. 57: 109.

11. Morel, G. M. and Wetmore, R.H., 1951, Fern callus tissue culture, Am. J. Bot. 38: 141.

12. Murashige , T. and Skoog, F., 1962, A revised medium for rapid growth and bioassays with tobacco tissue cultures, Physiol. Plant. 15: 473.

13. Nwankwo, B. A. and Krikorian, A., 1986, Morphogenetic potential of embryo- and seedling-derived callus of Elaeis guianeensis Jacq. var. pisifera Becc., Ann. Bot. 51: 65.

14. Pannetier, C. and Buffard-Morel, J., 1986, Coconut palm (Cocos nucifera L.) in "Biotechnology in Agriculture and Forestry, 1. Trees I, Y. P. S. Bajaj, ed., Springer-Verlag, Berlin.

15. Rabéchault, H. and Martin, J. P., 1976, Multiplication végétative du palmier à huile (Elaeis guianeensis Jacq.) à l'aide de culture de tissus foliares. C. R. Acad. Sci. Paris 283: 1735.

16. Reynolds, J. F., 1982, Vegetative propagation of palm trees, in "Tissue Culture in Forestry", J. M. Bonga and D. J. Durzan, eds, Martinus Nijhoff Publ., Dordrecht.

17. Reynolds, J. F. and Murashige, T., 1979, Assexual embryogenesis in callus cultures of palms, In Vitro 5: 383.

18. Smith, W. K. and Thomas, J. A., 1973, The isolation and in vitro cultivation of cells of Elaeis guianeensis, Oléagineux 28: 123.

19. Tisserat, B., 1979, Propagation of date palm (Phoenix dactylifera L.) in vitro, J. Exp. Bot. 30: 1275.

20. Tisserat, B., Esan, B. B. and Murashige, T., 1979, Somatic embryogenesis in angiosperms, Hortic. Rev. 1: 1.

21. Tisserat, B., 1984, Clonal propagation palms, in "Cell Culture and Somatic Cell Genetics of Plants", I. K. Vasil, ed., Academic Press, Orlando.

22. Tisserat, B., 1987, Palms, in "Cell and Tissue Culture in Forestry", J. M. Bonga and D. J. Durzan, eds, Martinus Nijhoff Publ., Dordrecht.

23. Valverde, R., Arias, O. and Thorpe, T. A., 1987, Picloran-induced somatic embryogenesis in pejibaye palm (<u>Bactris gasipaes</u> H. B. K.), <u>Plant Cell Tissue</u> and <u>Organ</u> Culture 10: 149.

GENETIC AND MORPHOGENIC STABILITY OF PLANTLETS REGENERATED

FROM SOMATIC EMBRYOS OF NORWAY SPRUCE (Picea abies)

L.H. Mo and S.von Arnold

Department of Forest Genetics, Swedish University of Agricultural
Sciences, Box 7027, S-750 07 Uppsala, Sweden

ABSTRACT

Embryogenic cultures were initiated from mature zygotic embryos of Picea abies.
The somatic embryos in the embryogenic cultures were first stimulated to mature and then
either to develop further into plantlets or to differentiate new embryogenic cultures. The
procedure was repeated three times during two years. The ability to give rise to new
embryogenic cultures or to develop into plantlets was similar for all somatic embryos
irrespective of how long they had been cultured in vitro. The nuclear DNA content,
measured in a flow cytometer, was estimated to 32 pg / G1 nuclei in seedlings developed
from zygotic embryos. Nuclei isolated from embryogenic cultures and from plantlets
regenerated from somatic embryos had the same DNA content as those isolated from
seedlings.

INTRODUCTION

There have been numerous reports showing that plantlets derived from tissue
culture are not identical to the original mother plant. In general, more aberrations are
induced after long term callus growth. Tissue culture-associated variation can for example
be observed as changes in plant morphology, growth requirements of the cultures and
DNA content and structure. Although it has generally been considered that gymnosperms
are more stable during tissue culture than angiosperms, there are reports showing that
some coniferous species for example polyploidize when cultured in vitro (2, with
references).

We have previously shown that root tip cells from Picea abies plantlets regenerated
from adventitious buds in vitro are diploid (1) and that the DNA profile for adventitious
buds and shoots does not deviate from that of resting buds in situ (3). Similar results have
been reported for adventitious shoots of Pinus taeda (7). In contrast, cells from
adventitious buds of Pinus coulteri showed a progressive increase in DNA level over
time; after 42 days in culture, the buds contained large populations of cells at the 8C level
and a considerable number of cells with DNA levels above 8C (6).

The only studies regarding stability in ploidy level in somatic embryos of conifers
are those by Schuller (8) and Lelu (4). Their results indicate that somatic embryos of
Abies alba are diploid whereas some of the somatic embryos of Picea abies are tetraploid.

RESULTS

Within one month, zygotic embryos cultured on callus medium (half concentrated
LP-medium containing 30 mM sucrose, 15 mM NH_4NO_3, 9.0 µM 2,4-dichlorophenoxy-
acetic acid and 4.4 µM benzyladenine, pH 6.1, gelled with 0.4% gellan gum, Kebo,
Merck) gave rise to embryogenic cultures. When the embryogenic cultures were

transferred to maturation medium, the somatic embryos were stimulated to mature. The maturation medium had the same composition as the callus medium except that it lacked 2,4-dichlorophenoxyacetic acid and benzyladenine, and contained 90 mM sucrose and 7.5 μM abscisic acid. The maturation process took about one month. Then the mature somatic embryos were transferred to full strength LP medium containing 60 mM sucrose but lacking plant growth regulators, and in a low frequency they developed into plantlets. However, isolated mature somatic embryos could also produce new embryogenic cultures in a similar way as zygotic embryos when transferred to callus medium. This procedure was repeated in four cycles (Fig 1).

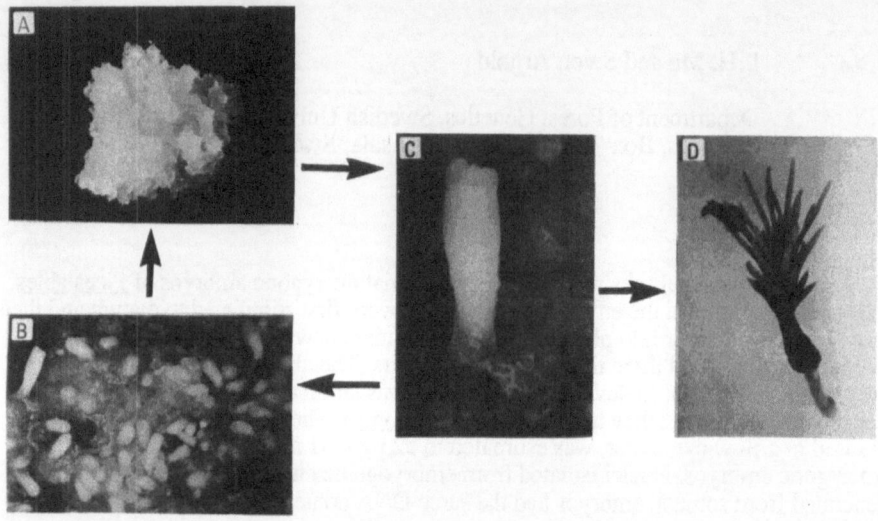

Fig.1. Development of somatic plantlets from embryogenic cultures of _Picea abies_. A) Embryogenic culture initiated from a zygotic embryo. B) Somatic embryos in various developmental stages one month after being transferred to medium containing 7.5 μM ABA. C) Higher magnification of a somatic embryo in B. D) Early somatic plantlet.

We analysed propidium iodide stained protoplastnuclei, from three types of material, in a flow cytometer, 1) plantlets developed from zygotic embryos, 2) first cycle embryogenic cultures and 3) plantlets developed from the fourth cycle mature somatic embryos. The DNA histogams of all _Picea abies_ samples contained a single G1 peak. To estimate the DNA content of the G1 peak, a part of each sample was also analysed with _Vicia faba_ as internal standard. The DNA content per nuclei in _Vicia faba_ has been reported to be 23.8 ± 0.66 pg(9). From that value we calculated that the G1 nuclei from seedlings of Picea abies contained 32 ± 2 (mean ± SE) pg DNA. Nuclear DNA content in plantlets regenerated from the fourth cycle mature somatic embryos was similar to that of plantlets regenerated from zygotic embryos. The experiment was repeated 4 times and the data obtained confirmed that the DNA content of plantlets regenerated from somatic embryos is in the same range and has the same distribution as that of plantlets regenerated from zygotic embryos. Nuclei isolated from the first cycle embryogenic cultures had the same DNA content as nuclei isolated from plantlets regenerated from zygotic embryos.

The results presented show that embryogenic cultures of _Picea abies_ retain the capacity to regenerate mature somatic embryos and plantlets after several cycles in _vitro_ and that plantlets regenerated from at least up to the fourth cycle somatic embryos remain at the diploid stage (5).

REFERENCES

1. Arnold von S.,1982, Factors influencing formation, development and rooting of adventitious shoots from embryos of Picea abies (L.) karst. Plant Sci. Lett. 27:275-287.
2. Bayliss M.W.,1980, Chromosomal variation in plant tissues in culture. Int. Rev. Cytol. 11A:113-144.
3. Hakman I., von Arnold S. and Bengtsson A.,1984, Cytofluorometric measurements of nuclear DNA in adventitious buds and shoots of Picea abies regenerated in vitro. Physiol. Plant. 60:321-325.
4. Lelu M.A.,1987, Variations morphologiques et génétiques chez Picea abies obtenues aprés embryogenése somatique. In: Annales de Recherches Sylvicoles, AFOCEL, 35-47.
5. Mo L.H., von Arnold S. and Lagercrantz U. ,1989, Morphogenic and genetic stability in longterm embryogenic cultures and somatic embryos of Norway spruce (Picea abies L. Karst). Plant Cell Reports 8:375-378.
6. Patel K.R. and Berlyn G.P.,1982, Genetic instability of multiple buds of Pinus coulteri regenerated from tissue culture. Can. J. For .Res. 12:93-101.
7. Renfroe M.H. and Berlyn G.P. ,1984, Stability of nuclear DNA content during adventitious shoot formation in Pinus taeda L. Am. J. Bot. 71 (2):268-272.
8. Schuller A., Reuther G. and Geier T., 1989, Somatic embryogenesis from seed explants of Abies alba. Plant Cell,Tissue and Organ Culture 17:53-58.
9. Ulrich I., Fritz B. and Ulrich W. ,1988, Application of DNA fluorochromes for flow cytometric DNA analysis of plant protoplasts. Plant Science 55: 151-158.

SOMATIC EMBRYOGENESIS AND ORGANOGENESIS OF EMBRYONIC EXPLANTS

OF ABIES ALBA AND ACER PSEUDOPLATANUS

Anton Grahsl, Josef Schmidt, and Eva Wilhelm

Austrian Research Centre Seibersdorf Ltd.
A-2444 Seibersdorf, AUSTRIA

ABSTRACT

With Abies alba somatic embryos were induced from the hypocotyl of ex-cised well developed zygotic embryos. Extended growth of embryogenic cells was observed on auxin-free medium with 2.2 µM/l BAP $\pm$ 2.3 µM/l kinetin. Transplants from several of these cultures developed continuously numerous embryos with green cotyledons, but root development was generally dis-turbed. Direct organogenesis was observed on the same media. Wreaths of buds formed around the hypocotyls right beneath the base of the cotyledons. Little bud formation occurred on medium containing auxin.

Acer pseudoplatanus does not express the same vegetative regeneration capacity in vitro as in vivo. Nevertheless buds formed from the hypocotyl part of segmented zygotic embryos when grown on induction medium supple-mented with 0.01 or 0.05 µM thidiazuron. Nearly 40 % of cultures produced shoots up to 20 mm in length. Bud production and shoot growth continued on the original explants or larger pieces of organogenetic callus, when trans-ferred on the respective medium at monthly intervals.

INTRODUCTION

Abies alba is among the most endangered central european tree species threatened by forest decline. Its in vitro regeneration capacity has been investigated only very recently (2, 5). In a similar approach using mature embryos we could observe differential development of the explants.

Acer pseudoplatanus has a high capacity for vegetative regeneration, but it reacts poorly to micropropagation. Therefore embryonic material was tested for its in vitro reactivity to thidiazuron, a compound with reported cytokinin action on Acer x freemanii (4).

MATERIALS AND METHODS

The silver fir seeds were sterilized (NaOCl), the embryos excised and presoaked for 24 h. After final disinfection (ascorbic acid/ethanol) they were placed on half strength MCM (1) in the dark. Explants and/or callus pieces were subcultured at 4-weekly intervals.

Mature but still green seed of Acer pseudoplatanus was repeatedly disinfected (NaOCl/ethanol) and embryos prepared. After final disinfection (NaOCl) the embryos were disected into shoot and root segments. The cotyledons were truncated to 1/3 length. The hypocotyl for the most part remained at the root segment. The segments were separately explanted on GD-media (3) supplemented with 1 µM BAP + 0.01 or 0.05 µM thidiazuron to induce organogenesis.

RESULTS AND DISCUSSION

1. Abies alba

1.1. Embryogenesis

The cotyledonal and hypocotyl parts of the embryo were the most reactive ones. Callus proliferation started primarily from the hypocotyl within two weeks. Auxin-induced callus growth was fast in the beginning, but ceased later without producing any embryogenic material. Extended growth of embryogenic cells was observed on auxin-free medium supplemented with 2.2 µM/l BAP ± 2.3 µM/l kinetin. Up to 20 % of embryo cultures from several provenances started to develop this type of cells after 8 weeks, but only few cultures proliferated over a longer period of time (including one from immature seed). Transplants from several of these cultures developed continuously numerous embryos with green cotyledons in the dark. After transfer to hormone-free medium embryo production was enhanced appreciably, but only for a limited time. After several weeks embryos were separately placed on hormone-free medium in a 16 h day (65 µE/m^{2}s fluorescent), but their further development was generally disturbed. In many cases the shoot rested with well developed cotyledons on top of suspensory cells, which either kept dividing without producing a functional root or died.

1.2. Organogenesis

We obtained direct organogenesis from the hypocotyls of freshly prepared embryos of different seed lots. This occured on the same media, which induced embryogenic cells. Buds formed like a wreath around the hypocotyls right beneath the base of the cotyledons. Little bud formation occurred on medium containing auxin. Shoots did not develop further when being separated from the embryo and transferred to liquid or solidified medium without hormones.

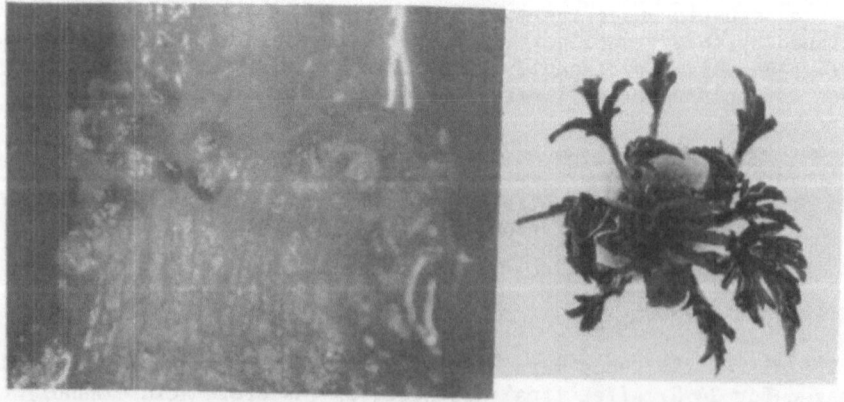

Fig. Multiple bud formation from the hypocotyl of Abies alba (left), shoots arising from callus of an Acer pseudoplatanus embryo (right).

2. Acer pseudoplatanus

Callus formation started within a few days from the hypocotyl end of the explants followed by the induction of shoots. Two weeks later, when first bud induction was observed, explants were transferred upon GD-media with various combinations of BAP (0 or 1.0 μM) with thidiazuron (0.01 or 0.05 μM). Thidiazuron effectively enhanced the induction of adventitious buds, especially at a concentration of 0.05 μM. There was only little additive effect when this concentration of thidiazuron was combined with BAP.

Nearly 40 % of cultures produced shoots up to 20 mm in length. Shoots did not continue to grow on hormone-free medium, when separated from their basal callus at an early stage. Nevertheless bud production and shoot growth continued on the original explants or larger pieces of organogenetic callus, when transferred on the respective medium at monthly intervals.

LITERATURE

1. Bornman, C.H., and Jansson, E., 1981, Regeneration of plants from the conifer leaf, Colloque sur la culture in vitro des essences forestieres, AFOCEL, Nangis, France, p. 41-55.
2. Gebhardt, K., Weisgerber, H., and Fröhlich, H. I., 1988, In vitro germination and production of embryogenic callus from liquid suspension cultures of Abies alba - silver fir, Poster Abstract, 4th Int. Conifer Tissue Culture Work Group, Saskatoon.
3. Gresshof, P.M., and Doy, C.H., 1972, Development and differentiation of haploid Lycopersicon esculentum (tomato), Planta, 17:161-170.
4. Kerns, H. R., and Meyer, M. M. Jr., 1986, Tissue culture propagation of Acer x freemanii using thidiazuron to stimulate shoot tip proliferation, HortScience, 21:1209-1210.
5. Schuller A., Reuther, G., and Geier, Th., 1989, Somatic embryogenesis from seed explants of Abies alba, Plant Cell, Tissue and Organ Culture, 17:53-58.

GENE TRANSFER AND EXPRESSION

GENETIC TRANSFORMATION OF TREES VIA DIRECT GENE TRANSFER

Brent H. McCown

Department of Horticulture
University of Wisconsin-Madison
Madison, WI 53706

ABSTRACT

The recent successes in genetically-engineering trees have in large part been due to advances in both the microculture and the gene transfer technologies. Besides regeneration, a critical problem with microculture has been developing efficient techniques for selecting cells and tissues that are expressing the inserted foreign genes. One solution has been the use of liquid overlays of the selective agent. In regards to gene transfer technologies, both vector-mediated and direct gene transfer techniques have been successful, particularly with *Populus*. However, the versatility and efficiency of particle bombardment appears to be a major advantage that may lead to the rapid adoption of such methods in biotechnology programs utilizing tree species.

INTRODUCTION

Being able to discuss the genetic transformation of trees at this Placerville meeting has special historical significance for me. It was at the first meeting on forest biotechnology held here in 1984 that a project was formulated by 3 separate research groups to transform poplar and insert genes coding for tolerance to the herbicide glyphosate. The project proved highly successful [2,8]. Now, 5 years later, we have gained enough experience with genetic engineering of trees to begin to evaluate the alternative methodologies of transferring genes into tree crops.

Any gene transfer system, whether utilized for trees or any other crop, has 4 requirements that must be fulfilled before it will be generally successful:
1. A microculture (tissue culture) system for the plant.
2. A methodology to transfer the DNA to the cultured tissues.
3. Genes, preferably isolated and characterized.
4. Control of the gene function in vivo.

Woody Plant Biotechnology, Edited by M.R. Ahuja
Plenum Press, New York, 1991

This paper will focus on the first two of these requirements. Other papers at this meeting are focussing on the other topics. I will restrict myself to the use of isolated genes and thus topics such as protoplast fusion will not be addressed.

MICROCULTURE SYSTEMS AND GENETIC TRANSFORMATION

Microculture has been and continues to be a necessity of all the most utilized genetic transformation methodologies:
  -Microcultured tissues are preferred sources of target tissues for gene insertion because of their sterility, reproducibility, and physiological responses.
  -Since all gene transfer occurs at the individual cell level, regeneration into normal plants is essential.
  -Selection of the transformed cells among the masses of living, non-transformed tissue is a critical step most efficiently conducted in vitro.
  -Micropropagation of the regenerated shoots aids immensely in subsequent testing and verification.

Developing an effective and efficient selection system has been one of the most difficult parts of the microculture system for us to perfect. After some years of using Kanamycin as a selective agent, a number of points have emerged as important:
  1. Dosage response curves must be determined using the exact tissue that will be selected in the transformation experiments. Seemingly minor changes such as the mass of the tissue being cultured can negate the effectiveness of the selection pressure.
  2. We have had the most success using levels of kanamycin that inhibit the rapid growth of the tissue without killing it in the short term. The idea is to provide an environment where the transformed cells have a marked growth advantage over the rest of the tissue without creating an inhibitory microenvironment of diing cells.
  3. Media and environment should be optimal for growth of the tissue being selected. Tissues that are not attempting rapid growth have been difficult to use in selective experiments.
  4. Kanamycin appears to not be readily transported through tissue masses. Thus cell that lie on the upper surfaces of callus, stems or leaves are often not highly effected by the kanamycin in the medium.

We now routinely use a liquid kanamycin overlay as the selection scheme [4]. Using a level of kanamycin that slows the growth of the parent tissue, kanamycin is added both to the base medium and as a water/kanamycin overlay that partially submerges the tissue mass. This technique provides uniform contact between the most all the tissue surfaces and the kanamycin and allows for periodic refreshing of the selective agent in long-term experiments typical of woody plants.

The proven alternative methods for transferring isolated DNA into tree cells can be grouped into 2 broad areas:
1. Biological methods
   -Utilization of biological vectors
2. Physical methods
   -Direct gene transfer

The only vector that has proven useful for tree cell transformation is *Agrobacterium* [2,5,6]. The method itself has become a rather trivial exercise and is the most successful of any gene transfer technology. However, the process of host/pathogen interaction is inherently complex and thus limits the amount of control the researcher has over the process. As with most plants, there is a strong host/vector genotype interaction which may necessitate extensive testing of alternative Agrobacterium strains for effectiveness on a specific genotype [8]. Even so, effective vectors that allow both gene transfer and plant regeneration from the transformed cells may not be readily found. Other problems we have encountered are difficulties in freing the infected tissue of the vector and ready acceptance of the transformed product by regulatory agencies since it may contain genes from a plant pathogen.

To avoid some of the problems with vector system and to gain better control and reproducibility over the transformation process, direct gene transfer techniques can be considered. For trees, two techniques, electroporation and particle bombardment, have been evaluated with a number of species [1,4,7]. Electroporation does not appear to offer a major advance over *Agrobacterium* since it depends on protoplast systems which are in themselves very complex. In our laboratory, the trade-off of extensive cell injury and death with adequate rates of poration and DNA transfer was severely limiting to the development of a highly useable system. However, accelerated particle bombardment has shown some very encouraging results.

The technique of bombardment which we use was developed at Agracetus company in Middleton, Wisconsin [3]. Briefly, DNA is coated onto gold particles about 1 micron in diameter, accelerated at high velocities by an electrically-generated shock wave, and used to impact the target tissue where the particle along with the DNA penetrate the target cells. The major advantages of the method are that it is primarily physical, thus it can be readily controlled and is reproducible. Most any tissue will serve as a suitable target, thus one can choose the best tissue in relation to regeneration capacity and recovery of plants.

A critical variable in bombardment success has been pretreatment of the tissue so that it is in a 'receptive' stage. The exact definition of this physiological state is not known yet, however tissues that are in growth that closely resembles meristems (cytoplasmically dense and actively dividing) appear to be the most efficient.

Table 1 summarizes the efforts using these gene transfer methodologies with *Populus*.

Table 1. Summary of progress in utilizing three methods of gene transfer for the genetic transformation of *Populus*.

Gene transfer method	Target tissue	Transformation rate (maximum)	Transformed plants recovered?
Agrobacterium vector	Leaves Roots	?	Yes (Multiple genotypes)
Electroporation	Leaf protoplasts	0.0003% of viable plated cells	No
Particle bombardment	Isolated cells, stems nodules	0.8% of viable target cells	Yes (multiple genotypes)

CONCLUSIONS

The specific process of gene transfer in the transformation of trees is rapidly becoming a trivial part of the total process of genetic engineering. Now, two proven and effective methods, Agrobacterium mediated transformation and particle bombardment, are available. The major problems remain regeneration, particularly with conifers, and selection efficiency.

REFERENCES

1. Bekkaoui, F., Pilon, M., Laine, E., Raju, D.S.S., Crosby, W.L., and Dunstan, D.I. 1988. Transient gene expression in electroporated *Picea glauca* protoplasts. Plant Cell Reports 7:481-484.
2. Fillatti, J.J., Sellmer, J., McCown, B., Haissig, B. and Comai, L. 1987. *Agrobacterium* mediated transformation and regeneration of poplar. Mol. Gen. Genet. 206:192-199.
3. McCabe, D.E., Swain, W.F., Martinell, B.J. and Christou, P. 1988. Stable transformation of soybean (*Glycine max*) by particle acceleration. Biotechnology 6:923-926.
4. McCown, B.H., McCabe, D.E., Russell, D.R., Robison, D.J., Barton, K.A., and Raffa, K.F. IN PRESS. Stable transformation of *Populus* and incorporation of pest resistance by electric discharge particle acceleration. Plant Cell Reports.

5. McGranahan, G., Leslie, C., Uratsu, S., Martin, L. and Dandekar, A. 1988. *Agrobacterium*-mediated transformation of walnut somatic embryos and regeneration of transgenic plants. Bio/Technology 6:800-804.
6. Pythoud, F., Sinkar, V.P., Nester, E.W. and Gordon, M.P. 1987. Increased virulence of *Agrobacterium rhizogenes* conferred by the vir region of pTiBo542: application to genetic engineering of poplar. Bio/Technology 5:1323-1327.
7. Seguin, A. and Lalonde, M. 1988. Gene transfer by electroporation in betulaceae protoplasts: *Alnus incana*. Plant Cell Rep. 7:367-370.
8. Sellmer, J.C. and McCown, B. H. 1989. Transformation in *Populus* spp. In: Y.P.S. Bajaj (ed), Plant Protoplasts and Genetic Engineering II. Springer Verlag, Berlin. pp. 155-172.

AGROBACTERIUM-MEDIATED TRANSFORMATION OF APPLE (MALUS PUMILA

MILL.)

David J. James

British Society for Horticultural Research
East Malling
Maidstone, Kent , ME19 6BJ
U.K.

ABSTRACT

Transgenic apple plants can now be produced via a leaf-disc
transformation procedure involving the use of disarmed binary
vectors in Agrobacterium . However despite having a highly
efficient regeneration protocol for non-transformed tissues the
production of transgenic apple plants still occurs at low
frequency. This paper discusses a strategy for making
improvements in the current procedures.

INTRODUCTION

Whole plants of many cultivated apple varieties can now be
regenerated from in vitro grown tissues via de novo organogenesis
or somatic embryogenesis (James, 1987). Most of this work has
involved the use of explants such as leaf pieces and in many
varieties high frequencies of regeneration have now been reported
(James et al. 1988; Fasolo et al. 1989 ; Predieri and Fasolo
1989; Welander, 1988).

We have used this character in the variety Greensleeves to
produce several transgenic apple clones (James et al. 1989). The
method used was based on the Agrobacterium-mediated leaf disc
transformation procedure as detailed by Horsch et al.(1985) and
used a prototype disarmed binary vector pBIN6 in A.tumefaciens
(Bevan 1984). This plasmid contains both a scorable marker gene
nos, encoding the enzyme nopaline synthase (Nos) and a selectable
marker, npt II, a chimaeric gene composed of the regulating
sequences of a nopaline synthase gene and the coding sequences of
a bacterial neomycin phosphotransferase gene. Although these
reports open up exciting future prospects for the genetic
improvement of apple our sucess is tempered by the findings that

the procedure has so far worked at only a low frequency over the past 4 years and is still less than 1 per cent on a per explant basis (James et al. 1990a). Failure to increase the frequency of transformation in an existing, highly regenerable cultivar makes the prospect of regenerating transformants of future, horticulturally superior cultivars with unknown regenerability a dim and distant one.

Our aim therefore is to make the transformation of apple and other trees a simple, reproducible and high frequency procedure where clonal fidelity is retained and somaclonal variation (Larkin and Scowcroft, 1983) avoided or even excluded. This paper will attempt to assess what strategies need to be developed and employed to achieve this aim.

THE CURRENT PROCEDURE FOR PRODUCING TRANSGENIC APPLE PLANTS

In essence the present methods for producing transgenic apple clones are much the same as those originally described for the leaf-disc transformation procedure (Horsch et al. 1985). There are four main components to this procedure;
Initiation, Infection, Selection, and Regeneration

INITIATION - the choice of the source plant tissue prior to infection and transformation

INFECTION - the genetic background of the bacteria used in transformation experiments and factors controlling their virulence

SELECTION - the selection of transformed cells that lead to the regeneration of transgenic plants

REGENERATION - all factors pertaining to the regeneration of transformed plants after the selection period. This requires frequent observations of explants and growing tissue during what can be a very lengthy period (as long as 9 months) before a regeneration event occurs.

1.INITIATION - Choice of the source tissue and cultivar variation

The first commercial apple cultivar to be regenerated from leaf discs was the roostock M.27 (James et al. 1984). The frequency was however rather low leading us to examine regeneration capacity in several other cultivars (James et al. 1988). After some searching the cultivar Greensleeves (James Grieve x Golden Delicious) was chosen since it reproducibly gave 100% adventitious shoot regeneration on a per explant basis and several (6 or more) shoots per explant. Consequently thousands of adventitious shoots may be regenerated under non-transforming conditions.

Micropropagation systems offer the possibility of using a range of tissues for transformation but leaves have been the organ of choice because of their abundance and ease of manipulation for infection with Agrobacteria. However leaves grown under strictly controlled tissue culture conditions still exhibit marked differences in their regeneration ability at different stages of growth and significant improvements in

214

performance of regeneration can be obtained by using leaves of the 'correct' age (Table 1) and stage of leaf development (James et al. 1990a). For this reason we regularly use rooted plantlets rather than those from micropropagating cultures where inter-shoot competition for light and nutrients will maximise variations in the response of excised tissues. Other work has also shown that discs from rooted shoots performed better than discs taken from leaves of unrooted plants (James et al. 1988). Other more recent work using the same cultivar has confirmed this (Viss, pers.comm. 1990).

Table 1. Influence of age of leaf from rooted plants on regeneration ability from 50 excised leaf discs of cv. Greensleeves.

Weeks on rooting media before disc excision	Number of shoots	% regen- eration	Shoots per regenerating disc
8	12	18	1.3
7	86	58	3.0
6	85	60	2.9
5	133	85	3.2
4	231	82	5.6

Clearly innate regeneration capacity is extremely high in this cultivar and would not appear to be a limiting factor for the regeneration of transformants. Moreover other apple cultivars with a very dissimilar parentage to Greensleeves are also capable of high levels of regeneration e.g. M.26 (Fasolo et al 1989; Predieri and Fasolo 1989).

However despite the improvements in regeneration frequencies 'optimising' a regeneration system is clearly relative and the ability to produce six shoots per explant could be replaced by another finding that 12 shoots or 20 shoots per explant can be produced by altering a biochemical or physiological variable. A good case in point is the difficult to regenerate but highly important commercial apple rootstock M.9. Using BAP as a cytokinin gave very low levels of regeneration but substitution with the diphenylurea thidiazuron (TDZ) markedly improved regeneration. Similar beneficial effects of this compound have been noted by Fasolo et al. (1989) in other apple cultivars.

Other physiological and biochemical factors influencing apple regeneration have been discussed elsewhere (James et al. 1988). More recently Kobayashi (1989) showed that one week of etiolation at the end of the subculture period and before excising the tissue was the most effective method of inducing adventitious shoot formation in Malus prunifolia Borkh.

There are reports that preculturing plant tissues prior to bacterial infection increases the frequency of transformants that could be produced (McHughen et al. 1989) and there are suggestions that cells actively undergoing division are more easily infected and transformed than older tissues (Chyi and Phillips 1987). In apple however when leaf discs were exposed to regeneration media for times from 0 - 8 days prior to inoculation any preculture treatment approximately halved the number of explants that eventually produced kanamycin resistant calli (Table 2) . These results need confirmation with biochemical and molecular analyses but if validated suggest that wounding factors may be lost by preculture, at least in the case of apple.

Table 2. Effect of a preculture period of leaf discs on regeneration media C81. Percentage discs with kanamycin resistant calli after 21 days selection on 50ug/ml kanamycin and three months growth on regeneration media (C81 ; 2mg/l BAP, 0.5 mg/l NAA). One hundred discs used per treatment.

Days of preculture on regeneration media	Per cent discs with kanamycin-resistant calli
8	35
4	27
2	34
0	63

Leaf material can be wounded in a variety of ways for infection and transformation but we found the use of discs of a known diameter (7mm) enabled us to quantify the response of the tissue to given variations in the infection and transformation schedules. However cutting discs can be a somewhat slow and laborious task using a cork borer and the use of strips or leaf halves is often much faster. It should be borne in mind however that the magnitude and dimension of the forces exerted when cutting a disc as compared to cutting strips with a scalpel are very different. The torque applied during the cutting of a disc using a cork borer will impart far greater shearing forces than those encountered during a simple cut across the surface of a leaf and the 'bruising zone' is much more obvious on the periphery of a cut disc than on a cut strip due to the greater downward pressure required for disc cutting. The importance of this could be considerable when considering the number and type of wounded plant cells within the damaged area. The implications for bacterial infection and subsequent transformation are not yet clear but must be borne in mind when using a leaf explant transformation procedure. It might be added that all the transgenic apple clones produced by us are of leaf disc origin.

2. INFECTION - Agrobacteria and their virulence

For anyone attempting to use agrobacteria-mediated transformation the nature of bacterial virulence is a key issue. The factors that determine virulence have been a subject of intensive study in the past few years and the detailed nature of the process of T-DNA transformation is now beginning to be unravelled (see e.g. Zambryski, 1988). It appears that the vir genes that are responsible for determining this character lie within both plasmid and chromosomal DNA. This should emphasise the need to use strains of agrobacteria that are known to infect the host plant or its close family relatives naturally. For example Dandekar et al. (1990) have recently shown that the Agrobacterium strains A281 and C58 are highly virulent on the apple cultivar Greensleeves. They also showed that virulence of the A281 strains could be further enhanced if the strain contained extra copies of the Ti-encoded vir genes A,B and G. The problem of matching strain with host is, of course, that wild type strains are of little use in producing regenerated plants with a normal phenotype and disarmed strains of Agrobacteria are still relatively few in number. Moreover disarming bacteria of their hormone-regulating genes is by no means a simple task.

In many of the experiments that have been carried out on the nature of virulence the phenomenon was assessed in prokaryotic systems by assaying promoter activities of various vir units fused to reporter genes (e.g. Rogowsky et al. 1987). Just such a system was used to discover that the plant phenolic compound, acetosyringone (AS), is a key factor in inducing virulence (Stachel et al. 1985). For this reason it is now regularly included in media during transformation experiments both during infection of plant tissues and during the co-cultivation period. There appears to be no doubt that this compound has extremely important properties as a signal molecule. What is not clear is the timing of the events for inducing virulence and how these affect subsequent ability of the bacteria to transform plant cells. We have used this compound in apple transformation at a range of concentrations from 0.1 to 1mM but the results so far are equivocal (James, unpublished data). Certainly our transgenic apple clones were produced without using AS as were all (almost 100) transgenic strawberry clones (James et al. 1990b). This is not to say it does not have a role in improving the efficiency of transformation in these species.
The investigations of Alt-Moerbe et al. (1989) and other groups on the role of external and controllable variables on the expression of vir genes may have important consequences for those attempting to transform plant species. These workers showed that

temperatures below 28C and the presence of sucrose were essential
for expression of these genes and that such expression was
stimulated by AS and acidic pH. Furthermore there were important
lag phases for the appearance of the encoded proteins which,
depending on the gene, were from 2 to several hours duration .
When employing the leaf explant transformation procedure the time
between transferring bacteria from their overnight bacterial
growth medium at pH 7.2 to an acid pH medium for subsequent
infection is not usually stated. In our work (and presumably in
others too) the two are carried out sequentially. In the studies
of Alt-Moerbe et al. (1989) however it was clearly shown that an
acid pH, sucrose at 2-3% and a period of between 2-5 h were
necessary to incite the virulence of virD and virB genes.
Although AS was not essential it could considerably enhance
virulence. The implications for those attempting to transform
plant cells with Agrobacteria are considerable. Usually bacterial
cells are resuspended in Murashige-Skoog medium at pH 5.5 with
sucrose. However MS media have low buffering capacity and a fall
in the pH below 5.2 drastically reduced virulence in their
assays. They consequently recommended the use of a simplified
induction medium (SIM) based on 20mM sodium citrate buffer at
pH5.5 containing sucrose and acetosyringone at 0.1mM. They also
showed that ammonium chloride considerably reduced induction of
the vir genes. Ammonium ions are present at high concentration in
the form of ammonium nitrate in MS medium, and this salt that has
already been shown to be suboptimal for the growth and
regeneration of some apple cultivars in vitro (Fasolo et al.
1989).

The efficacy of changes in the induction medium and the
period needed for induction before infecting plant tissues now
needs to be assessed using a suitable scorable marker such as
B-glucuronidase (GUS ; Jefferson et al. 1987).

3.SELECTION

a)   The length of the co-cultivation period

The period immediately after infection is usually of 1-3 days
duration and is the time during which transformation occurs. This
period is something of a 'black box' and the cellular and
molecular events as they relate to plant cell transformation are
virtually unresearched. From an empirical standpoint the time
required is something of a compromise between allowing enough
time for transformation to occur and not letting agrobacteria
overgrow the plant tissue. For apple explants we have found 3
days to be slightly better than 1 or 2 as assessed by the number
of kanamycin resistant calli formed in a 3 month period following
infection. Again the use of the GUS assay procedure for both a
quantitative and qualitative assessment of the effects of
co-cultivation period on transformation would be useful here.

Feeder layers of plant cells usually in the form of cell
suspensions are often used during the co-cultivation period to
enhance transformation and restrict the growth of agrobacteria
(Horsch et al. 1985). We have found in apple that neither potato
or apple endosperm cell suspensions affect the frequency of
transformed callus or shoots that subsequently develop
(unpublished data).

b) Antibiotic selection

One possible explanation for a low number of transformants in
an otherwise highly efficient regeneration system is that the

antibiotic is inhibitory to the process of regeneration from transformed cells. This has already been claimed in sunflower (Everett et al. 1987) and may be operative in apple. Ways of obviating this are to lower the contact time, the concentration or both. Certainly with Brassica napus efficiency was markedly increased by using only 15 rather than 50µg/ml kanamycin in the selection procedure (Moloney et al. 1989) . Although there is a greater risk of permitting 'escapes' to regenerate it may allow some true transformants to develop that would otherwise not have done so at higher concentrations of the antibiotic.

Although our transgenic apple clones were produced at 60µg/ml we are now experimenting with 25µg/ml since we know that this permits no root formation in control untransformed Greensleeves material that normally gives very high rooting levels (James et al. 1989). The use of the different selection strategies and their effects on the regeneration of apple transformants have been discussed elsewhere (James et al. 1989, 1990a). Although we have obtained transgenic apple plants with selection periods as short as 5 days this only slightly reduces the time taken for transgenic shoots to regenerate compared to shoots regenerated after much longer periods of exposure to the antibiotic (Table 3). At the same time short selection times often produce large number of 'escapes' which take considerable time to screen for the scorable marker and even longer to screen for kanamycin resistance since several subcultures are required to produce enough shoots to test for their ability to root on kanamycin (see below). We have now compromised on using 21 days on kanamycin as a suitable time for screening out untransformed cells without subjecting the tissues to overlong exposures to the antibiotic should it interfere with the regeneration process. There are reports that other antibiotics related to kanamycin and equally effective at selection are less inhibitory to the regeneration process viz. paromomycin (Chupeau et al. 1989) and phleomycin (Perez et al. 1989).

4. REGENERATION - Organogenesis from leaf explants

If we assume that the majority of cells within a leaf explant are not transformed immediately after the transformation process they will be either dead or dying after exposure to the antibiotic. The necrotic products could then have an inhibitory effect on growth of nearby viable transformed cells. De Block (1988) was able to regenerate transformants of potato cultivars at a much higher frequency by including silver nitrate at 10mg/l in the regeneration medium. Silver nitrate has long been known to inhibit the production of ethylene during senescence and other age related phenomenon (Aharoni et al. 1979; Beyer 1976). More recently it has been shown to stimulate shoot regeneration in tissue cultures of both monocots and dicots (Purnhauser et al. 1987).

Higher regeneration frequencies using altered tissue culture media for apple have already been mentioned in the context of measuring adventitious bud formation per se. Media manipulations during the antibiotic selection phase have so far not been reported in apple. It is likely that frequencies of regeneration from apple leaf explants can be much higher than those already recorded. Recent unpublished work by Viss (pers. comm. 1990) showed that leaf explants of the apple cultivar Greensleeves could regenerate on average more than 100 shoots per explant using TDZ combined with other media changes. The question here is whether all these buds were of adventitious origin and a moments reflection should question whether it is in fact physically possible for this many shoots to arise on a surface area of approximately 2 square centimetres (adaxial and abaxial surfaces) It needs to be determined then how many of these were true de novo regeneration events and how many were 'condensed' axillary shoots laid down on initial de novo primordia. This question can only be resolved by a detailed histological analysis based on the time course of regeneration. Whatever the answer to this question observations such as this and others mentioned above serve to show that the role of media manipulations in increasing the efficiency of transformation may yet be paramount.

THE CURRENT SITUATION AND THE NATURE OF TRANSGENIC APPLE TREES

Not all apple shoots that regenerate after kanamycin selection have proven to be transgenic and escapes are screened out with the nopaline synthase assay (Otten and Schilperoort, 1978). Moreover shoots that have proven positive for Nos activity do not always go on to produce micropropagating cultures. The reasons for this are unknown. Once a clone is established however it can take several months to produce sufficient material for rooting, biochemical and molecular analyses. To produce sufficient apple shoots for the rooting assay is critical since regenerates can show a reduced rooting capacity in the absence of kanamycin compared to the control parent line. Consequently we routinely produce a minimum of 30 shoots for rooting assays on kanamycin before subjecting the clone to molecular analyses (Table 3) .One of our clones ,B, has consistently rooted at a significantly lower level than other transgenic clones irrespective of the presence of kanamycin in the rooting media (James et al. 1989).

To date we have generated 8 transgenic shoots that have successfully adapted to micropropagation procedures and have been cloned for rooting and callus bioassays, for biochemical assay and for molecular analyses (Table 4) .Six of these are of sufficient age to have spent at least one year growing under greenhouse conditions. After this time there were no apparent morphological differences between any of the clones and the micropropagated untransformed, non-regenerated controls.

The oldest two clones, A and B however have now spent two years under greenhouse conditions. Preliminary observations of the growth of clone B suggest that changes in physiology may have taken place during the second year. When grafted onto the parent control rootstock (also tissue culture propagated and of the same age) at the beginning of the second growth year they were healthy 5 months after grafting.

In replicated reciprocal grafts i.e. using the transgenic clone as the rootstock and the untransformed parent as the scion, growth of the scion was inhibited and death ensued after 4-6 weeks. Transgenic plants of the same clone grown under glass for the same time but not grafted, shed their leaves 4 weeks before the untransformed parent. Advanced leaf fall was not observed in one year old clones. Reciprocal grafting experiments are now being duplicated in other clones to verify how general is the effect.

CONCLUSIONS

It should be clear that all four aspects of the leaf disc

Table 3. Rooting of some of the pBin6 transgenic apple clones in presence/absence of 50 µg/ml kanamycin. At least 30 shoots used for each rooting treatment.

Clone	Kanamycin	%shoots rooting	No. roots/ rooted shoot	Nopaline synthase activity	Southern Blot*
Control	+	0	0	–	–
Control	–	96	5.3	–	–
A	+	85	4.6	–	–
	–	90	5.9	–	–
B	+	61	2.5	+	+
	–	83	3.9	+	+
D	+	83	3.7	+	+
	–	79	4.5	+	+
E	+	94	5.7	+	+
	–	96	5.8	+	+
F	+	0	0	–	–
	–	87	5.8	–	–
H	+	76	4.1	+	+
	–	90	4.7	+	+

* A 7.5kbp Bgl II fragment from pBIN1, a precursor of pBIN6, which spans the latters entire T-DNA region was used as a probe.

Table 4. Regenerated Bin 6 clones of apple. Selection was at 50 or 60µg/ml kanamycin.

Clone	expt/ code	Nos assays	Southerns blots*	Root-ing on kanamycin	Times to regenerate (weeks)	Days on kanamycin
A	860806	-VE	(-VE) rearranged)	YES	20+	>150
B	860806	+VE	+VE	YES	20+	>150
D	870814/1	+VE	+VE	YES	20+	>150
E	871030/1	+VE	+VE	YES	20+	>150
H	880223	+VE	+VE	+VE	10	5
J	880223	+VE	+VE	+VE	16	5
K	871030/1	+VE	+VE	+VE	36	>150
M	890207	+VE	ND	IN PROGRESS	14	21
P	890609	+VE	ND	ND	32	21

ND ; not done
* A 7.5kbp Bgl II fragment from pBIN1, a precursor of pBIN6, which spans the latters entire T-DNA region was used as a probe.

transformation procedure,'initiation, infection, selection and regeneration, still need investigation. All suffer from a lack of understanding of the basic physiology and biochemistry that governs host susceptibility, bacterial virulence, cell senescence, cell dedifferentiation and shoot regeneration and there can be no doubt that progress in fundamental studies will bring us closer to making transformation of apple and other tree species a routine procedure. The work on virulence is an obvious example of this and the role of signal molecules a particularly good example of applying fundamental knowledge to solve a practical problem. On the other hand considerable progress has been made in developing very efficient regeneration protocols using an empirical approach. The progress has been rapid enough to make remarks made a few years ago concerning the intractability of regeneration from woody species now seem rather lame and it is interesting to note that what was assumed to be very difficult at one time is routine just a few years later. This appears to have happened without parallel fundamental studies of the underlying mechanisms of how plant growth regulators control cellular dedifferentiation and organogenesis and/or somatic embryogenesis.

One of the major problems in plant cell transformation work is the inability to identify quickly the effects of the several variables that are being examined in the assessment of transformation efficiency . For this recourse to the GUS assay (Jefferson et al. 1987) and other techniques such as in situ hybridisation may permit a more rapid assessment of the role of the variables discussed above. Implementation of some of these ideas may soon see the appearance of apple trees carrying economically important traits such as insect resistance conferred through such genes as Bt (Bacillus thuringiensis) insecticidal proteins (Fischhoff et al. 1987 ; Vaeck et al. 1987) or the cow pea trypsin inhibitor (Hilder et al. 1987). It is not such a great leap from here to the implementation of anti-sense gene technology (Van der Krol et al. 1988) where genes concerned with the ripening process of climacteric fruit such as tomato (Smith et al. 1988) may already be regulated in order to alter shelf-life. On the other hand isolation and cloning of the single gene determining the 'Columnar habit' in apple (Lapins and Watkins, 1973) may be farther off but it too will ultimately be dependent on solving the problems of plant cell transformation that have been outlined here.

REFERENCES

Aharoni, N., Anderson, J. D., and Lieberman, M., 1979, Production and action of ethylene in senescing leaf discs. Effects of indoleacetic acid, kinetin, silver ion, and carbon dioxide. Plant Physiol. 64:805.

Alt-Moerbe, J., Kuhlmann,H., Schroder,J., 1989, Differences in induction of Ti plasmid virulence genes virG and virD and continued control of virD expression by four external factors, Molecular Plant-Microbe Interactions, 2: 301.

Bevan, M. W., 1984, Binary Agrobacterium vectors for plant transformation, Nuc. Acids Res., 12: 8711.

Beyer, E. M., 1976, A potent inhibitor of ethylene action in plants. Plant Physiol. 58:268.

Chyi, Y-S., and Phillips, G., C., 1987, High efficiency Agrobacterium mediated transformation of Lycopersicon based on conditions favourable for regeneration, Plant Cell Reports, 6: 105.

Chupeau, M-C., Bellini, C., Guerche, P., Maisonneuve, B., Vastra, G., and Chupeau, Y., 1989, Transgenic plants of lettuce obtained through electroporation of protoplasts. Bio/Technology, 7: 503.

Dandekar, A. D., Uratsu, S. L., and Matsuta, N., 1990, Agrobacteria- mediated transformation of apple . Factors influencing virulence. Acta Hort., (In Press).

De Block, M., 1988, Genotype independent leaf disc transformation of potato (Solanum tuberosum) using Agrobacterium tumefaciens, Theor. Appl. Genet., 76:767.

Everett, N. P., Robinson, K. E. P., and Mascarenhas, D., 1987, Genetic engineering of sunflower (Helianthus annuus L.), Bio/technology, 5:1201.

Fasolo, F., Zimmerman, R. H., and Fordham, I., 1989, Adventitious shoot formation on excised leaves of in vitro grown shoots of apple cultivars, Plant Cell Tissue Organ Culture, 16: 75.

Fischhoff, D. A., Bowdish, K. A., Perlak, J., Marrone, P. G., McCormick, S. M., Niedermeyer, J. G., Dean, D. A., Kusano-Kretzer, K., Mayer, E. J., Rochester, D. E., Rogers, S. G., and Fraley, T., 1987, Insect tolerant transgenic tomato plants, Bio/technology, 5: 807.

Hilder, V. A., Gatehouse, M. R., Sheerman, S. E., Barker, R. F., and Boulter, D., 1987, A novel mechanism of insect resistance engineered into tobacco, Nature, 330: 160.

Horsch, R. B., Fry, J. E., Hoffman, N. L., Eicholtz, D., Rogers, S. G., and Fraley, R. T., 1985, A simplified and general method of transferring genes into plants, Science, 227: 1229.

James, D. J., Cell and tissue culture technology for the genetic manipulation of temperate fruit trees. in: Biotechnology and Genetic Engineering Reviews 5:33, G. Russell, ed., Intercept Ltd., Newcastle-upon-Tyne, 1987.

James, D. J., Passey, A. J., and Malhotra, S. B., 1984, Organogenesis in callus derived from stem and leaf tissues of apple and cherry rootstocks. Plant Cell Tissue Organ Cult., 3: 333.

James, D. J., Passey, A. J., and Rugini, E., 1988, Factors affecting high frequency plant regeneration from apple leaf tissues cultured in vitro. J. Plant Physiol., 132:148.

James, D. J., Passey, A. J., Barbara, D. J., and Bevan, M. W., 1989, Genetic transformation of apple (Malus pumila Mill.) using a disarmed Ti-binary vector, Plant Cell Rep., 7: 658.

224

James, D. J., Passey, A. J., and Barbara, D. J. 1990a, Regeneration and transformation of apple and strawberry using disarmed Ti-binary vectors. in: "Genetic Engineering of Crop Plants", 49th Nottingham Easter School, University of Nottingham Sutton Bonington, G. Lycett and D. Grierson, eds., Butterworths, (In press).

James, D. J., Passey, A. J., and Barbara, D. J. 1990b, Agrobacteria-mediated transformation of the cultivated strawberry (Fragaria x anannassa Duch.) using disarmed Ti-binary vectors. Plant Science (In Press).

Jefferson, R. A., Kavanagh, T. A., and Bevan, M. W., 1987, GUS fusions: B-glucuronidase as a sensitive and versatile gene fusion marker in higher plants, The EMBO J., 6:3901.

Kobayashi A., 1989, Studies on cell culture and plant regeneration in apple. III. Influence of etiolation on adventitious shoot formation from leaf discs, Bull. Fruit Trees Res. Stn. C., 16:15.

Lapins, K.O. and Watkins, R. (1973). Genetics of compact growth habit. Rep. East Malling Res. Stn. for 1972. p.136.

Larkin, P. J., and Scowcroft, W. R. 1983, Somaclonal variation and crop improvement, in: "Genetic Engineering of Plants: An Agricultural Perspective," T. Kosuge, C.P. Meredith, and A. Hollaender, eds., p 289. , Plenum Press, New York.

McHughen, A., Jordan, M., and Feist, G., 1989, A preculture period prior to Agrobacterium inoculation increases production of transgenic plants. J. Plant Physiol., 135:245.

Moloney, M. M., Walker, J. M., and Sharma, K. K., 1989, High efficiency transformation of Brassica napus using Agrobacterium vectors, Plant Cell Rep., 8:238.

Otten, L. A. B. M. and Schilperoort, R. A., 1978, A rapid microscale method for the detection of lysopine and nopaline dehydrogenase activities. Biochem. Biophys. Acta. 527:497.

Perez, P., Tiraby, G., Kallerhoff, J., Perret, J., 1989, Phleomycin resistance as a dominant selectable marker for plant cell transformation, Plant Mol. Biol., 13:365.

Predieri, S., and Fasolo Fabbri Malavasi, F., 1989. High frequency shoot regeneration from leaves of the apple rootstock M.26 (Malus pumila Mill.), Plant Cell Tissue Organ Cult., 17:133.

Purnhauser, L., Medgyesy, P., Czako, M., Dix, P. J., and Marton, L. , 1987, Stimulation of shoot regeneration in Triticum aestivum and Nicotiana plumbaginifolia Viv. tissue cultures using the ethylene inhibitor AgNO3. Plant Cell Rep. 6:1.

Rogowsky, P. M., Close, T. J., Chimera, J. A., Shaw, J. J., and Kado, C. I., 1987, Regulation of the vir genes of Agrobacterium tumefaciens plasmid pTiC58. J. Bacteriol., 169:5101.

Smith, C. J. S., Watson, C. F., Ray, J., Bird, C. R., Morris, P. C., Schuch, W., and Grierson, D., 1988, Antisense RNA inhibition of polygalacturonase gene expression in transgenic tomatoes. Nature 334:726.

Stachel, S. E., Messens, E., Van Montague, M., and Zambryski, P. C., 1985, Identification of the signa molecules produced by wounded plant cells that activate T-DNA transfer in Agrobacterium tumefaciens, Nature 318:624.

Vaeck, M., Reynaerts, A., Hofte, H., Jansens, S., De Beuckeleer, M., Dean, C., Zabeau, M., Van Montagu, M., and Leemans, J., 1987, Transgenic plants protected from insect attack, Nature, 328:33.

Van der Krol, A. R., Mol, J. N. M., and Stuitje, A. R., 1988, Antisense genes in plants: an overview, Gene, 72:45.

Welander, M., 1988, Plant regeneration from leaf and stem segments of shoots raised in vitro from mature apple trees, J. Plant Physiol., 132:738-744.

Zambryski, P. C., 1988, Basic Processes underlying Agrobacterium mediated DNA transfer to plant cells, Ann. Rev. Genet., 22:1-30.

TRANSFER OF FOREIGN GENES INTO YELLOW-POPLAR (<u>LIRIODENDRON TULIPIFERA</u>)

H.D. Wilde[1], R.B. Meagher[2], and S.A. Merkle[1]

[1]School of Forest Resources and [2]Department of Genetics
University of Georgia, Athens 30602 USA

ABSTRACT

Yellow-poplar can be regenerated from tissue culture by somatic embryogenesis. The transformation of embryogenic cultures should allow the production of transgenic yellow-poplar plants. Direct gene transfer methods have been used to introduce the gene encoding ß-glucuronidase (GUS) into protoplasts (polyethyene-glycol mediated uptake, electroporation) and intact cells (microprojectile bombardment) from suspension cultures. In addition to other parameters, the physiological age of the protoplasts was found to be an important factor in the transient expression of GUS. Intact cells, bombarded with DNA-coated tungsten particles, were found to express the marker gene individually and in cell clusters.

INTRODUCTION

Gene transfer technology can be applied to forest species for which regeneration systems have been developed. Yellow-poplar (<u>Liriodendron tulipifera</u>) is an ideal candidate for genetic transformation because it can be regenerated from cultured cells by somatic embryogenesis (6). Protoplasts from embryogenic suspension cultures will form callus that retains developmental potential (7). Methods involving the direct DNA transformation of protoplasts and intact cells can be used as an alternative to <u>Agrobacterium</u>-mediated transformation. This approach avoids the host-range limitations of <u>Agrobacterium</u>.

Protoplasts can be induced to take up foreign DNA with a brief exposure to a high intensity electric field (electroporation) or polyethylene glycol (PEG). These techniques must be optimized for each species and tissue source. DNA can be introduced into intact cells by microprojectile bombardment (10). The efficiency of DNA uptake can be measured by the degree of transient expression of a reporter gene.

Protoplast Preparation

Protoplasts were isolated from two embryogenic suspension culture lines by a modification of the procedures of Merkle and Sommer (7). Briefly, approximately one gram of tissue was placed in 10 ml of filter-sterilized digestion medium containing 1% Cellulysin and 0.5% Macerase. After 24 hours of incubation at 30^0 C, the cells were filtered through Miracloth, washed twice by sedimentation, and sieved through a 25 μm pore stainless steel mesh. The length of time between subculture of the cell suspensions and protoplast preparation was varied. The average number of protoplasts/g after 5-day and 20-day intervals was 6.4×10^6 and 1.7×10^7, respectively.

Transformation of Protoplasts

The protoplast concentration was adjusted to 1×10^6 protoplasts/ml for electroporation and 2×10^6 protoplasts/ml for PEG-mediated transformation. Uncut plasmid DNA was added at a final concentration of 25 μg/ml. Either pBI221, which carries the CaMV 35S/GUS/NOS construct, or pBI121, which carries the NOS/NPTII/NOS construct in addition to the same GUS construct, was used (4). A GUS construct with a soybean heat shock promoter was a gift from M. A. Ainley, Botany Dept., UGA. Protoplasts were electroporated with a Promega Model 450 in a buffer containing 100 mM NaCl, 4 mM $CaCl_2$, 8 % PEG, and 500 mM mannitol. The voltage, capacitance, and pulse length were varied. The pulse decay constant was calculated as the product of the capacitance and resistance (3). Since the initial voltage (V_0), capacitance (C), and pulse length (t) are set and the voltage remaining after the pulse (V_t) is observed, the resistance (R) could be calculated from the following equation:

$$V_t = V_0 e^{-t/RC}.$$

For PEG-mediated transformation, the protoplast/DNA suspension was mixed with one volume of 40% PEG-6000 and incubated at room temperature for 45 minutes. PEG-treated and electroporated protoplasts were washed with yellow-poplar regeneration medium I (7) and incubated for 18 hours at 30^0 C in a thin layer of the same medium. Controls without DNA were always included. Protoplast viability was determined by exclusion of 0.5% Evan's Blue dye. After transformation with pBI121, a fraction of the protoplasts was plated in agarose droplets (7) and were selected with 100 μg/ml kanamycin after two weeks.

ß-Glucuronidase Assay

Protoplasts were sedimented at 100 x g in 15 ml Falcon tubes for 5 minutes and the pellet was resupended in 50 μl of GUS extraction medium (4). The suspension was transferred to eppendorf tubes, homogenized briefly with a glass pestle, and spun for 5 minutes in a microcentrifuge. The supernatant was removed and a 10 μl aliquot was taken for protein quantitation by Bradford assay (Bio-Rad kit procedures). An equal volume of extraction buffer containing 2 mM 4-methyl umbelliferyl glucuronide was added to the protein extracts and incubated at 37^0 C. The reaction was stopped at different timepoints with the addition of 0.5 ml of 0.2 M Na_2CO_3. ß-Glucuronidase hydrolyzes the substrate, producing methyl umbelliferone (MU), which is fluorescent when excited by long-wave UV light. The fluorescence of the samples was quantitated with a Hoefer TKO 100 Mini-Fluorometer and the MU concentration was determined from a standard curve. Background fluorescence, determined from the controls without DNA additions, was subtracted and GUS activity was normalized on a per mg protein basis.

Microprojectile bombardment

Two-week old suspension cultures (0.5g) were collected on 4.25 cm filter paper discs using a Buchner funnel. The discs were transferred to 60 mm Petri plates containing agar-solidified medium. Tungsten particles (1.1 μm diameter) were coated with plasmid DNA (pBI 121 or pBI 221) using the calcium-spermidine preciptation procedure (5). There was approximately 4 μg DNA/mg tungsten and the final particle concentration was 63 mg tungsten/ml. A DuPont "Gene Gun" was used to accelerate 2 μl of this suspension toward a target placed 10 cm below the end of the barrel. Each plate was bombarded twice.

For the histochemical GUS assay, the cultures were transferred after two days to new petri dishes and incubated with 0.5 ml of GUS assay solution containing 2 mM 5-bromo-4-chloro-3-indoyl-ß-D-glucuronic acid (X-Gluc; ref.5). For antibiotic selection, cultures were moved after 7 days to plates containing agar-solidified medium with 200 μg/ml kanamycin.

RESULTS AND DISCUSSION

Electroporation Parameters

Sucessful transformation by electroporation requires a degree of membrane permeablization that balances gene transfer efficiency with cell viability. Generally, this involves electroporation conditions that have a decay time constant of 5-15 ms and lead to a reduction of

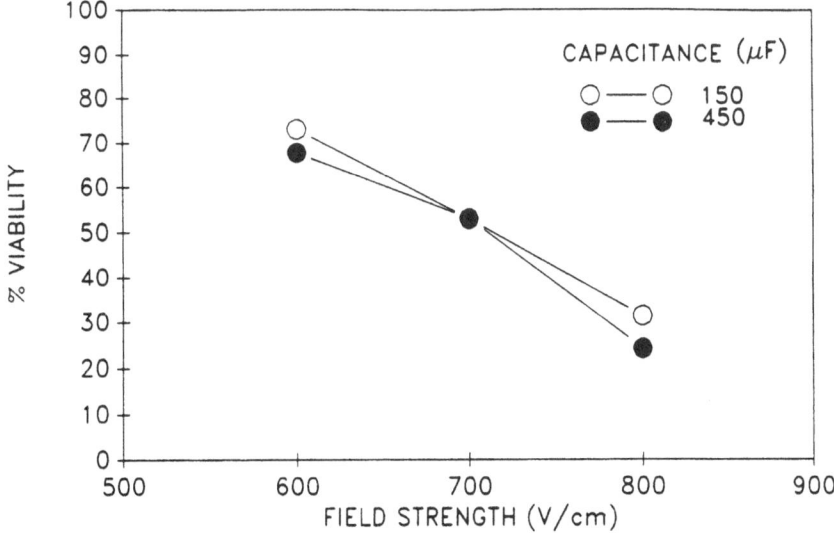

Figure 1. Effect of field strength on protoplast viability. Protoplasts were electroporated and viability was assessed the following day by dye exclusion.

Table 1. Electroporation parameters of protoplasts from suspension cultures with regenerative potential.

	Field strength	Capacitance	Decay constant
yellow-poplar	700 V/cm	400 μF	12 ms
rice	750	22	4
carrot	875	980	20
maize	500	245	4

50% in protoplast viability (3). The decay time constant is a function of the capacitance of the electroporator, the voltage to which it is charged, and the resistance of the buffer containing the protoplasts.

Figure 1 shows the effect of field strength (V/cm) on protoplast viability. With a capacitance of 450 microfarads (μF), cell viability was reduced to 24% when a field strength of 800 V/cm was applied. A capacitance of 450 μF was used in further experiments because the higher capacitance did not reduce viability significantly. With a field strength of 700 V/cm, protoplast viability was 53%, which is nearly optimal. The resistance of the electroporation buffer, determined by its ionic strength, was held constant. When the pulse length was set to 30 ms, and with a capacitance of 450 μF and a field strength of 700 V/cm, the decay time constant was calculated to be 12 ms.

At a field strength of 600 V/cm, protoplast viability remained near 70%. It is typical for small, embryogenic cells to require higher field strengths for membrane permeabilization than larger, vacuolated cells from, for example, mesophyll tissue. Yellow-poplar protoplasts isolated from embryogenic cell suspensions are small (10-20 μm diameter) and densely cytoplasmic. Electroporation conditions for these protoplasts are comparable to those for other protoplasts with regenerative potential. Table 1 shows electroporation parameters for yellow-poplar, rice (11), carrot (1), and maize (2).

Influence of culture cycle on transient expression

Yellow-poplar suspension cultures were subcultured into fresh medium every two weeks. Protoplasts were routinely isolated one week after subculture, but GUS expression was never detected after these protoplasts were electroporated with either pBI121 or pBI221. GUS was expressed transiently, however, in protoplasts isolated two and three weeks after subculture. To examine this variation in more detail, protoplasts were isolated at five-day intervals after a subculture and were incubated with pBI221 and 20% PEG. After washing and overnight incubation, the protoplasts were lysed and assayed for GUS activity. The results are shown in Figure 2.

PEG-mediated transformation of protoplasts isolated from culture line TP 14x108 at 5, 10, and 25 days after subculture produced low levels of GUS activity. The identical treatment of protoplasts isolated 15 and 20 days after subculture led to a 2-fold and 6-fold increase, respectively, in GUS activity. A similar, though less well pronounced, pattern of GUS expression was evident in protoplasts from tissue culture line TP 4x12. This suggests that the physiological age of the protoplasts may affect their ability to take up or express foreign DNA. The differential expression of the chaemeric gene does not appear to be related to the CaMV 35S promoter, because the same results were observed when a soybean heat shock promoter-GUS construct was used. Transformation efficiency was found to be affected by the mitotic state of the protoplasts in tobacco (8,9). Yellow-poplar suspension cultures are asynchronous, however, so the contribution of a single mitotic state would be minimal.

When the PEG concentration was increased from 8 to 20%, PEG-mediated DNA uptake led to levels of GUS activity in the range previously obtained by electroporation. This method of transformation was less deleterious to protoplasts, as evidenced by improved colony production when these protoplasts were plated in agarose droplets. Kanamycin (100 µg/ml) was used to select colonies derived from protoplasts after PEG-mediated uptake of pBI121. This plasmid carries a gene encoding neomycin phosphotransferase II, which confers antibiotic resistance. While cells not exposed to pBI121 died, some colonies which had been treated with the plasmid survived for up to three months. Similarly treated protoplasts from the same experiment showed transient GUS activity. The loss of kanamycin resistance may be related to a low rate of DNA integration, limited promoter activity, or gene inactivation.

Microprojectile bombardment

The plasmid pBI121 could be introduced into intact cells penetrated by DNA-coated tungsten particles. These particles were accelerated toward a thin layer of embryogenic culture by the force of a gunpowder charge. Expression of the GUS gene could be detected after two days using the indigogenic substrate X-Gluc. GUS was found to be expressed both in single cells and in cell clusters. Continued growth of the callus produced sectors of blue cells, suggesting that stable integration of the plasmid had taken place. However, expression of the linked gene NPT II could not be detected.

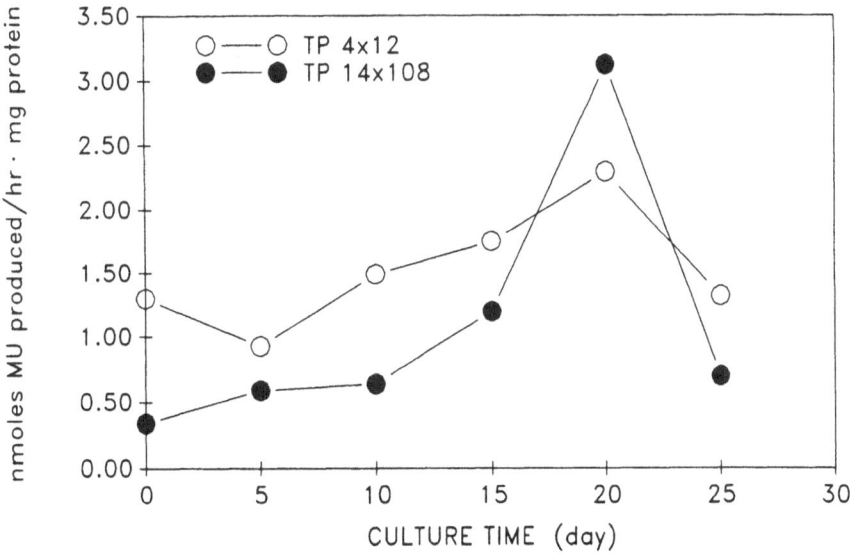

Figure 2. Transient GUS expression in protoplasts isolated at different timepoints after subculture. GUS activity shown is the average of two transformations per timepoint per experiment. The experiment was repeated twice for each culture line.

Bombarded tissue, selected with kanamycin after one week, responded no differently than untreated controls.

CONCLUSIONS

Plasmid DNA was introduced into yellow-poplar protoplasts by electroporation and PEG-mediated uptake and into intact cells by microporojectile bombardment. Electroporation parameters were adjusted to give a pulse decay constant of 12 ms and protoplast viability of 53%. Conditions for electroporation of yellow-poplar protoplasts were in the range of those for small, developmentally active protoplasts from other species. Transient expression of the GUS gene showed that DNA uptake by PEG treatment alone was as effective as electroporation when the PEG concentration was raised to 20%. The physiological age of the protoplasts influenced the amount of GUS activity observed after the uptake of pBI221. In contrast to the GUS gene, the NPT II construct was poorly expressed in yellow-poplar. Alternative promoters or different selectable markers may improve screening for stable integration of foreign genes.

LITERATURE CITED

1. Fromm, M., Taylor, L.P., and Walbot, V., 1985, Expression of genes transferred into monocot and dicot cells by electroporation. Proc. Natl. Acad. Sci. USA 82: 5824-5828.
2. Fromm, M.E., Taylor, L.P., and Walbot, V., 1986, Stable transformation of maize after gene transfer by electroporation. Nature 319: 791-793.
3. Fromm, M., Callis, J., Taylor, L.P., and Walbot, V., 1987, Electroporation of DNA and RNA into plant protoplasts. Methods in Enzymol. 153: 351-366.
4. Jefferson, R.A., Kavanaugh, T.A., and Bevan, M.W., 1987, GUS fusions: ß-Glucuronidase as a sensitive and versatile fusion marker in higher plants. EMBO 6: 3901-3907.
5. Klein, T.M., Gradziel, T., Fromm, M.E., and Sanford, J. C., 1988, Factors influencing gene delivery into Zea mays by high velocity microprojectiles. Biotechnology 6: 559-563.
6. Merkle, S.A., and Sommer, H.E., 1986, Somatic embryogenesis in tissue cultures of Liriodendron tulipifera. Can. J. For. Res. 16: 420-422.
7. Merkle, S.A., and Sommer, H.E., 1987, Regeneration of Liriodendron tulipifera (Family Magnoliaceae) from protoplast culture. Amer. J. Bot. 74: 1317-1321.
8. Meyer, P., Walgenbach, E., Bussmann, K., Hombrecher, G., and Saedler, H., 1985, Synchronized tobacco protoplasts are efficiently transformed by DNA. Mol. Gen. Genet. 201: 513-518.
9. Okada, K., Takebe, I., and Nagata, T., 1986, Expression and integration of genes introduced into highly synchronized plant protoplasts. Mol. Gen. Genet. 205: 398-403.
10. Sanford, J.C., 1988, The biolistic process. Trends in Biotech. 6: 299-302.
11. Toriyama, K., Arimoto, Y., Uchimaya, H., and Hinata, K., 1988, Transgenic rice plants after direct gene transfer into protoplasts. Biotechnology 6: 1072-1075.

EVIDENCE FOR AGROBACTERIUM-MEDIATED GENETIC TRANSFORMATION IN LARIX

DECIDUA

Yinghua Huang, Dong-Ill Shin, and David F. Karnosky

Michigan Technological University
School of Forestry and Wood Products
Houghton, Michigan USA 49931

ABSTRACT

Our research efforts have demonstrated that European larch (Larix decidua) is susceptible to Agrobacterium infection, that agrobacterial plasmid transfer into larch cells occurs, and that subsequent expression of the plasmid-harbored genes takes place in the transformed larch tissues. These results provide documented proof of Agrobacterium-mediated gene transfer in this softwood species.

INTRODUCTION

We have been working to develop an Agrobacterium-mediated genetic transformation in larch (Larix spp.) for the past five years. Larches are important conifer species which have shown rapid growth and good wood and pulp quality in northern climates (3,8,16). In addition, larches are highly amenable to in vitro tissue culture, including micropropagation (5,12), embryogenesis (17,23), and organogenesis (13). However, the problems that face the increased use of larch trees are their extreme sensitivity to herbicides (19) and their susceptibility to serious insect pests (14,20,21).

Although Agrobacterium-mediated genetic transformation has been proven in many agronomic and horticultural crop plants, reported successes have been few in hardwood trees and yet more rare in commercially valuable softwood tree species (1,4). The purpose of our research is to develop methods for genetic transformation of conifer trees via Agrobacterium-mediated gene transfer systems and to use this technology to improve conifers for reforestation in northern regions.

Initially, our artificial inoculations of European larch with Agrobacterium were successful either in vitro or in the greenhouse. Sterile seven-day-old seedlings of larch were linearly wounded on their upper hypocotyls and aseptically inoculated by placing a scrape from the bacterial "lawn" at the wound site. Agrobacterium tumefaciens strains Bo542 and ATCC15955 and A. rhizogenes strain ATCC11325 were used. Inoculated seedlings were then placed on modified Gresshoff and Doy medium (15) and incubated in a growth chamber to allow the bacteria to transform the plant cells. Pathological symptoms such as swollen tumors, multiple bud clusters and adventitious hairy roots developed from the inoculated sites about three to five weeks after inoculation. In the greenhouse, stems of actively growing three-year-old European

larch seedlings were wounded with 18-gauge or 22-gauge needles and inoculated
with fresh Agrobacterium suspensions of the same strains as used in the in
vitro inoculations. At various times after infection, seedlings were scored
for reaction to Agrobacterium. The formation of swelling at inoculation
sites began first within one month after inoculation and then the formation
of callus was visible on the swollen inoculation sites within two months.
Under both cases, artificial inoculation of larch seedlings resulted in a
high frequency of infection (e.g. tumorous growth) by both A. tumefaciens
and A. rhizogenes. However, no symptoms were seen on the wounded but unin-
oculated controls.

The symptomatic tissues excised from the wounds of inoculated seedlings
were cultured on a callus medium (2) containing appropriate antibiotics for
disinfection of Agrobacterium. The calli derived from the transformed tissues were assayed for detection of T-DNA-specified opines by high voltage
paper electrophoresis described by Otten and Schilperoort (18). The electro-
phoretograms showed the phenanthrenequinon-positive spots that extracts of
tissue samples co-migrated with synthetic nopaline. The extracted substance
had the same electrophoretic mobility as standard nopaline. This substance
was not detected in untransformed larch tissues.

The calli initiated from symptomatic tissues inoculated with either
A4pARC8 or WAG11pWB101, which are engineered plasmids bearing a marker gene
(npt II) for kanamycin resistance, were subjected to selection on the callus
medium supplemented with 40 mg/l kanamycin. Callus lines resistant to kanamycin were recovered from 25% of the symptomatic tissues, and maintained on
kanamycin-containing medium. It is believed that the survival of the callus
lines resulted from the expression of the npt II gene for biosynthesis of
neomycin phosphotransferase (NPT II), which enabled the larch callus cultures
to tolerate kanamycin. These results from the opine assay and kanamycin resistant selection experiments provide evidence for gene transfer to and expression of bacterial genes in the transformed larch cells.

The most convincing evidence for demonstration of genetic transformation
of callus lines is provided by Southern blot analysis (22). Calli developed
from two lines each of WAG11pWB101-transformed and A4pARC8-transformed larch
tissues were submitted to Southern blotting. When probed with the entire
plasmid vectors, the probes hybridized to a few bands in the lanes containing
the high molecular weight (plant) DNA isolated from the transformed callus
lines of European larch. The positive result from the Southern analysis confirmed the physical insertion of foreign DNA sequences into larch genome.

Recently, we investigated the interaction of agrobacteria with cells of
Larix decidua using scanning electron microscopy. Larch tissues were wound-
inoculated with both Agrobacterium tumefaciens strain ATCC1595 and Agrobacterium rhizogenes strain ATCC11325, and co-cultivated on the GD medium (10) in
Petri dishes for periods of six to 48 hours. The infected tissues were prepared for scanning electron microscope study by a modification of Graves'
procedure (9). The electron-micrographs showed the virulent agrobacteria
attached to larch cells at wound sites in a similar manner to that described
for cells of susceptible dicot angiosperms (11). The observations from this
experiment have verified our previous inoculation tests (6,7) in which young
seedlings of European larch were successfully inoculated with Agrobacterium.
This is one more piece of evidence to indicate that it is possible to infect
larch species and to promote gene transfer into larch cells with virulent
Agrobacterium strains.

Our results pave the way for Agrobacterium-mediated gene transfer in
larch trees with improved disease resistance, increased herbicide tolerance
and modified lignin production as goals.

LITERATURE CITED

1. Ahuja, M.R. 1988. Gene transfer in forest trees. In: Hanover J. and D. Kiethley (Eds.) Genetic Manipulation of Woody Plants. Plenum Press, New York. Pp. 25-41.

2. Brown, C.L. and R.H. Lawrence. 1968. Culture of pine callus on a defined medium. For. Sci. 14:62-64.

3. Chiang, V.L., G.D. Mroz, S.M. Shaler and D.D. Reed. 1988. Pulp production of a 16-year-old larch stand. Tappi J. 71:179-180.

4. De Cleene, M. and J. De Ley. 1976. The host range of crown gall. Bot. Rev. 42:389-446.

5. Diner, A.M., A. Strickler and D.F. Karnosky. 1986. Initiation, elongation and remultiplication of Larix decidua micropropagules. New Zealand J. For. Sci. 16:306-318.

6. Diner, A.M. and D.F. Karnosky. 1986. Agrobacterium: Potential for genetic engineering in Larix. Tappi Proc. Sept., 1986. Pp. 93-94.

7. Diner, A.M. and D.F. Karnosky. 1987. Differential responses of two conifers to in vitro inoculation with Agrobacterium rhizogenes. Eur. J. For. Path. 17:211-216.

8. Einspahr, D.W., G.W. Wyckoff and M.H. Fiscus. 1984. Larch: A fast-growing fiber source for Lake States and the Northeast. J. For. 82:104-106.

9. Graves, A.E., S.L. Goldman, S.W. Banks and A.C.F. Graves. 1988. Scanning electron microscope studies of Agrobacterium tumefaciens attachment to Zea mays, Gladiolus sp., and Triticum aestivum. J. Bacteriol. 170:2395-2400.

10. Gresshoff, P.M. and C.H. Doy. 1972. Development and differentiation of haploid Lycopersicon esculentum (tomato). Planta 107:161-170.

11. Huang, Y. and D.F. Karnosky. 1989. Interaction of Larix decidua tissues with Agrobacterium rhizogenes and Agrobacterium tumefaciens as depicted by scanning electron microscopy. (Submitted for Publication).

12. Karnosky, D.F. and A. Mulcahey. 1988. Explant orientation and media manipulation affect Larix decidua adventitious bud quality and quantity. In Vitro 24(3). Part II:51.

13. Laliberte, S. and M. LaLonde. 1988. Sustained caulogenesis in callus cultures of Larix x eurolepis initiated from short-shoot buds of a 12-year-old tree. Amer. J. Bot. 75:767-777.

14. Langor, D.N. and A.G. Raske. 1989. The eastern larch beetle, another threat to our forests (Coleoptera: Scolytiae). For. Chron. 65:276-279.

15. Mehra-Palta, A., R.H. Smeltser and R.L. Mott. 1978. Hormonal control of induced organogenesis: Experiments with excised parts of loblolly pine. Tappi J. 61:37-40.

16. Mroz, G.D., D.D. Reed and H.O. Lichty. 1988. Volume production of a 16-year-old European larch stand. Northern J. Appl. For. 5:160-161.

17. Nagmani, R. and J.M. Bonga. 1985. Embryogenesis in subcultured callus of Larix decidua. Can. J. For. Res. 15:1088-1091.

18. Otten, L.A. and R.A. Schilperoort. 1978. A rapid microscale method for the detection of lysopine and nopaline dehydrogenase activities. Biochem. Biophys. Acta 527:497-500.

19. Perala, D.A. 1982. Early release: Current technology and conifer release. In: Proc. Artificial Regeneration of Conifers in the Upper Great Lakes Region. Mich. Tech. Univ., Houghton, Mich. Pp. 396-410.

20. Robbis, K. 1985. Risks associated with growing non-native larches in eastern North America. North. J. Appl. For. 2:101-104.

21. Ryan, R.B., S. Tunnock and F.W. Ebel. 1987. The larch casebearer in North America. J. For. 85(7):33-39.

22. Southern, E.M. 1975. Detection of specific sequences among DNA fragments separated by gel electrophoresis. J. Mol. Biol. 98:503-517.

23. Von Aderkas, P. and J.M. Bonga. 1988. Formation of haploid embryoids of Larix decidua: Early embryogenesis. Amer. J. Bot. 75:619-628.

REGENERATION AND TRANSFORMATION EXPERIMENTS IN APPLE

Margareta Welander[1] and Gowri Maheswaran[2]

[1]Department of Horticultural Science
S-230 53 Alnarp, Sweden

[2]Horticultural Research Institute, Knoxfield
Ferntree Gully, Victoria 3156, Australia

ABSTRACT

Successful introduction of foreign genes using a leaf disc procedure and cocultivation with Agrobacterium tumefaciens requires a good regeneration protocol and tissue susceptible to infection. In our experiments, apple rootstocks M26 and M9 and the cultivar McIntosh Wijcik were used. A good regeneration system was developed for M9 whereas a suitable protocol for the others had been previously developed. Transformation experiments were carried out with three wild type strains and three disarmed vectors harbouring ß-glucuronidase and neomycin phosphotransferase genes. Infection with three wild type strains (C58, A348, A281) did not cause gall formation. However, the binary vector pBII21 (35SGUS) x LBA 4404 resulted in transformed shoots of M26 and transformed callus of M9. This was confirmed histochemically using the substrate 5-bromo-4-chloro-3-indolyl glucoronide (X-Gluc) staining. Cefotaxime, used to control growth of the bacteria, had an effect simalar to cytokinin by stimulating shoot formation in vitro.

INTRODUCTION

Application of traditional plant breeding methods to fruit tree species has been limited by their long generation times. Genetic manipulation techniques are therefore potentially useful for fruit trees because specific genetic changes could be made in a short period of time. An agriculturally important trait concerns pest resistant cultivars. In apples, sources for single genes for resistance to the major pests are available. Dominant genes giving resistance to apple scab (Venturia inaequalis) are found in Malus floribunda, M. prunifolia, M. atrosanguinea and M. micromalus (19). Single gene resistance for mildew (Podosphaera leucotrichia) is found in M. robusta and M. Zumi (11). The apple cultivar "Northern Spy" carries a single gene resistance for woolly aphid (Eriosoma lanigerum) and collar rot (Phytophtora cocotorum) (1). Another useful target for genetic engineering would be to transfer the compact gene found in the cultivar McIntosh Wijcik.

Development of successful DNA transfer systems has been largely based on exploitation of the <u>Agrobacterium tumefaciens</u> mediated DNA delivery system (10). <u>Agrobacterium</u> has a broad but discontinuous host range. Susceptibility depends upon interaction of specific genotypes of host and pathogen (2). For this reason we have tested different bacterial strains and constructs. Successful introduction of foreign genes into excised tissues also requires an adequate regeneration system. The aim of this investigation was to develop a successful regeneration and transformation procedure for some important apple rootstocks and McIntosh Wijcik.

MATERIAL AND METHODS

Plant material

<u>In vitro</u> grown shoots of the apple rootstocks M9 (clone 8172 from Belgium) and M26 and the cultivar McIntosh Wijcik were used in the experiments. Established shoot cultures were routinely subcultured every fourth week. The medium used for M26 consisted of MS (Murashige and Skoog, 15) salts and vitamins plus 5 uM BAP, 1uM IBA, 3% sucrose and 0.7% Difco Bacto agar at pH 5.8. The same medium was used for M9 and McIntosh Wijcik except that IBA was omitted and sucrose was replaced by sorbitol at the same concentration. The cultures were maintained in a growth room at 25^{0}C under 16h photoperiod at an irradiance level of 50 umol m^{-2} s^{-1} from cool-white fluorescent tubes.

Regeneration studies

The regeneration system developed for M26 and McIntosh Wijcik (20) was not sufficient for M9. In preliminary experiments we found marked differences between leaf age, basal media and gelling agent in the regeneration capacity. We also observed that if leaves from micropropageted shoots were pretreated in liqiud regeneration medium, the regeneration capacity was improved. Based on this information only the youngest leaves were utilized and classified in two age groups: 1) folded leaves 2) the first unfolded ones. Each leaf was cut in three segments and placed in Petri dishes on the different media. Two basal media were selected : MS and B5 (Gamborg, 5). These media were used together with 10 uM BAP gelled with 0.25% gelrite and combined with different concentrations of auxin and sucrose. The pH of the media was 5.7. In the first experiment, the two basal media were used together with 2% sucrose in combination with 6 NAA concentrations, 0, 0.25, 0.5, 0.75, 1.0 and 2.0 uM. In the second experiment the two basal media including 1 uM NAA were combined with 6 sucrose concentrations, 0, 1, 2, 3, 4, 5%.

Host range survey

<u>In vitro</u> cultured shoots 2-3 cm in length and 1 mm

thick leaf segments of young expanded leaves were inoculated with three wild type strains of Agrobacterium tumefaciens, C58, A348 and A281. The different strains were kind-
ly provided by Paul Ebbert, Department of Botany, Mel-
bourne. The bacteria cultures were maintained at 4^0C on YEP
agar medium (1% yeast extract, 1% peptone, 1% NaCl and 1.5%
Difco bacto agar at pH 7). Bacteria for inoculation were
cultured for 2 days on the YEP medium at 28^0C, scraped off
into 50 ml of liquid medium and agitated in a water bath
overnight at 28^0C. The bacteria were used at an concentration of $5x10^8$ bacteria per ml. The shoots were inoculated at
different positions on the stem and petioles using a hypo-
dermic syringe and needle. The leaf segments were soaked
in the bacterial solution for 5 minutes and blotted dry on
a filter paper. The medium used for both shoots and leaf
segments consisted of MS basal medium, 3% sucrose 0.7% agar
and 300 mg/l cefotaxime. Noninfected shoots and leaves
placed on the same medium with and without cefotaxime were
used as controls. Tumor formation was scored within 8 weeks
after inoculation.

Transformation experiments with disarmed strains

Transformation experiments were carried out with the
genetically engineered Agrobacterium tumefaciens strain C58
containing either the binary vector pBI121 (35SGUS) x LBA
4404 or the co-integrated vector 1103(35SGUS).pGV3850 and
the strain A281 containing the binary vector pCGN 257 x EHA
101. All vectors harboured neomycin phoshotransferase NPT
II and -glucoronidase (GUS). The NPT II gene is under
control of the nopaline-synthase (NOS) promoter and the GUS
gene under 35S promoter in pBII21 and 1103 whereas both
genes are under the control of the 35S promoter in pCGN
257. The strains were a generous gift from Danny Llewellyn,
CSIRO, Div. of Plant Industry, Canberra and Calgene Pacific.

The transformation procedure was performed as described
below. Only folded and the first unfolded leaves were used.
The leaves were harvested and placed in the regeneration
medium for 2-3 h. Five leaves were simultaneously placed on
a filter paper, soaked in the bacteria solution, wounded
and cut into 3 segments and left in the bacteria solution
for 3 min. The bacteria solution was from an overnight
liquid culture consisting of LB medium plus 25 mg/l kanamycin at a concentration of $5 x 10^8$ bacteria/ml. The leaf
segments were blotted on a filter paper and transferred to
a petri dish containing the regeneration medium, N6 (Chu et
al.,3) or MS. After two days of cocultivation, the segments
were transferred to the regeneration medium containing
cefotaxime (200-500 mg/l) and kanamycin (100mg/l). Noninfected leaves placed on the regeneration media with
either kanamycin or cefotaxime were used as controls. The
Petri dishes were placed in dark at 25^0C until visible shoots were observed. Thereafter, cultures were transferred to
low light intensity. Subculturing was performed every
second week as cefotaxime or by-products from cefotaxime
disintegration seem to inhibit growth and development of
the explants.

Assay for transformants

The GUS-assay according to Jefferson (9) was used to measure gene expression and test for putative transformants. The GUS enzyme activity was located histochemically in freehand sections using the substrate 5-bromo-4-chloro-3-indolyl glucoronide (X-Gluc). Five mg of the substrate was dissolved in 50 ul dimethylformamide and diluted to 5 ml in 50 mM phosphate buffer at pH 7. The sections were incubated at 37°C for 1-3 h depending on the intensity of the staining.

RESULTS

Regeneration experiments

The effect of basal media, NAA concentration and leaf age on adventitious shoot regeneration in leaf segments of the apple rootstock M9 is shown in Table 1. MS medium was clearly superior to B5 in the percentage of explants forming shoots. Responses to NAA concentration differed between the media as well as the age groups. For MS medium the optimum regeneration for age 1 was obtained between 0.75 and 1.0 uM NAA with a reduction at 2 uM. For age 2, no reduction was observed at the highest concentrations. For B5 medium the optimum regeneration was obtained at lower NAA concentrations. Percentage of explants forming shoots was higher for leaf age 1 than 2 at the optimum NAA concentration.

Table 1. The effect of leaf age, basal media and different NAA concentrations on per cent explants forming shoots.

Regeneration medium: MS + 10 uM BAP + 2% sucrose

Leaf age	uM NAA					
	0	0.25	0.5	0.75	1.0	2.0
1	7	53	61	78	73	58
2	4	42	52	42	60	60

Regeneration medium: B5 + 10 uM BAP + 2% sucrose

Leaf age	0	0.25	0.5	0.75	1.0	2.0
1	3	22	40	24	14	12
2	0	14	8	14	2	8

Leaf age:1= folded leaves 2= first unfolded ones
For comparison within a row the least significant difference (P=0.05) = 11 and within a column = 12

The same level of shoot regeneration could be obtained for age 2 as age 1 by increasing the sucrose concentration to 4% in the MS medium (Table 2). Average number of shoots formed per explant was 3.5 + 1.1. Neither NAA nor the sucrose concentration significantly affected shoot number. Thus by using the MS medium together with 10 uM BAP, 1uM NAA, 4% sucrose and 0.25% gelrite, a suitable medium was obtained for both leaf groups in M9 in the transformation experiments. Shoot regeneration for M9 is shown in Fig.1.

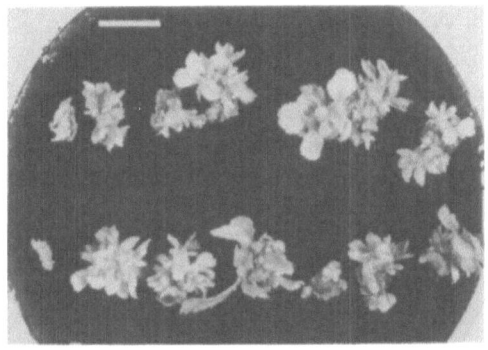

Fig. 1. Shoot proliferation in leaf segments of M9 using the optimum regeneration medium, scale bar = 9 mm.

Host range survey

None of the apple cultivars formed tumours from infection with any of the various <u>Agrobacterium</u> strains (Table 3). However, cefotaxime, added to the medium in order to control the growth of the bacteria, enhanced the axillary shoot production. Leaf segments responded in the same way with increased callus and shoot formation with the presence of cefotaxime (data not shown). Thus from these experiments it was not possible to select the most suitable strain for transformation.

Table 2. The effect of leaf age, basal media and different sucrose concentrations on per cent explants forming shoots.

Regeneration medium: MS + 10uM BAP + 1uM NAA

Leaf age	sucrose concentration (%)				
	1	2	3	4	5
1	40	73	68	78	78
2	51	60	67	80	69

Regeneratiom medium: B5 + 10 uM BAP + 1uM NAA

1	12	14	6	10	20
2	4	2	8	19	12

For definition of leaf age see Table 1. For comparison within a row the least significant difference (P=0.05)=11 and within a column =14

Table 3.   The response of the different apple
           cultivars to inoculation with Agro-
           bacterium tumefaciens wild type strains.

Strain	No. of shoots	% shoots forming tumors	No of ax.shoots per explant mean + SD
Cultivar M9			
A281	20	0	13.3 + 3.1
A348	20	0	7.0 + 2.1
C58	20	0	9.3 + 2.5
cont.-cef	20	0	1.0 + 0
cont.+cef	20	0	7.7 + 5.5
Cultivar M26			
A281	20	0	7.1 + 1.3
A348	20	0	10.0 + 1.2
C58	20	0	13.3 + 2.8
cont.-cef	20	0	1.0 + 0
cont.+cef	20	0	7.8 + 3.4
Cultivar McIntosh Wijcik			
A281	20	0	5.3 + 1.2
A348	20	0	5.3 + 3.0
C58	20	0	12.7 + 2.9
cont.-cef	20	0	1.3 + 0.5
cont.+cef	20	0	12.5 + 3.3

Transformation experiments

     Effect of kanamycin and cefotaxime on shoot regeneration of
the three apple cultivars is shown in Table 4. All of the
kanamycin concentrations tested inhibited callus and shoot
formation. However, the tissue was not killed, and the segments
remained green for several months as long as they were kept in
dark. When segments were transferred to light they were completely bleached. Since it is has been observed that tissues
after cocultivation with bacteria seem to tolerate higher
concentration of kanamycin, 100 mg/l was chosen which would
minimize the number of nontransformed shoots. The low concentration of cefotaxime did not affect shoot regeneration
whereas 500 mg/l reduced both per cent shoot regeneration and
number of shoots per explant. However for some strains of
bacteria, the high concentration of cefotaxime was necassary in
order to control the growth of the bacteria.

Table 4. The effect of kanamycin and cefotaxime on shoot regeneration from leaves of different apple cultivars.

Cultivar	M9	M26	McWj
Variat	% explants forming shoots (No. of shoots per explant)		
Regeneration medium (RM)	78 (3.5)	88 (13.0)	75 (4.3)
RM + 50 mg/l kan.	0	0	0
RM + 75 " "	0	0	0
RM + 100 " "	0	0	0
RM + 125 " "	0	0	0
RM + 200 " cef.	78 (3.7)	84 (12.5)	73 (4.5)
RM + 500 " "	62 (2.3)	78 (5.7)	61 (3.2)
RM + 100 " kan. + 200 " cef.	0	0	0

In the transformation experiments, positive results were obtained only with the binary vector 4404 and with the cultivars M26 and M9. Transformed cells were observed 1 week after cocultivation which was confirmed by X-Gluc staining. After 5 weeks (3 weeks in dark and 2 weeks in light), transformed shoots were obtained in M26 (Fig.2B) and transformed callus in M9. We also observed that nontransformed shoots survived in the vicinity of a transformed shoot therefore each shoot must be examined carefully. Figure 2A shows a shoot cluster of putative transformed shoots of M26. The material is now being multiplied for the final test with the southern blot assay.

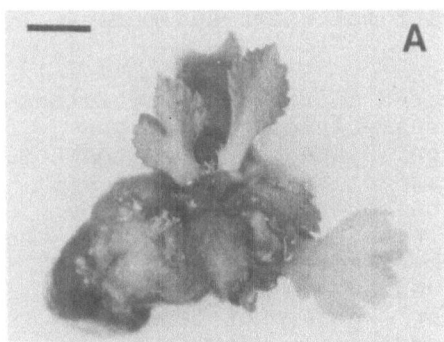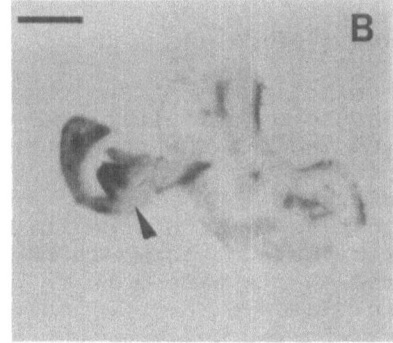

Fig. 2. A. Putative transformed shoots of M26 selected on a medium containing 100 mg/l kanamycin 8 weeks after cocultivation, scale bar = 3.5 mm. B. Transformed shoot (arrow) confirmed by X-Gluc staining 3 weeks after cocultivation, scale bar = 1.9 mm.

DISCUSSION

The effect of gelling agent on growth _in vitro_ has been reported previously (18,16,4). We found a strong interaction between gelling agent and composition of the inorganic nutrients. Modified N6 medium together with agar resulted in a high regeneration capacity for M26 and McIntosh Wijcik but not for M9 whereas MS medium together with gelrite was excellent for M9. Gelrite together with N6 medium and MS with agar resulted in poor regeneration capacity for all cultivars (data not shown). There are several possible explanations for the differences between the gelling agents. All agar types contain impurities at different concentrations (17). Gelrite causes vitrification _in vitro_ to a higher degree than agar (16) indicating differences in the water holding capacity. This may also influence diffusion rates of nutrients into the tissue.

Although members of _Agrobacterium_ are considered wide host-range pathogens, the susceptibility to infection varies widely among both species and cultivars (2). In our experiments, none of the cultivars formed tumors with the wild type plasmids tested. Despite this T-DNA transfer was obtained in M26 and M9 with one of the disarmed Agrobacterium strain. James (7) using the wild type plasmid T37 obtained tumor formation in several apple cultivars although at a low frequency. However, transformation was only reported for the cultivar Greensleeves (8). Monocotyledonous plants are not susceptible to tumor formation, and for a long time it was generally believed that the Ti plasmid could not be used as a vector for monocots. However, later reports (6) indicate that T-DNA can be transferred to monocotyledonous plants without formation of tumors. Thus, it can be concluded that the absence of tumor formation does not always exclude infection.

Cefotaxime at a concentration of 300 mg/l stimulated the axillary shoot production in the absence of cytokinin in the apple cultivars tested. This effect seem to be general for several species since improved growth and regeneration by cefotaxime has been reported earlier for wheat and barley callus (14,13).

Selection for transformants based on kanamycin resistance proved to be difficult. All the kanamycin concentrations tested inhibited shoot regeneration. Since the leaf segments did not show any chlorophyll impairment as long as the explants were kept in dark, selection could not be based on green tissue. After cocultivation with the Agrobacterium, the sensitivity to kanamycin seem to be altered and nontransformed shoots were able to grow even at 100 mg/l of kanamycin. James et al. (8) showed that exposure of leaf discs to kanamycin (60 mg/l) longer than 4 d greatly reduced shoot regeneration but increased the proportion of transformed callus. However, the percentage of shoots from callus was extremely low. Mathews (12) showed that the kanamycin concentration for inhibition of growth varied considerably between species as well as different explants of the same species. However, the GUS gene was a very good marker in apples. The GUS activity was easily detected without interfering inhibitors or background activity. Infected cells could be detected already after 4-5 days. Histochemical assays can be made in a part of an explant while retaining the remaining part for further growth.

REFERENCES

1. Alston, F.H., 1970, Resistance to collar rot. Phy-tophthora cactorum (Leb.and Cohn) Schroet.,in apple. Rep. E.Malling Res. Stn. for 1969: 143-145.

2 . Byrne, M.C., McDonnel, R.E., Wright, M.S., and Carnes, M.G., 1987, Strain and cultivar specifi-city in the Agrobacterium-soybean interaction. Plant Cell Tiss. Org. Cult. 8: 3-15.

3. Chu, C.C., Wang, C.C., Sun, C.S., Hsü, C., Yin, K.C., Chu, C.Y., and Bi, F.Y., 1975, Establishment of an efficient medium for anther culture of rice through comparative experiments on the nitrogen sources. Sci. Sin. 18: 659-668.

4. Debergh, P., Harbaoui, Y., and Lemeur, R., 1981, Mass propagation of globe artichoke (Cynara scolymus): Evaluation of different hypothesis to overcome vitri-fication with special reference to water potential. Physiol. Plant. 53: 181-187.

5. Gamborg, O.L., Miller, R.A., and Ojima, K., 1968, Nutrient requirements of suspension cultures of soybean root cells. Exp. Cell Res. 50: 151-158.

6. Hooykaas-Van Slogteren, G.M.S., Hooykaas, P.J.J., and Schilperoort, R.A., 1984, Expression of Ti plasmid genes in monocotyledonous plants infected with Agrobacterium tumefaciens. Nature vol 311 (5988): 763-764.

7. James, D.J., 1987, Cell and tissue culture technology for the genetic manipulation of temperate fruit trees. Biotechnology and Genetic Reviews vol.5: 33-79.

8. James, D.J., Passey, A.J., Barbara, D.J.,and Bevan, M., 1989, Genetic transformation of apple (Malus pumila Mill) using a disarmed Ti-binary vector. Plant Cell Reports 7: 658-661.

9. Jefferson, R.A., Burgess, S.M., and Hirsh, D., 1986, ß-Glucuronidase from E.coli as a gen fusion marker. Proc. Natl. Acad. Sci.USA 83: 8447-51.

10. Klee, H., Horsch, R., and Rogers, S., 1987, Agrobac-terium-mediated plant transformation and its further applications to plant biology. Ann. Rev. Plant. Phy-siol. 38: 467- 486.

11. Knight, R.L., and Alston, F.H., 1969, Developments in apple breeding. Rep. E.Malling Res. Stn. for 1968: 125-132.

12. Mathews, H., 1988, _In vitro_ responses of _Brassica juncea_ and _Vigna radiata_ to the antibiotic kanamycin. Ann. Bot. 62: 671-675.

13. Mathias, R.J., and Boyd, L.,1986, Cefotaxime stimulates callus growth, embryogenesis and regeneration in hexaploid bread wheat (_Triticum aestivum L EM. Thell_). Plant Sci. 46: 217-223.

14. Mathias, R.J., and Mukasa, C., 1987, The effect of cefotaxime on the growth and regeneration of callus from four varieties of barley (_Hordeum vulgare L._). Plant Cell Reports 6: 454-457.

15. Murashige, T., and Skoog, F., 1962, A revised medium for rapid growth and bioassays with tobacco tissue cultures. Physiol. Plant. 15: 473-497.

16. Pasqualetto, P.L., Zimmerman R.H., and Fordham, I., 1986, Gelling agent and growth regulator effects on shoot vitrification of " Gala" apple in vitro. J. Amer. Soc. Hort. Sci. 111(6): 976-980.

17. Pierik, R.L.M., 1971, Plant tissue culture as motivation for the symposium. Misc. Pap. Landbouwhogesch. Wageningen. 9: 3-13.

18. Romberger, J.A., and Tabor, C.A., 1971, The _Picea abies_ shoot apical meristem in culture. I. Agar and autoclaving effects. Amer. J. Bot. 58(2): 131-140.

19. Shay, J.R., Dayton, D.F., and Hough, L.F., 1953, Apple scab resistance from a number of _Malus_ species. Proc. Soc. Hort. Sci. 62: 348-356.

20. Welander,M., 1988, Plant regeneration from leaf and stem segments of shoots raised _in vitro_ from mature apple trees. J. Plant. Physiol. 132: 738-744.

PRODUCING HERBICIDE TOLERANT POPULUS USING GENETIC TRANSFORMATION

MEDIATED BY AGROBACTERIUM TUMEFACIENS C58: A SUMMARY OF RECENT RESEARCH

Don E. Riemenschneider
and Bruce E. Haissig

USDA Forest Service
North Central Experiment Station
Forestry Sciences Laboratory
P.O. Box 898
5985 Highway K
Rhinelander, Wisconsin 54501, USA

ABSTRACT

We tested the hypothesis that genetic transformation mediated by Agrobacterium tumefaciens strain C58 will produce Populus with better than normal tolerance of the herbicide Roundup (glyphosate). The basic strategy was so-called target enzyme protection. Using a binary C58-based vector, T-DNA from two engineered plasmids (pPMG 85/587 or pCGN 1107) was separately inserted into Populus alba x P. grandidentata cv. 'Crandon.' Both these plasmids contained a bacterial gene (aroA) that encodes a chimeric 5-enol-pyruvylshikimate-3-phosphate (EPSP) synthase that tolerates Roundup better than normal enzyme. Compared to plasmid pPMG 85/587, pCGN 1107 contained a better gene promoter and added coding to help movement of chimeric EPSP synthase from cytoplasm to chloroplasts. Transformations with plasmid pPMG 85/587 yielded normal greenhouse plants with aroA and increased Roundup tolerance. This was the first transformation and regeneration of a woody plant with an important gene. Southern blots of greenhouse plants from tissues transformed with pCGN 1107 also showed that aroA was present; hence, plants were produced wherein chimeric EPSP synthase was targeted for chloroplast expression. Roundup tolerance tests of those plants are underway. A host range study revealed that C58 transforms Populus of diverse ancestry. But, parental genotype governs whether a progeny is susceptible or resistant to C58.

INTRODUCTION

In 1984, we began testing the general hypothesis that gene insertion will modify economically valuable traits of forest tree productivity, especially traits that are difficult to achieve by only conventional selection and breeding (cf. 20, 40). At that time no economically important gene had been inserted into any woody plant species and no genes at all had been inserted into any forest tree species. Hence, lack of background knowledge suggested the need for careful planning (21), as discussed below.

Woody Plant Biotechnology, Edited by M.R. Ahuja
Plenum Press, New York, 1991

Our selection of the final genetic trait and target tree species was based on the following rationale (21). We felt that progress in genetic modification would be most rapid when knowledge about the trait could be transferred from agronomy to forestry without modification and when enough knowledge about the tree species existed. Much fundamental knowledge is needed to modify higher plants by gene insertion, basically knowledge of "trait components" and "species components." Many trait components such as the molecular biology and biochemistry of herbicide resistance differ little among higher plant species, whereas species components such as in vitro regeneration ability (a prerequisite for in vitro genetic transformations) are highly variable, even within given plant genera (18). Unskilled selection of either trait or species would perhaps stymie our research because of insufficient knowledge to meet or exceed the critical threshold for progress (Fig. 1).

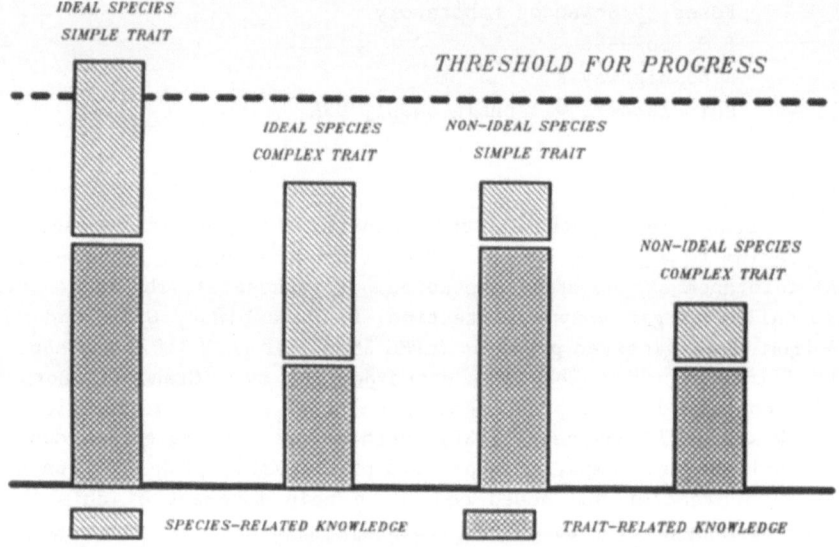

Figure 1. The magnitudes of species- and trait-related knowledge determine the probability of successfully modifying genetic traits (21).

We theorized that an appropriate target trait for trees would achieve a characteristic that cannot be equalled or exceeded by manipulation of other organisms in the ecosystem and (or) that would result in such high added value that economic opportunity would drive deployment of the genetically different trees. In addition, we felt that the most suitable target trait would have four characteristics: 1) definition based on biochemistry rather than morphology; 2) commonness to various plant species, thus having a substantial knowledge base; 3) importance in either maturation or secondary vascularization (i.e., wood formation); and 4) association with tree response (resistance) to biotic or abiotic damaging agents (stresses).

In 1984, characteristic (3) could not be readily targeted and even now is highly challenging (24). However, herbicide tolerance satisfied all the other characteristics (21, 33).

Herbicide tolerance can be achieved in at least six ways: decreased uptake, decreased translocation, intracellular sequestration, detoxification (metabolism), target enzyme overproduction, and target enzyme alteration (33). Of the aforementioned, detoxification (metabolism), overproduction of the target enzyme, and (or) alteration of the target enzyme of the herbicide seemed to be the most promising for achieving herbicide tolerances in forest trees (33). We chose the strategy of target protein alteration because knowledge, materials, and methods associated with an effective forestry herbicide (Roundup, a.k.a. glyphosate) were available (33). Thus we tested the specific hypothesis that genetic transformation mediated by <u>Agrobacterium</u> <u>tumefaciens</u> will produce <u>Populus</u> that tolerate Roundup. We selected transformation mediated by <u>A. tumefaciens</u> because it was potentially the most effective of alternatives in 1984 (13, 15, 16, 33). The targeted genetic basis for Roundup tolerance was expression in <u>Populus</u> of a mutant bacterial <u>aro</u>A gene. Roundup kills plants by blocking aromatic amino acid biosynthesis by inhibiting EPSP synthase, the product of the <u>aro</u>A gene. A mutant <u>aro</u>A gene, which encodes an EPSP synthase that better tolerates Roundup than normal enzyme, had been isolated from <u>Salmonella</u> <u>typhimurium</u> and incorporated into <u>A. tumefaciens</u> vectors (6, 7), which we used. The mutant <u>aro</u>A gene encodes a single amino acid substitution in EPSP synthase (49).

We elected to insert the bacterial <u>aro</u>A gene into the genus <u>Populus</u> for several reasons. <u>Populus</u> is an important genus in world forestry; however, establishment of intensively cultured <u>Populus</u> plantations (i.e., like agronomy; 22) is limited by a lack of herbicides that control weed competition but do not damage the trees (23, 33). Of even greater importance, we believed that, compared to many other forest tree species, some <u>Populus</u> genotypes would satisfy the prerequisite species component of robust <u>in</u> <u>vitro</u> regeneration ability. We specifically selected hybrid <u>Populus</u> NC-5339 cv. 'Crandon' (<u>Populus</u> <u>alba</u> x <u>Populus</u> <u>grandidentata</u>) because we were able to develop efficient <u>in</u> <u>vitro</u> culture and regeneration protocols for its production (29, 46) and it is a productive clone for woody biomass.

As described below, our research has been divided into three phases, each of which was designed to obtain knowledge about genetic transformation of <u>Populus</u> NC-5339 for Roundup tolerance [phase I (13, 15, 42) and phase II (unpublished)], or of <u>Populus</u> spp. for many desired genetic traits [phase III (in press)]. In other research we have discovered that <u>in</u> <u>vitro</u> (i.e., "somaclonal") selection can achieve Roundup and Oust (sulfometuron methyl) tolerances in <u>Populus</u> spp. (30), but that research will be considered here only in discussion because of subject matter and space limitations.

<u>Abbreviations and Terminology</u>

<u>aro</u>A, a bacterial gene that encodes EPSP synthase; BA, benzyladenine; CaMV 35S, a gene promoter from cauliflower mosaic virus; EcoRI, a restriction endonuclease that cleaves DNA at a specific base pair sequence; EPSP synthase, 5-enolpyruvylshikimate-3-phosphate synthase (E.C. 2.5.1.19), a chloroplast enzyme of the shikimic acid pathway needed for synthesis of aromatic amino acids; IAA, indoleacetic acid; kb, kilobase pairs; LPI, leaf plastochron index (26); MAS (<u>mas</u>), mannopine synthase; MS, Murashige and Skoog (32) basal nutrient medium; NC-, a permanent plant accession record of North

Central Forest Experiment Station; NPT II' (npt II'), neomycin phosphotransferase; OCS (ocs), octopine synthase; oncogenes, genes of the Agrobacterium tumefaciens Ti plasmid T-DNA that encode auxin and cytokinin
syntheses and cause gall formation when expressed by a host plant; PGR,
plant growth regulator; RUBISCO, ribulose-1,5-bisphosphate carboxylase (E.C.
4.1.1.39); subclone, a clone derived from a parent (control) clone by
regeneration from a cell or group of cells having genetic properties (markers)
not shared by the parent clone. ["Subclone" is a proper extension of the
term "substrain" (43). A subclone is putative until some genetic difference
from the parent clone has been unequivocally demonstrated.]; T-DNA, DNA that
is transferred from A. tumefaciens to higher plant cells during crown gall
disease infection (which involves genetic transformation); Ti, A. tumefaciens
plasmid that contains T-DNA and a virulence coding region which is responsible
for the excision, packaging and transfer of T-DNA; TML (tml), a DNA sequence
that provides a polyadenylation signal during aroA transcription.

MATERIALS AND METHODS

All experimentation utilized the hybrid Populus NC-5339 cv. 'Crandon'
(P. alba x P. grandidentata) and Agrobacterium tumefaciens strain C58. The
C58 binary vector included an engineered plasmid and the wild-type C58 Ti
plasmid. The engineered plasmids differed between phase I and II research,
as described below (Fig. 2).

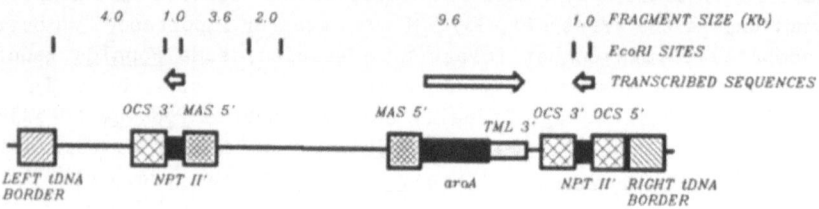

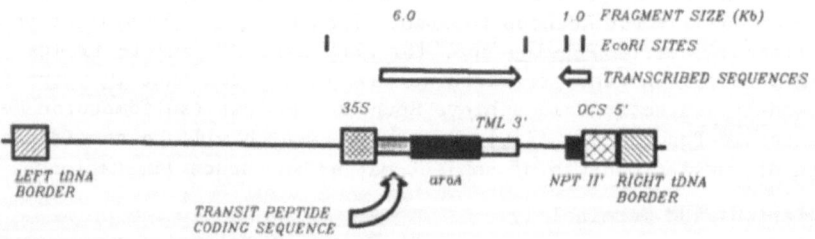

Figure 2. Diagrammatic representations of plasmids pPMG 85/587 and
pGCN 1107 harbored in Agrobacterium tumefaciens strain C58. Plasmid
pPMG 85/587 includes: two selectable markers (NPT II') that impart
resistance to the antibiotic kanamycin; and the mutant aroA gene
fused to a mannopine synthase promoter (MAS 5'). Plasmid pCGN 1107
includes: one NPT II' coding sequence; the mutant aroA gene fused
to a 35S cauliflower mosaic virus promoter; and transit and mature
RUBISCO sequences from Pisum sativum and Glycine max.

Phase I - Insertion of T-DNA from Plasmid pPMG 85/587

Transformation and regeneration protocols were described in detail (15, 45, 46). Leaves of <u>Populus</u> NC-5339 were obtained from continuous shoot cultures (29, 46) and prepared for cocultivation by aseptically excising the leaf margins, which resulted in irregularly shaped pieces. These wounded leaf pieces were then cocultivated with a liquid culture of <u>A. tumefaciens</u> C58/85/587 ($2 \cdot 10^8$ mL^{-1} bacteria containing plasmid pPMG 85/587) (Fig. 2) for 48 to 96 h over a tobacco feeder layer (14, 25). Adventitious shoots were regenerated from the cocultivated leaf discs on solidified (0.6% Phyto-agar) MS medium that contained sucrose (30 g L^{-1}), BA (1.0 mg L^{-1}), zeatin (1.0 mg L^{-1}), kanamycin (60 mg L^{-1}, a selective agent), and carbenicillin (500 mg L^{-1}, a bacteriostat). Individual regenerated shoots were excised, uniquely identified, and propagated <u>in vitro</u>, yielding subclones. Subclones were divided into two visually obvious morphological types, normal (type A) or abnormal (teratomous, type B). Control plants were also propagated <u>in vitro</u>, except the culture medium did not contain kanamycin or carbenicillin.

NPT II' enzyme activity assay (39), nopaline assay (36), and Western blot assay for production of bacterial EPSP synthase (7) were conducted on extracts of leaves and shoots taken from <u>in vitro</u> cultures of the subclones and control. Nuclear DNA for Southern analysis (27) was extracted from leaves of the subclones and control (15). DNA was digested with EcoRI restriction endonuclease, separated by gel electrophoresis, transferred to nitrocellulose, and hybridized separately to ^{32}P labeled <u>aro</u>A, NPT II', and C58 nopaline synthase probes.

Herbicide tolerance of the subclones and control was estimated by spraying greenhouse plants with 0.07 or 0.28 kg ha^{-1} of the active ingredient (N-phosphonomethylglycine, a.k.a. glyphosate) of the Roundup herbicide formulation. To propagate transformed subclones and control plants, 3- to 5-cm-long shoots were excised from <u>in vitro</u> shoot cultures, rooted <u>ex vitro</u> without chemical treatment in sterile potting medium, and transplanted to 10.2 cm dia. plastic pots. Plants were sprayed with Roundup when they were about 25 cm in height. Plant height (cm) was measured on the day of spraying and at 3-d intervals for 15 d.

Phase II - Insertion of T-DNA from Plasmid pCGN 1107

Leaves were obtained from shoot cultures as before (15), but 1-cm dia. leaf discs, centered on the leaf midvein, were used. Leaf discs were cocultivated with strain C58, containing plasmid pCGN 1107 (Fig. 2). Again, a tobacco cell feeder layer was used during cocultivation. Various cocultivation protocols were tested to determine optimum conditions for shoot regeneration. First, the effect of time was estimated by conducting thrice replicated trials to compare 24-, 48-, and 72-h-long cocultivations. Second, the effect of bacterial titer was tested in thrice replicated trials that compared five bacterial concentrations, ranging from 10^7 to 10^9 mL^{-1} bacteria.

Compared to phase I, an improved protocol for adventitious shoot regeneration after cocultivation was developed and used in phase II (C. Michler, unpublished). After cocultivation, leaf discs were plated on solidified (0.6%, w/v, Difco Bacto Agar) MS medium containing 5.0 mg L^{-1} IAA and 0.01 mg L^{-1} BA. Cultures were initially placed in darkness for 72 h. Then, the leaf discs were plated on MS medium containing 0.10 mg L^{-1} BA as the sole PGR and cultured in light (25 °C, 25-30 μmol m^{-2} s^{-1}, General Electric cool white F20T12-CW and F40-CW, 16 h photoperiod). In addition, all regeneration and shoot maintenance media contained kanamycin (50 mg L^{-1}, a selective agent) and cefotaxime (300 mg L^{-1}, a bacteriostat). After regeneration, subclones were separated according to morphological type A or B and uniquely identified, as described for phase I. Regenerated shoots of type A were propagated _in vitro_ for 3 to 4 passages, then shoots were rooted _ex vitro_ and used to establish greenhouse stock plants.

For Southern blots, nuclear DNA was extracted from 1 g (fresh weight) of LPI 3-6 greenhouse stock plant leaves (15). Ten μg of nuclear DNA from each of ten subclones was digested with EcoRI restriction endonuclease (Promega) according to manufacturer's instructions. Digested DNA was separated electrophoretically on a 0.7% agarose gel and transferred to nylon membrane (Fisher Magnagraph) (2). _AroA_ and NPT II' probes were labeled with ^{32}P using the random primer method (11, 12) and hybridized separately to nylon membrane-bound _Populus_ DNA (2).

Phase III - Host Range of Agrobacterium tumefaciens C58

Controlled hydridizations were made between females of a single species (_P. deltoides_ Bart. ex. Marsh.) and males of four species (_P. maximowiczii_ Henry, _P. balsamifera_ L., _P. nigra_ L., and _P. deltoides_) (Table 1). Seedlings were grown in a greenhouse (ca. 25 °C days and 20 °C nights, 18 h photoperiod) for one growing season. Starting in September, growth cessation and dormancy were induced in the seedlings by lowering greenhouse temperature (finally to 4 °C), utilizing natural photoperiod, and decreasing nutrient fertilization. After 5 months, dormant seedlings were pruned to a uniform height of 10 cm and reflushed in a warm greenhouse (25 °C, 18 h photoperiod).

Table 1. Total number of crown galls produced by _Populus_ hybrids after inoculation with _Agrobacterium tumefaciens_ C58. Data were obtained after two separate inoculations of each of 6 seedlings per family. An asterisk indicates that gall tissue produced nopaline and proliferated callus _in vitro_ in the absence of plant growth regulators. Female parents were all _Populus deltoides_ (D1-D10); male parents were _P. maximowiczii_ (M1-M6), _P. balsamifera_ (B1-B6), _P. nigra_ (N1-N2), or _P. deltoides_ (D11-D15). Blank cells (-) indicate failed crosses.

Female	Male									
	M1	M2	M3	B1	B2	B3	N1	D11	D12	D15
D1	2*	0	0	1	0	0	0	-	-	0
D2	1*	1	1		1*	0	3	0	0	
D3	0	0	1*	0	1	0	0	0	0	0
D4	0	0	0	0	-	0	0	0	1	-
D5	0	0	2	4	8*	1*	0	0	1	0
	M4	M5	M6	B4	B5	B6	N1	N2	D13	D14
D6	0	0	0	0	0	0	0	0	0	0
D7	2	0	2	1	0	0	0	0	2	0
D8	3	1*	1	0	1	0	0	0	0	1
D9	0	0	0	0	0	0	0	0	0	0
D10	0	2*	0	0	0	1	1	0	0	0

Strain C58 cultures were maintained on AB minimal medium (53) and transferred to fresh medium every 30 d. Five mL of MG/L broth (4) in a 25 mm x 150 mm test tube was inoculated with a single bacterial colony on the day prior to seedling inoculations. The liquid cultures were grown overnight in an incubated orbital shaker (30 °C, 200 rpm), then diluted to $2 \cdot 10^8$ mL^{-1} bacteria, based on light absorption by cultures (550 nm). Seedlings were inoculated with strain C58 by piercing each stem at different locations with a 0.5 mm dia. needle that was dipped in the overnight culture of strain C58. In one experiment, six seedlings from each of 95 hybrid families were inoculated at two different stem positions with strain C58; six seedlings from the same families were similarly double-inoculated but with sterile bacterial medium alone (Table 1). In a second experiment, 20 seedlings from six families were each inoculated at four different stem positions. Those six families formed a 2 x 3 factorial mating design using P. deltoides females and P. balsamifera males (clones D4 and D5, and B1, B2, and B3, resp.; Table 1). Gall formation was evaluated in both experiments after 60 d.

All galls were excised from one seedling in each of the eight families that formed galls in the first experiment (Table 1); seedlings from which galls were excised were used as controls for subsequent nopaline assays, as described below. Galls were surface-sterilized in 10% Clorox (v/v) for 3 min, then washed three times in sterile deionized water. The surface-sterilized galls were then sectioned aseptically into 3- to 5-mm-thick discs and plated on solidified (0.7%, w/v, Difco Bacto Agar) MS medium that contained 20 g L^{-1} glucose and 300 mg L^{-1} carbenicillin (without any PGR). Cultures were maintained by subculture at 30-d intervals under environmental conditions identical to those used for Phase II transformations.

Nopaline assays were conducted on two types of tissue: 1) 5 to 10 mm pieces of callus from each of the eight in vitro gall tissue cultures (second and fifth passages); and 2) gall-free stem apex (3 to 5 cm, with leaves) from the original seedlings whose galls were excised as a basis for the in vitro cultures. Samples were individually homogenized (Virtis VertiShear, 25,000 rpm, 5 s, 3 mL 80% ethanol, 5 °C). Then, homogenates were transferred to 1.5 mL Eppendorf tubes and centrifuged (ca. 2,000 xg, 5 min). Supernatants were stored (5 °C) up to one week before electrophoretic separation and visualization of nopaline (36). The whole test was performed twice.

RESULTS AND DISCUSSION

Phase I

Overall, results of the present tests with plasmid pPMG 85/587 indicated that morphologically normal NC-5339 plants with increased tolerance to Roundup were produced using A. tumefaciens C58-mediated transformation. This research was the first unequivocal demonstration of transformation and regeneration of a woody plant with an economically important gene (13, 15, 16).

On average, 32% ± 2.1% of cocultivated leaf pieces regenerated adventitious shoots on selective medium, with a mean of 4.4 ± 0.90 shoots per explant. Of shoots that regenerated on selective medium, analyses revealed that type A shoots produced NPT II' enzyme (38 of 40 tested) and bacterial EPSP synthase (2 of 2 tested), but not nopaline. Type B shoots also produced NPT II' enzyme (18 of 20 tested). But, in contrast to type A, type B shoots produced nopaline, had wild-type C58 T-DNA in their nuclear genomes (2 of 2 tested), and did not regenerate adventitious roots under conditions that resulted in adventitious rooting by type A or control shoots (cf. 15). Based on Southern analysis, shoots of both types had a 1 kb NPT II' DNA fragment in their nuclear DNA. The two tested type A shoots, which expressed bacterial EPSP synthase activity, also had a 9.7 kb aroA-containing fragment in their nuclear DNA. However, shoots with the NPT II' DNA fragment did not always contain the expected aroA fragment.

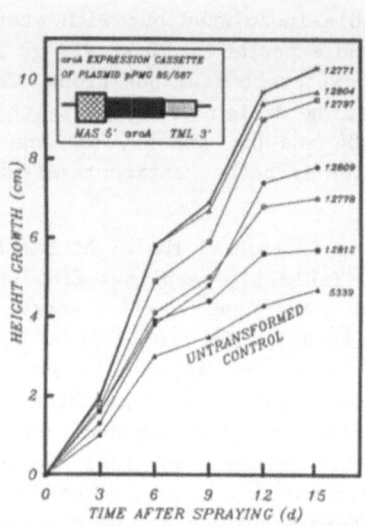

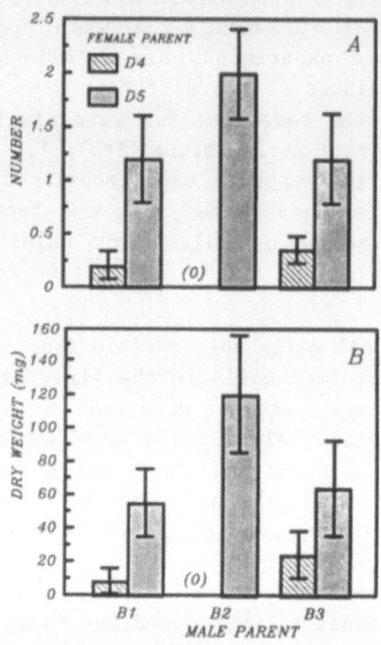

Figure 3. Mean height growth (cm) of plants from six (numbered) subclones and the control (parent clone NC-5339) after overspray with the herbicide Roundup (0.07 kg ha^{-1}). Subclones contained T-DNA from Agrobacterium tumefaciens C58, plasmid pPMG 85/587, including a mutant bacterial aroA gene fused to a mannopine synthase promoter [MAS 5'; see Fig. 2 and (42)].

Figure 4. Influence of host Populus genotype on mean number of crown galls produced per tree (A) and mean total gall dry weight (mg) per tree (B) after inoculation with Agrobacterium tumefaciens C58, plasmid pPMG 85/587 (see Fig. 2). Variance among families for both variables was due only to female parent (see Table 3). I-shaped symbols represent ± S.E. mean.

The foregoing results indicated that type A shoots received DNA from only the engineered plasmid of the binary vector. Type B shoots received DNA from the engineered plasmid (and could thus regenerate shoots on selective kanamycin medium) and also received T-DNA from the wild-type C58 Ti plasmid (resulting in abnormal morphology and non-rooting). Reduced rooting would be expected for shoots that contained oncogenes because applied cytokinins usually inhibit adventitious rooting (8). Our observation that some type A shoots possessed NPT II' DNA but not aroA DNA suggested that sometimes only part of the engineered plasmid's DNA was incorporated into the host genome, either because of recombination between homologous sequences within the T-DNA or because of only partial DNA transfer.

Successful insertion of the aroA expression cassette into Populus NC-5339 almost always resulted in increased Roundup tolerance compared to control plants. For example, in greenhouse tests subclones grew taller than controls after application of 0.07 kg ha^{-1} (active ingredient) Roundup (Fig. 3). Indeed, four of six subclones grew as well as unsprayed plants of the same subclones. At the end of the test, mean height growth of plants of all but one subclone treated with 0.07 kg ha^{-1} Roundup was significantly greater than plants of the identically treated control (Fig. 3). Subclones often survived treatment with 0.07 kg ha^{-1} Roundup, but damage to shoot apices, prolific axillary shooting, and poor plant form were common (42). In comparison with 0.07 kg ha^{-1} Roundup, subclones and the control grew less after spray treatment with 0.28 kg ha^{-1} Roundup. However, growth of subclones still exceeded the control. Nonetheless, treatment with 0.28 kg ha^{-1} Roundup eventually resulted in death of all subclones and the control (42).

Lack of tolerance of phase I aroA-containing subclones to the high dosage of Roundup is perhaps attributable to two specific features of plasmid pPMG 85/587 (Fig. 2). First, the bacterial aroA gene was fused to the MAS 5' promoter, which is a weaker promoter in higher plants than, for example, CaMV 35S. Second, the aroA gene product was not specifically directed to chloroplasts, the main site of action of higher plant EPSP synthase (31). Thus the structure of the aroA-containing plasmid was modified for phase II research (Fig. 2). However, lack of tolerance may also be due to biochemical perturbations caused by Roundup, other than inhibition of EPSP synthase [reviewed in (33)].

Phase II

For phase II research, the aroA expression cassette was like that used in phase I but contained the aroA gene fused to the CaMV 35S promoter, and also to a peptide transit sequence from the gene that encodes the small subunit of RUBISCO in Pisum sativum L. and 23 codons from the mature RUBISCO gene of Glycine max (L.) Merr. (Fig. 2). The peptide transit sequence should result in transport of the aroA gene product from the cytoplasm to chloroplasts, which are the normal site of action of EPSP synthase (31). Previous research has shown that RUBISCO transit peptide coding sequences can direct the translocation of NPT II' into chloroplasts in vivo (44, 51) and in vitro (52). The 35S promoter has been shown to direct high level expression of foreign genes in higher plants (35). Hence, the plasmid that we used for phase II should result in Populus subclones with greater and chloroplast-based bacterial EPSP synthase activity and, therefore, greater Roundup tolerance than achieved in phase I.

Thus far, phase II tests with plasmid pCGN 1107 unambiguously demonstrated insertion of the mutant bacterial aroA gene into Populus NC-5339, with the gene product specifically targeted for expression in chloroplasts. However, herbicide tolerance of those greenhouse plants has yet to be determined. We also learned that strain C58 concentration during cocultivation and the cocultivation time greatly varied the frequency of adventitious shoot regeneration by Populus NC-5339 leaf pieces on selective medium. The previous conclusions were based on the following evidence.

Southern analysis verified insertion of the aroA gene in 7 of 10 subclones tested (Table 2). The aroA probe hybridized to a 6.0 kb EcoRI nuclear DNA fragment in the seven aroA-positive subclones. A 6.0 kb fragment was expected based on a restriction endonuclease map of the pCGN 1107 plasmid (Fig. 2). The 6.0 kb fragment contained the 35S promoter, transit peptide coding sequence, and mutant bacterial aroA gene, but not the NPT II' selectable marker gene. Neither the aroA nor NPT II' probes hybridized with nuclear DNA from subclone PT2 or from the control, which suggested that PT2 probably originated from an untransformed shoot that escaped kanamycin selection. The NPT II' probe, but not the aroA probe, hybridized with DNA from subclones PT4 and PT5, which suggested that PT4 and PT5 resulted from recombination or partial insertion of pCGN 1107 T-DNA that eliminated the aroA gene. Hence, absence of aroA DNA in some subclones was sometimes the outcome of otherwise successful transformation with both A. tumefaciens C58/587/85 and C58/1107.

Table 2. Positive (+) and negative (-) Southern blots for subclones of hybrid Populus NC-5339 that were transformed with Agrobacterium tumefaciens C58/pCGN 1107.

Subclone No.	Hybridization Probe and Size of Genomic EcoRI Fragment (kb)			
	aroA	Size	NPT II'	Size
PT1a	+	6.0	+	1.0
PT1b	+	6.0	+	1.0
PT2	-	---	-	---
PT3	+	6.0	+	1.0
PT4	-	---	+	1.0
PT5	-	---	+	1.0
PT6	+	6.0	+	1.0
PT7	+	6.0	+	1.0
PT9	+	6.0	+	1.0
PT13	+	6.0	+	1.0
5339 (control)	-	---	-	1.0

The most effective cocultivation protocol in phase II ($1 \cdot 10^8$ mL^{-1} bacteria, 48 h cocultivation) was about the same as that used in phase I (15), even though the phase I protocol was developed for tomato (14). In both phases I and II about the same regeneration frequencies were obtained, as expected based on the similar cocultivation and regeneration protocols. In phase II, 45% of leaf discs regenerated callus and 31% regenerated shoots after cocultivation for 48 h ($2 \cdot 10^8$ mL^{-1} bacteria). Only 5% of leaf discs regenerated shoots after cocultivation for 24 or 72 h. Also, 61% of leaf discs cocultivated with $1 \cdot 10^8$ mL^{-1} bacteria (48 h cocultivation) regenerated callus whereas 39% regenerated shoots. Increasing bacterial concentration to $5 \cdot 10^8$ or decreasing concentration to $5 \cdot 10^7$ significantly reduced frequency of callus and shoot regeneration.

Phase III

Our host range study provided much previously unknown information, for example, that A. tumefaciens C58 transforms (infects) Populus of diverse ancestry. In addition, results indicated that, within any parental species combination, parental genotype determines whether a progeny is highly susceptible or highly resistant to transformation by strain C58.

For instance, in the first experiment, crown galls formed on seedlings from 27 of 95 hybrid families inoculated with strain C58 (Table 1). No galls formed on seedlings that were inoculated with sterile MG/L broth alone. Gall-forming families produced a mean of 1.8 ± 0.3 galls per 12 inoculation sites (six trees, two inoculations per tree). Families D5 x B1 and D5 x B2 produced the most galls (four and eight, resp.). Crown galls formed on at least one family from each parental species (but not clone) combination in the crossing scheme (Table 1). Crown gall tissue from all tested plants proliferated callus on PGR-free medium for 18 months and continues to do so. Nopaline was detected in all crown gall callus cultures but never in uninfected control tissue.

Table 3. Analyses of variance of total gall dry weight per tree and total number of galls per tree (four inoculation sites) on seedlings from the six full-sib families in phase III, experiment 2 (see Fig. 4). Data were transformed as $\sqrt{X+1}$ before analysis. (** = $\underline{P}$ < 0.01, ns = not significant)

Source	df	Number		Dry Weight (mg)	
		MS	f	MS	f
Families	5	10.8	7.03 **	0.035	3.52 **
Female	1	48.0	31.2 **	0.130	13.4 **
Male	2	0.672	0.440 ns	0.007	0.710 ns
F x M	2	2.39	1.55 ns	0.016	1.66 ns
Error	106	1.54		0.010	

Strain C58 was virulent over a broad range of intra- and inter-specific Populus hybrids (Table 1). Insertion and expression of wild-type T-DNA was suggested by three lines of evidence. First, none of the control seedlings inoculated with sterile MG/L broth produced galls. Thus, gall formation was most probably due to interaction of the host plant with strain C58, and not due to wound healing, interaction with bacterial medium, or infection by a secondary pathogen. Second, growth of crown gall-derived callus tissue in vitro on PGR-free medium required expression in Populus cells of auxin (50) and cytokinin (1) synthesis genes (oncogenes) introduced by bacterial T-DNA. Last, nopaline synthesis was found in gall-derived callus tissue from all eight genotypes tested (Table 1). Nopaline synthesis by gall-derived callus tissue required host expression of the nopaline synthase gene also introduced on bacterial T-DNA (34). Furthermore, gall-free stem and leaf tissue from the eight gall-forming seedlings had no detectable nopaline content. Hence, production of opines by untransformed plants (5), which may yield false evidence of transformation, was not found in our study.

Results of the second experiment indicated that the Populus host's susceptibility to infection by strain C58 is significantly influenced by intra-specific genetic variation in the female parent. In the second experiment, six families differed significantly in number of galls formed per tree (Table 3). Variation in mean number of galls per tree and total gall dry weight per tree (Fig. 4) was attributable only to female parents (Table 3). No significant variation in gall number or weight was attributable to female x male interaction or differences among males.

CONCLUSIONS

Our research has established that hybrid _Populus_ NC-5339 can be genetically transformed by using _A. tumefaciens_ C58-based binary vectors. Two different genetic constructs have been inserted by using leaf explant cocultivation. Plant regeneration frequencies as high as 39% resulted. Multiple foreign genes were incorporated into the nuclear genome of the host _Populus_ cells and expressed. However, the gene that was subjected to early selection pressure (NPT II') was incorporated more reliably than the gene that was not subjected to selection (_aroA_). After transformation, subclones with normal morphology were easily produced in the greenhouse and carried through repeated vegetative propagation cycles. Tolerance to normal field application rates of Roundup was not achieved to date. But, Roundup tests of subclones that contain the potentially very effective _aroA_ construct from plasmid pCGN 1107 have not been completed.

Varying success has been achieved by other investigators who have attempted to transform _Populus_ genotypes other than NC-5339 with _A. tumefaciens_ C58/587/85 (our Phase I construct). For instance, successful transformation and regeneration were obtained with the Leuce (taxonomic section) clonal cultivar _P. tremula_ 'Erecta' (45). In contrast, the previous authors could not transform NC-5331 (_P. nigra betulifolia_ x _P. trichocarpa_), NC-5326 (_P. deltoides_ x _P. nigra_ cv. 'Eugenei'), NC-5272 (_P. nigra_ x _P. laurifolia_ 'Strathglass'), or NC-11390 (_P. maximowiczii_ x _P. trichocarpa_ cv. 'Androscoggin'). Such inability to transform _Populus_ hybrids from the sections Aigeiros and Tacamahaca may be due to two factors. First, there are no protocols for robust regeneration of shoots from tissues of many _Populus_ and other tree species (18), especially after cocultivation with _A. tumefaciens_. As one example, Sellmer and McCown (45) could not regenerate shoots from cocultivated _P. tremula_ 'Erecta' leaf pieces on selective medium, even though transformation occurred, as demonstrated by post-regeneration selection experiments. Therefore, the stress of cocultivation, combined with stresses induced by the normally lethal selective agent, may preclude shoot regeneration by genotypes for which _in vitro_ culture protocols are suboptimal. Second, some genotypes may not be susceptible to _A. tumefaciens_ C58. In one instance, transformed clones were not recovered either when selection was applied during regeneration or afterward (45). Furthermore, our Phase III studies clearly demonstrated host genotype-related constraints to transformation with _A. tumefaciens_ C58. _Agrobacterium_-mediated transformation of _Populus_ hybrids of Aigeiros and Tacamahaca parentage has been accomplished, but other strains or another species were used, for example, _A. tumefaciens_ A281 and A348 (37) and _A. rhizogenes_ R1600 and R1601 (38).

During our research similar experiments with different plant species have been successfully conducted by others (17, 28). For example, the mutant bacterial _aroA_ gene used in the present tests has been incorporated into _Nicotiana tabacum_ L. (7) and _Lycopersicon esculentum_ L. (14), yielding plants with increased herbicide tolerance. In other similar research, a _Petunia hybrida_ L. EPSP synthase gene that conferred glyphosate resistance through amplification was cloned and used to transform _Petunia_ leaf discs (47). That chimeric gene included the _Petunia_ EPSP synthase transit peptide coding sequence and thus its product was translocated to chloroplasts (47). The _Petunia_ EPSP synthase transit coding sequence was also fused to a mutant _E. coli_ EPSP synthase gene. As a result, the chimeric pre-enzyme was imported into chloroplasts _in vitro_, where it was active in the presence of glyphosate (10). The chimeric _E. coli_ gene was also inserted by means of an _A. tumefaciens_ vector into <u>Nicotiana</u> <u>tabacum</u>, where its product was also translocated into chloroplasts and active in the presence of glyphosate (10).

Our other previously stated objective (33) of obtaining herbicide tolerant <u>Populus</u> by using <u>in</u> <u>vitro</u> (somaclonal) selection also yielded positive results. In that research, tissues or cells of several <u>Populus</u> genotypes were challenged directly <u>in</u> <u>vitro</u> with an herbicide (Roundup or Oust), then caused to regenerate adventitious shoots and greenhouse plants. Subclones with increased tolerance to Roundup and Oust resulted (30); tolerance to field dosages of Oust was obtained and Roundup tolerance was greatly increased but not to field dosage levels. We continue to study the bases of herbicide tolerances in those subclones. Somaclonal variants with high tolerance (or resistance) to an herbicide might provide a source of genes for subsequent insertion into <u>Populus</u> or other species. Indeed, that was the strategy utilized by Shah et al. (47) to produce glyphosate tolerant transgenic <u>Petunia</u>.

We are continuing research related to the phase II subclones. Molecular tests are appraising expression of NPT II' and bacterial <u>aro</u>A. Tests of <u>aro</u>A expression on cytoplasmic and chloroplastic cell fractions are determining functioning, if any, of the transit peptide coding sequence of plasmid pCGN 1107. Tests of greenhouse plants are establishing Roundup tolerance levels (cf. 42) and biochemically and physiologically based productivity traits, with and without Roundup challenge. Successful greenhouse tolerance tests are prerequisite to field tests of Roundup tolerance and plantation pro- ductivity characteristics.

Our results have also suggested that research is needed to determine whether genotypes of contrasting <u>ex</u> <u>vitro</u> susceptibility to a given <u>A.</u> <u>tumefaciens</u> strain retain the same susceptibility when exposed to a binary vector <u>in</u> <u>vitro</u>. Such studies must consider genotypic differences in regeneration ability that might confound determination of transformation efficiency (19, 30, 41). Research is also needed to understand the molecular and biochemical bases for varying susceptibility of <u>Populus</u> hybrids to <u>A.</u> <u>tumefaciens</u> (e.g., 3, 9, 48). Such research might facilitate genetic transformation of non-susceptible genotypes by discovering now-missing adjuvants for cocultivation and regeneration.

ACKNOWLEDGMENTS

We thank Dr. B.H. McCown and Mr. J.C. Sellmer (University of Wisconsin-Madison) for their significant contributions to phase I research, and Ms. J.J. Fillatti and Dr. L. Comai (Calgene, Inc., Davis, CA) for their major contributions to research in phases I and II. We also thank Dr. C.H. Michler for related <u>Populus</u> <u>in</u> <u>vitro</u> culture research data. Finally, we thank Dr. R.E. Dickson, Dr. D. Ellis , Dr. D. Karnosky, and Dr. C.H. Michler for critical review of the manuscript. This research was partially supported by the Department of Energy under Interagency Agreement DE-A105-800R20763.

Roundup and Oust are registered trademarks of Monsanto Co. and Du Pont Co, respectively. Mention of tradenames does not constitute endorsement by the U.S. Forest Service.

PESTICIDE PRECAUTIONARY STATEMENT

This publication reports research involving pesticides. It does not contain recommendations for their use, nor does it imply that the uses discussed here have been registered. All uses of pesticides must be registered by appropriate State and (or) Federal agencies before they can be recommended.

Caution: Pesticides can be injurious to humans, domestic animals, desirable plants, and fish or other wildlife--if they are not handled properly. Use all pesticides selectively and carefully. Follow recommended practices for the disposal of surplus pesticides and pesticide containers.

REFERENCES

1. Akiyoshi, D.E., Klee, H., Amasino, R.M., Nester, E.W., and Gordon, M.P., 1984, T-DNA of Agrobacterium tumefaciens encodes an enzyme of cytokinin biosynthesis, Proc. Natl. Acad. Sci. USA 81:5994-5998.

2. Ausubel, F.M, Brent, R., Kingston, R., Moore, D., Smith, J., Seidman, J., and Struhl, K., 1987, "Current Protocols in Molecular Biology," John Wiley & Sons, New York.

3. Bolton, G.W., Nester, E.W., and Gordon, M.P., 1986, Plant phenolic compounds induce expression of the Agrobacterium tumefaciens loci needed for virulence, Science 232:983-985.

4. Chilton, M-D., Currier, T.C., Farrand, S.K., Bendich, A.J., Gordon, M.P., and Nester, E.W., 1974, Agrobacterium tumefaciens DNA and PS8 bacteriophage DNA not detected in crown gall tumors, Proc. Natl. Acad. Sci. USA 71:3672-3676.

5. Christou, P., Platt, S.G., and Ackerman, M.C., 1986, Opine synthesis in wild-type plant tissue, Plant Physiol. 82:218-221.

6. Comai, L., Sen, L., and Stalker, D., 1983, An altered aroA gene product confers resistance to the herbicide glyphosate, Science 221:370-371.

7. Comai, L., Facciotti, D., Hiatt, W.R., Thompson, G., Rose, R.E., and Stalker, D.M., 1985, Expression in plants of a mutant aroA gene from Salmonella typhimurium confers tolerance to glyphosate, Nature 317:741-744.

8. Davis, T.D., and Haissig, B.E., 1990, Chemical control of adventitious root formation by cuttings, Plant Growth Regul. Soc. Amer. Quart. (in press),

9. DeCleene, M., 1988, The susceptibility of plants to Agrobacterium: A discussion of the role of phenolic compounds, FEMS Microbiol. Rev. 54:1-8.

10. della-Cioppa, G., Bauer, S.C., Taylor, M.L., Rochester, D.E., Klein, B.K., Shah, D.M., Fraley, R.T., and Kishore, G.M., 1987, Targeting a herbicide-resistant enzyme from Escherichia coli to chloroplasts of higher plants, Bio/Technology 5:579-584.

11. Feinberg, A.P., and Vogelstein, B., 1983, A technique for radiolabeling DNA restriction endonuclease fragments to high specific activity, Anal. Biochem. 132:6-13.

12. Feinberg, A.P., and Vogelstein, B., 1984, A technique for radiolabeling DNA restriction endonuclease fragments to high specific activity, Addendum Anal. Biochem. 137:266-267.

13. Fillatti, J.J., McCown, B.H., Sellmer, J., and Haissig, B., 1986, The introduction and expression of a gene conferring tolerance to the herbicide glyphosate in Populus NC-5339, pp. 83-84, in: Proc. TAPPI Res. and Dev. Conf., Raleigh, NC.

14. Fillatti, J.J., Kiser, J., Rose, R., and Comai, L., 1987a, Efficient transfer of a glyphosate tolerance gene into tomato using a binary Agrobacterium tumefaciens vector, Bio/Technology 5:726-730.

15. Fillatti, J.J., Sellmer, J., McCown, B., Haissig, B., and Comai, L. 1987b. Agrobacterium-mediated transformation and regeneration of Populus, Mol. Gen. Genet. 206:192-199.

16. Fillatti, J.J., Haissig, B., McCown, B., Comai, L., and Riemensch-neider, D., 1988, The development of glyphosate-tolerant _Populus_ plants through expression of a mutant _aroA_ gene from _Salmonella typhimurium_, pp. 243-249, _in_: "Genetic Manipulation of Woody Plants," J.W. Hanover and D. Keathley, eds., Plenum Press, New York.

17. Gasser, C.S., and Fraley, R.T., 1989, Genetically engineering plants for crop improvement, _Science_ 244:1293-1299.

18. Haissig, B.E., 1989, Status of forest tree vegetative regeneration for biotechnology, _Amer. Biotech. Lab._ 7(1):48-51.

19. Haissig, B.E., and Riemenschneider, D.E., 1988, Genetics of adven-titious rooting, pp. 47-60, _in_: "Adventitious Root Formation by Cuttings," T.D. Davis, B.E. Haissig, and N. Sankhla, eds., Advances in Plant Sciences Series, Vol. 2, Dioscorides Press, Portland, OR.

20. Haissig, B.E., Nelson, N.D., and Kidd, G.H., 1987, Trends in the use of tissue culture in forest improvement, _Bio/Technology_ 5:52-56 and 59.

21. Haissig, B.E., and Riemenschneider, D.E., 1990, Strategic planning for application of biotechnology in woody plant genetics and breeding, _in_: "Micropropagation of Woody Plants", M.R. Ahuja, ed., Kluwer Academic Pubs., Dordrecht, Boston, Lancaster (in press).

22. Hansen, E.A. (Compiler), "Intensive Plantation Culture: 12 Years Research," Gen. Tech. Rept. NC-91, USDA Forest Service, North Central Forest Experiment Station, St. Paul, MN, (1983).

23. Hansen, E.A., and Netzer, D.A., "Weed Control Using Herbicides in Short-Rotation Intensively Cultured Poplar Plantations," Res. Pap. NC-260, USDA For. Serv., North Central For. Exp. Sta., St. Paul, MN, (1986).

24. Harry, D.E., and Sederoff, R.R., "Biotechnology in Biomass Crop Production. The Relationship of Biomass Production and Plant Genetic Engineering," Pub. No. 3411, Oak Ridge Nat. Lab., Environmental Sci. Div., Oak Ridge, TN, (1989).

25. Horsch, R.B., Fry, J.B., Hoffman, N.L., Wallroth, M., Eichholtz, D., Rogers, S.G., and Fragley, R.T., 1985, A simple and general method for transferring genes into plants, _Science_ 227:1229-1231.

26. Larson, P.R., and Isebrands, J.G., 1971, The plastochron index as applied to developmental studies of cottonwood, _Can. J. For. Res._ 1:1-11.

27. Maniatis T., Fritsche, E.F., and Sambrook, J., "Molecular Cloning: A Laboratory Manual," Cold Spring Harbor Laboratory, Cold Spring Harbor, New York, (1982).

28. Mazur, B.J., and Falco, S.C., 1989, The development of herbicide resistant crops, _Annu. Rev. Plant Physiol. Plant Mol. Biol._ 40:441-470.

29. McCown, B.H., 1985, From gene manipulation to forest establishment: shoot cultures of woody plants can be a central tool, _TAPPI J._ 68:116-119.

30. Michler, C.H., and Haissig, B.E., 1988, Increased herbicide tolerance of _in vitro_ selected hybrid poplar, pp. 183-188, _in_: "Somatic Cell Genetics of Woody Plants," M.R. Ahuja, ed., Kluwer Academic Pubs., Dordrecht, Boston, London.

31. Mousdale, D. N., and Coggins, J.R., 1985, Subcellular localization of the common shikimate-pathway enzyme in _Pisum sativum_, _Planta_ 163:241-249.

32. Murashige, T., and Skoog, F., 1962, A revised medium for rapid growth and bioassays with tobacco tissue culture, _Physiol. Plant._ 15:473-497.

33. Nelson, N.D., and Haissig, B.E., 1986, Herbicide stress: use of biotechnology to confer herbicide resistance to selected woody plants, in: "Stress Physiology and Forest Productivity," Hennessey T.C., Dougherty P.M., Kossuth, J.D. Johnson, eds., Martinus Nijhoff/Dr. W. Junk Pubs., The Hague, The Netherlands.

34. Nester, E.W. and Kosuge, T., 1981, Plasmids specifying plant hyperplasias, Annu. Rev. Microbiol. 35:531-565.

35. Odell, J.T., Nagy, F., and Chua, N.H., 1985, Identification of DNA sequences required for activity of cauliflower mosaic virus 35S promoter, Nature 313:810-812.

36. Otten, L.A.B.M., and Schilperoort, R.A., 1978, A rapid micro scale method for the detection of lysopine and nopaline dehydrogenase activities, Biochim. Biophys. Acta 527:497-500.

37. Parsons, T.J., Sinkar, V.P., Stettler, R.F., Nester, E.W., and Gordon, M.P., 1986, Transformation of poplar by Agrobacterium tumefaciens, Bio/Technology 4:533-536.

38. Pythoud, F., Sinkar, V.P., Nester, E.W., and Gordon, M.P., 1987, Increased virulence of Agrobacterium rhizogenes conferred by the vir region of pTiBo542: Application to genetic engineering of poplar, Bio/Technology 5:1323-1327.

39. Reiss, B., Sprengle, R., Will, H., and H. Schaller, 1984, A new sensitive method for qualitative and quantitative assay of neomycin phosphotransferase in crude cell extracts, Gene 30:211-217.

40. Riemenschneider, D.E., Haissig, B.E., and Bingham, E.T., 1988a, Integrating genetic manipulation into plant breeding programs, pp. 433-451, in: "Genetic Manipulation of Woody Plants," J.W. Hanover and D. Keathley, eds., Plenum Press, New York.

41. Riemenschneider, D.E., Haissig, B., and Michler, C., 1988b, Genetic effects on adventitious rooting in vitro in Populus deltoides, In Vitro Part II 24(3):52A.

42. Riemenschneider, D.E., Haissig, B.E., Sellmer, J., and Fillatti, J.J., 1988c, Expression of an herbicide tolerance gene in young plants of a transgenic hybrid poplar clone, pp. 73-80, in: M.R. Ahuja, ed., Somatic Cell Genetics of Woody Plants, Kluwer Academic Pubs., Dordrecht, Boston, London.

43. Schaeffer, W.I., 1989, Terminology associated with cell, tissue and organ culture, molecular biology and molecular genetics, TCA Rept. 23(5):5-6,11-13.

44. Schreier, P.H., Seftor, E.A., Schell, J., and Bohnert, H.J., 1985, The use of nuclear-encoded sequences to direct the light-regulated synthesis and transport of a foreign protein into plant chloroplasts, EMBO J. 4:25-32.

45. Sellmer, J.C., and McCown, B.H., 1989, Transformation in Populus spp., pp. 155-172, in: "Biotechnology in Agriculture and Forestry", Vol. 9, "Plant Protoplasts and Genetic Engineering II," Y.P.S. Bajaj, ed., Springer-Verlag, Heidelberg, West Germany.

46. Sellmer, J.C., McCown, B.H., and Haissig, B.E., 1989, Shoot culture dynamics of six Populus clones, Tree Physiol. 5:219-227.

47. Shah, D.M., Horsch, R.B., Klee, H.J., Kishore, G.M., Winter, J.A., Tumer, N.E., Hironaka, C.M., Sanders, P.R., Gasser, C.S., Aykent, S., Siegel, N.R., Rogers, S.G., and Fraley, R.T., 1986, Engineering herbicide tolerance in transgenic plants, Science 233:478-481.

48. Stachel, S.E., Messens, E., Van Montagu, M., and Zambryski, P., 1985, Identification of the signal molecules produced by wounded plant cells that activate T-DNA transfer in Agrobacterium tumefaciens, Nature 318:624-629.

49. Stalker, D.M., Hyatt, W.R., and Comai, L., 1985, A single amino acid substitution in the enzyme 5-enolpyruvylshikimate-3-phosphate synthase confers resistance to the herbicide glyphosate, J. Biol. Chem. 260:4724-4728.

50. Thomashow, M.F., Hugly, S., Buchholz, W.G., and Thomashow, L.S., 1986, Molecular basis for the auxin-independent phenotype of crown gall tumor tissues, Science, 231:616-618.

51. Van den Broeck, G., Timko, M.P., Kausch, A.P., Cashmore, A.R., Van Montagu, M., and Herrera-Estrella, L., 1985, Targeting of a foreign protein to chloroplasts by fusion to the transit peptide from the small subunit of ribulose 1,5-bisphosphate carboxylase, Nature 313:358-363.

52. Wasmann, C.C., Reiss, B., Barlett, S.G., and Bohnert, H.J., 1986, The importance of the transit peptide and the transported protein for protein import into chloroplasts, Mol. Gen. Genet. 205:446-453.

53. Watson, B., Currier, T.C., Gordon, M.P., Chilton, M.-D., and Nester, E.W., 1975, Plasmid required for virulence of Agrobacterium tumefaciens, J. Bacteriol. 123:255-264.

WOUND-RESPONSIVE GENE EXPRESSION IN POPLARS

Harvey D. Bradshaw, Jr. and Milton P. Gordon

Dept. of Biochemistry SJ-70
University of Washington
Seattle, WA USA 91895

ABSTRACT

We have cloned cDNA copies of several transcripts specifically accumulating in the unwounded upper leaves of a hybrid poplar whose lower leaves were mechanically wounded. Of those cDNAs sequenced, two encode proteins with sequence similarity to plant endochitinases and another specifies a polypeptide similar to Kunitz trypsin inhibitors from legume seeds. Genomic clones of one of the chitinase gene families reveal a clustered organization of these genes on the poplar chromosomes.

INTRODUCTION

Trees, like other plants, protect themselves from insect herbivores and pathogenic microorganisms with biochemical defenses. Some of these defenses are induced throughout the tree in reaction to a local injury. This systemic response to herbivory or pathogen attack may require new gene expression in tissues remote from the pest invasion. Wound-responsive genes and their protein products potentially may be used to enhance pest and pathogen resistance in trees, either by providing guidance to conventional breeding programs or by direct genetic engineering.

The biochemical defenses of many herbaceous plants are mobilized only when the plant is attacked by herbivores or pathogens, or in response to environmental stress (reviewed in reference 3). These defensive compounds include low molecular weight substances such as the phytoalexins (10), cell wall components like the hydroxyproline-rich glycoproteins (2,4), hydrolytic enzymes such as chitinases

(5,11) and glucanases (9), inhibitors of insect gut
proteinases (13), and several proteins (some of unknown
function) called "pathogenesis-related" (PR) proteins (14).
Mechanical wounding of leaves, like herbivore chewing or
pathogen invasion of leaf tissue, elicits the defensive
response.

RESULTS

When the lower leaves of a hybrid poplar tree (Populus
trichocarpa x P. deltoides) are mechanically wounded, novel
mRNAs accumulate in the unwounded upper leaves (12). cDNAs
corresponding to several of these wound-responsive
transcripts have been cloned and their nucleotide sequences
determined. Two of the newly accumulated mRNAs, win6 and
win8 (wound-inducible) appear to encode chitinases, enzymes
which may be active in degrading the cell walls of invading
fungi or bacteria. Another wound-responsive transcript,
win3, codes for a polypeptide similar to Kunitz-type trypsin
inhibitors found in the seeds of legumes (1). Since other
classes of proteinase inhibitors (some of which are
transcriptionally induced by wounding in herbaceous plant
systems) are potent deterrents of insect feeding (6,8), it
seems probable that this poplar protein is part of a defense
against herbivory.

Poplar genomic DNA clones for win6 chitinases have been
isolated, mapped, and sequenced. win6 is a multigene family
with as many as a dozen members, and several of the copies
are clustered within the poplar genome. We have found up to
three family members on a single lambda phage clone, and
these are transcriptionally oriented in the same direction.
DNA sequences upstream from the start of translation of
several win6 genes have been fused to the B-glucuronidase
(GUS, 7) gene to enable us to study the spatial and temporal
expression of this poplar chitinase in transgenic plants.

DISCUSSION

By understanding the mechanisms of control of gene
expression in response to insect herbivory or pathogen
invasion, we hope to accelerate tree improvement by
genetically engineering hybrid poplars for enhanced
resistance to insect and microbial pests. Two approaches
are being investigated: 1) the levels of expression of the
poplar's own defense genes could be elevated by manipulating
transcription or translation, and 2) the transcriptional
control elements from wound-responsive genes might be used
to drive the expression of heterologous genes, such as that
for the insecticidal endotoxin from Bacillus thuringiensis.

ACKNOWLEDGEMENTS

This work was supported by grants from the Washington
Technology Center and the United States Department of
Agriculture (88-33520-4072) to MPG.

266

REFERENCES

1. Bradshaw, H.D., Jr., Hollick, J.B., Parsons, T.J., Clarke, H.R.G., and Gordon, M.P. (1989) Systemically wound-responsive genes in poplar trees encode proteins similar to sweet potato sporamins and legume Kunitz trypsin inhibitors. Plant Mol. Biol., 14: 51-59.

2. Chen, J. and Varner, J.E. (1985) An extracellular matrix protein in plants: Characterization of a genomic clone for carrot extensin. EMBO J., 4: 2145-2151.

3. Collinge, D.B. and Slusarenko, A.J. (1987) Plant gene expression in response to pathogens. Plant Mol. Biol., 9: 389-410.

4. Corbin, D.R., Sauer, N., and Lamb, C.J. (1987) Differential regulation of a hydroxyproline-rich glycoprotein gene family in wounded and infected plants. Mol. Cell. Biol., 7: 4337-4344.

5. Hedrick, S.A., Bell, J.N., Boller, T., and Lamb, C.J. (1988) Chitinase cDNA cloning and mRNA induction by fungal elicitor, wounding, and infection. Plant Physiol., 86: 182-186.

6. Hilder, V.A., Gatehouse, A.M.R., Sheerman, S.E., Barker, R.F., and Boulter, D. (1987) A novel mechanism of insect resistance engineered into tobacco. Nature, 330: 160-163.

7. Jefferson, R.A., 1987. Assaying chimeric genes in plants: The GUS gene fusion system. Plant Mol. Biol. Rep., 5: 387-405.

8. Johnson, R., Narvaez, J., An, G., and Ryan, C. (1989) Expression of proteinase inhibitors I and II in transgenic tobacco plants: Effects on natural defense against Manduca sexta larvae. Proc. Natl. Acad. Sci. USA, 86: 9871-9875.

9. Kauffmann, S., Legrand, M., Geoffroy, P., and Fritig, B. (1987) Biological function of pathogenesis-related proteins: Four PR proteins of tobacco have 1,3-B-glucanase activity. EMBO J., 6: 3209-3212.

10. Lawton, M.A., Dixon, R.A., Hahlbrock, K., and Lamb, C.J. (1983) Rapid induction of the synthesis of phenylalanine ammonia-lyase and of chalcone synthase in elicitor-treated plant cells. Eur. J. Biochem., 129: 593-601.

11. Legrand, M., Kauffmann, S., Geoffroy, P., and Fritig, B. (1987) Biological function of pathogenesis-related proteins: Four tobacco pathogenesis-related proteins are chitinases. Proc. Natl. Acad. Sci. USA, 84: 6750-6754.

12. Parsons, T.J., Bradshaw, H.D., Jr., and Gordon, M.P. (1989) Systemic accumulation of specific mRNAs in response to wounding in poplar trees. Proc. Natl. Acad. Sci. USA, 86: 7895-7899.

13. Ryan, C.A. (1983) Systemic response to wounding. In: R. Denno and M. McClure (Editors), Variable Plants and Herbivores in Natural and Managed Systems. Academic Press, New York.

14. van Loon, L.C. (1985) Pathogenesis-related proteins. Plant Mol. Biol., 4: 111-116.

A TRANSIENT ASSAY FOR HETEROLOGOUS PROMOTER ACTIVITY IN *PICEA GLAUCA*

David D. Ellis, Dennis McCabe[1], Dave Russell[1], Brent McCown and Brian Martinell[1]

Department of Horticulture, University of Wisconsin-Madison, Madison, WI and [1]Agracetus, Middleton, WI

ABSTRACT

One important aspect of the successful genetic engineering of trees is the proper and coordinated expression of introduced foreign genes in the tree. To date, few studies have focused on the controlled expression of foreign genes in forest trees. Using electrical discharge particle acceleration, the expression of several heterologous promoters were tested in *Picea glauca* (white spruce) embryos, seedlings and embryogenic callus. Promoters tested include: *Arabidopsis* and soybean ribulose-1,5-bisphosphate small subunit (rbcS) promoters, a maize phosphoenolpyruvate carboxylase (PEP) promoter, a soybean heat shock promoter, a maize alcohol dehydrogenase (ADH) promoter and a soybean auxin inducible promoter. All promoters were used to drive the β-glucuronidase marker gene and expression was determined both with and without the respective induction stimulus required for maximal promoter activity in the source plant. Furthermore, the induction stimuli were also tested to determine the effects of the stimuli on the expression of a cauliflower mosaic virus 35s promoter. While β-glucuronidase gene expression in white spruce embryos was detected from all promoters tested, inducible gene expression was observed only with the heat shock and ADH promoters. Expression in seedlings and embryogenic callus indicate that this method for transient assays allows testing not only of promoter activity but also of tissue and/or cell specific expression.

INTRODUCTION

When working with long-lived perennials, such as forest trees, the proper regulation of foreign genes could be instrumental in the successful deployment of transgenic plants. Any factor which places a particular tree at a selective disadvantage, such as a disruption in the energy balance due to the production of additional gene products, could have a profound long-term effect over the decades often needed for a tree crop to reach economic maturity. Such effects would not necessarily be obvious in an herbaceous crop, as accumulation of the selective pressures may not be measurable in a one or two year rotation. One strategy which could decrease any

Woody Plant Biotechnology, Edited by M.R. Ahuja
Plenum Press, New York, 1991

deleterious effects from the expression of foreign genes in woody perennials is the limited expression of introduced genes in only those specific cells or tissues which would directly benefit from the expression of the gene. Examples of this would be leaf specific expression of genes coding for herbicide or insect resistance.

The mechanisms governing such tight control of gene expression are poorly understood. Despite this, promoters, or regulatory sequences of DNA, from genes normally expressed in a tissue or organ specific manner can confer a similar level of specific gene expression to genes linked to them. Transgenic plants containing marker genes linked to specific promoters have been used to precisely regulate these introduced genes in either an environmentally inducible or tissue specific manner (Simpson et al. 1985, Kuhlemeier et al. 1987, Schoffl et al. 1988). Theoretically, by properly identifying and isolating a particular promoter, the expression of introduced genes should be able to be targeted to most any cell type, developmental stage or tissue within a plant.

Unfortunately very little work has been done with the identification of promoters from forest trees. All the above examples were concerned with the use of dicot or monocot promoters to regulate the expression of genes in other angiosperms. Therefore, two fundamental questions which exist with a large group of temperate forest trees, the conifers, are 1) can these angiosperm promoters be used to express introduced genes across such a wide taxonomic distance as exists between angiosperms and gymnosperms and 2) will these promoters confer the same tissue or organ specific gene expression that they do in angiosperms.

The testing of heterologous promoters (promoters from other species) in conifers has been limited by the lack of a reliable transformation system for these trees. While this still remains a limitation, electrical discharge particle acceleration provides a highly reliable means for testing the transient (short term) expression of genes in intact tissues. This technique bombards tissue with microscopic gold particles coated with DNA with such force as to drive the particles through the cell walls and membranes so that they penetrate and remain lodged in the plant cell (see McCown, this volume). Using *Picea glauca* (white spruce) embryos, we have developed a very reproducible (Figure 4) and quantitative assay to monitor transient gene activity. Although initial work centered on the manipulation of both biological and physical parameters which affect gene expression, subsequent research has focused on the use of this technique to test the expression of various inducible angiosperm promoters in a gymnosperm, white spruce.

MATERIALS AND METHODS

Open pollinated white spruce seed was collected from the Wisconsin Department of Natural Resources Lake Tomahawk seed orchard, Lake Tomahawk, WI and stored at 4°C. Prior to embryo excision, the seed was placed under running tap water for 16 hrs and then kept at 4°C for a minimum of two days to allow easy separation of the embryo from the megagametophyte. The seed was then surface sterilized in 20% clorox (5.25% sodium hypochlorite) for 20 min followed by rinsing three times in sterile distilled water.

For all studies using embryos, the embryos were aseptically excised from the seed and placed on woody plant medium (WPM) (Lloyd and McCown 1980) supplemented with 50 μM zeatin and 0.01 μM thidiazuron. Embryos were pretreated on bud induction medium for seven days prior to particle bombardment unless otherwise noted. In the seedling experiments, excised embryos were placed on hormone free WPM for seven days to allow the *in vitro* germination of the embryo prior to particle bombardment. Embryogenic callus was induced from immature zygotic embryos and cultured as previously described (Webb et al. 1989). All tissue was maintained at 27°C, with embryos and seedlings grown in continuous light and embryogenic callus in the dark.

For particle acceleration, 1 μg of plasmid DNA was adhered to 10 mg of 1-3 μm gold particles. The DNA coated gold particles were accelerated by vaporizing a drop of water at a discharge voltage of 20 kilovolt (KV) with a particle load of 0.05 mg/cm^2. Targets for particle bombardment consisted of fifteen embryos, ten seedlings or 200 mg of embryogenic callus placed in a 1.2 cm^2 area on a 15X60 mm petri dish containing 10 ml of the appropriate medium supplemented with 250 μg/ml carbenicillin. All embryos used for defining the parameters important for transient gene expression were histochemically stained for β-glucuronidase (GUS) enzyme activity with 5-bromo-4-chloro-3-indolyl glucuronide (x-gluc) (Jefferson et al. 1987) two days after bombardment. Tissue used in the promoter studies was histochemically stained for GUS activity six days after particle bombardment. The level of gene expression was quantified by counting the number of cells (Figure 1) expressing the GUS enzyme for a given parameter or promoter and using this number as a relative level of gene expression. 4-methyl umbelliferyl glucuronide (MUG), a fluorometric assay which is used to quantify GUS enzyme activity, was used to verify that the relative level of gene expression obtained by counting the GUS expressing spots was correlated to enzyme activity.

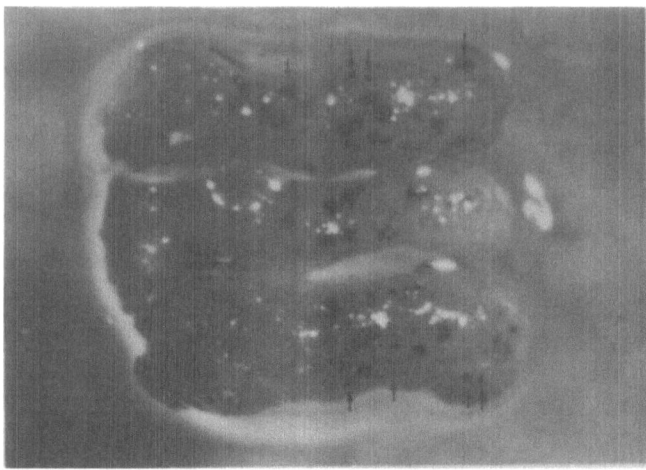

Fig. 1. White spruce embryos with blue spots (arrows) which represent transient expression in embryonic cells of a CaMV 35s-GUS construct two days following particle acceleration.

In all preliminary experiments which tested parameters affecting transient expression, a derivative of pTV4AMVBTS (Barton et al. 1987), pBTV41100, was used. This plasmid differs from pTV4AMVBTS mainly by the addition of a cauliflower mosaic virus 35s (CaMV 35s)-GUS construct. Constructs to test the expression and inducibility of angiosperm promoters in white spruce were made by fusing the various promoters to the GUS coding region and introducing these fusions into a pUC19 derived expression vector. The promoters used include: the light inducible promoters from maize phosphoenylpyryvate carboxylase (PEP) (W. Taylor, unpublished), soybean ribulose-1,5-bisphosphate carboxylase (rbcS) (Berry-Lowe et al. 1982) and *Arabidopsis* rbcS (Timko et al. 1988); an anaerobiasis induced promoter from maize alcohol dehydrogenase (ADH) (Walker et al. 1987); a soybean heat shock inducible promoter (Schoffl et al. 1984); and a soybean auxin inducible promoter (McClure et al. 1989). The size of the upstream fragments used are listed in Table 1. A construct containing a CaMV 35s promoter fused to the GUS coding region was used as a control to determine if the induction stimuli used had a general effect on other promoters introduced into white spruce. The protocol used for the induction of gene expression in tissue containing the various promoters is listed in Table 2. The induction stimulus was given five days after particle bombardment. Non-induced controls were included for all promoters.

Table 1. Inducible heterologous promoters used to test expression in white spruce, the plant source from where the promoter was originally isolated, the reference to the promoter sequence and the size of the DNA promoter fragment used.

INDUCIBLE PROMOTER	SOURCE	REFERENCE	PROMOTER SIZE
AUXIN	soybean	McClure et al. 1989	−830 to +1
HEAT SHOCK	soybean	Schoffl et al. 1984	−420 to +100
ADH	maize	Walker et al. 1987	−1090 to +106
rbcS	soybean	Berry-Lowe et al. 1982	−1550 to +45
rbcS	*Arabidopsis*	Timko et al. 1988	−320 to +1
PEP	maize	W. Taylor, unpublished	−990 to +60

Table 2. The induction protocol used to induce gene expression from the various promoters.

PROMOTER	INDUCTION PROTOCOL
AUXIN	Flood with 25 μM 2,4-D or place on medium containing 50 μM 2,4-D for 24 hr
HEAT SHOCK	1 hr at 42° + 3 hr at 25°
ADH	Flood tissue for 24 hr
PEP + rbcS	24 hr light

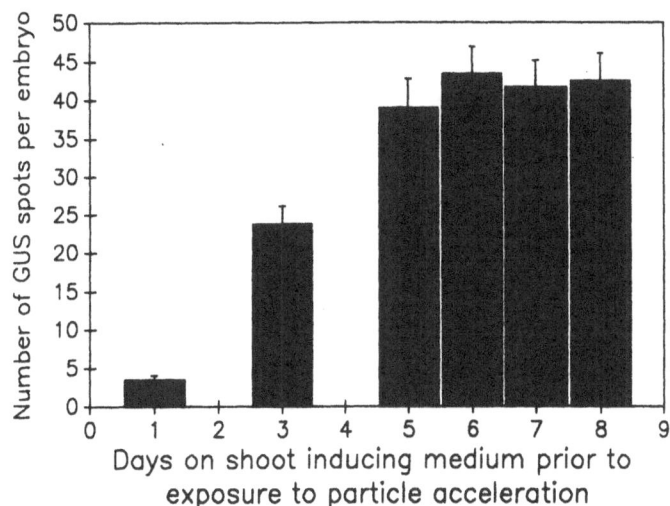

Fig. 2. The effect on transient GUS expression of pretreating
white spruce embryos on bud induction medium for
varying days prior to particle acceleration. Embryos
were assayed for GUS enzyme activity two days after
particle acceleration. Bars represent +/- SE.

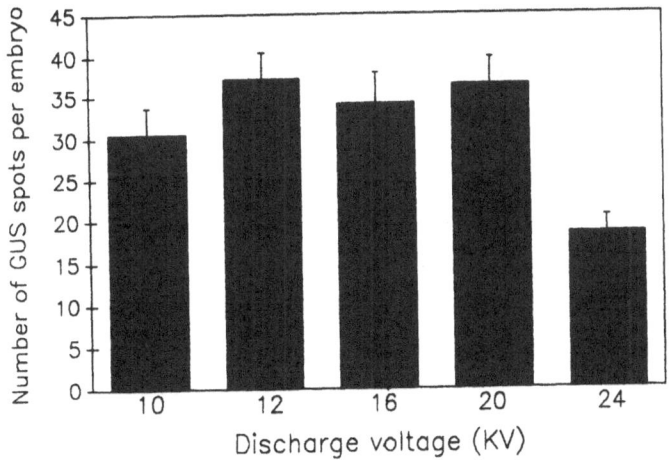

Fig. 3. The effect of varying the discharge voltage used with
particle acceleration on GUS expression in embryos of
white spruce. Embryos were assayed for GUS enzyme
activity two days after particle acceleration. Bars
represent +/- SE.

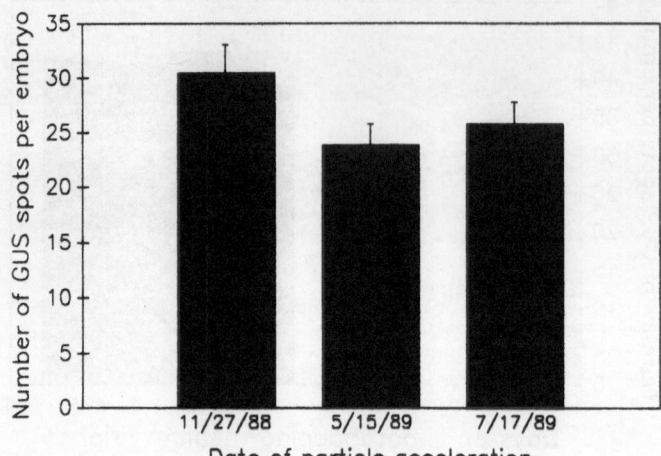

Fig. 4. Variation between experiments in the number of GUS
expressing spots per white spruce embryo. Discharge
voltage for these experiments was 12 KV and embryos
were assayed for GUS enzyme activity two days after
particle acceleration. Bars represent +/- SE.

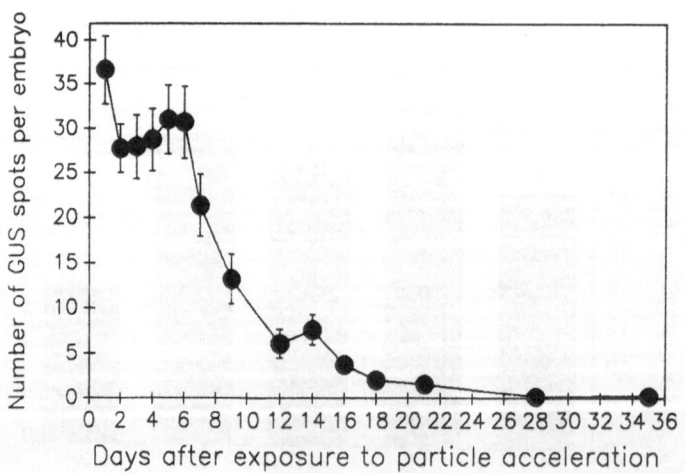

Fig. 5. The effect of time following particle acceleration on
the expression of GUS enzyme in embryos of white
spruce. Embryos were assayed for GUS activity at
varying days following particle acceleration. Bars
represent +/- SE.

RESULTS AND DISCUSSION

The developmental stage of the embryos had a profound affect on the level of transient gene activity (Figure 2). If the embryos were placed on bud induction medium for varying periods of time prior to particle acceleration, the level of transient GUS expression would increase for up to five days of pretreatment. This period correlates to the time when meristematic centers are first forming in the subdermal cells of the hypocotyl. This would suggest that either a specific cellular organization or cellular state is required for a high level of gene expression. Similar effects of the developmental stage on transient gene expression have also been noted with poplar (McCown, this volume, and Sellmer personal communication) and cranberry (Serres, personal communication).

Transient gene expression can also be affected by varying the physical parameters of particle acceleration. The discharge voltage used to vaporize the water droplet can be used to control the velocity of the particles. With a higher discharge voltage, the vaporizing water creates a larger explosive force which drives the carrier sheet containing the particles against the stopping screen with greater force and propels the particles into the tissue at an increased velocity. As the discharge voltage is increased, the depth that the particles penetrate the tissue increases. Therefore, by varying the discharge voltage, DNA can be targeted to particular cell layers. In addition to affecting particle penetration, the discharge voltage used can also affect transient expression (Figure 3). If low discharge voltage is used, below 10 KV, the number of particles penetrating the epidermis decreases and thus the level of transient expression declines. Between 12 and 20 KV the level of transient expression remains stable yet the number of particles penetrating the 2nd and 3rd cell layers increases as the voltage increases (data not shown). Above 20 KV, although the number of particles in the subdermal layers increase, the damage to the epidermal cells also increases and thus the level of transient gene expression decreases.

By fine tuning those parameters which maximize transient expression within defined limits, electrical discharge particle acceleration is highly reproducible. Figure 4 clearly shows that by maintaining both biological and physical parameters, the level of transient gene expression can be maintained and repeated over relatively long periods of time. Such repeatability has not been demonstrated with other transformation systems used with woody plants. Further, this reproducibility is well-suited for use in a study aimed at examining promoter activity.

An initial concern with the use of inducible heterologous promoters in white spruce was that the simple act of bombarding the cells with the particles could give a non-specific wound induction of some of these promoters. Although this possibility has not been ruled out, assays for transient expression were done as long after particle bombardment as possible to decrease such non-specific responses. By assaying the tissue for CaMV-GUS expression at varying days after particle bombardment, it was apparent that after seven days there is a sharp decline in gene expression (Figure 5). This decline continued until 28 days when the level of GUS expression in embryos was

275

diminished. To avoid this decrease in gene expression at seven days yet to assay for promoter activity as long after particle bombardment as possible, GUS expression for the promoter assays were done six days after particle acceleration.

It is clear from Figure 6 that angiosperm promoters will function in gymnosperms. Of the promoters tested, only the soybean heat shock promoter was strongly inducible. The maize ADH promoter was slightly induced with flooding yet the level of expression even with induction was quite low. Of the promoters where gene expression was not stimulated with an induction stimulus, the *Arabidopsis* rbcS and the maize PEP promoters were expressed at a relatively high level, roughly 1/3 that of CaMV 35s. The induction protocol used did not have a significant effect on the expression of CaMV 35s suggesting that the induction stimuli by themselves did not cause a non-specific stimulation on gene expression. The relative levels of gene expression obtained from counting the GUS expressing spots correlates relates well to GUS enzyme activity (Figure 7). It should be noted however, that because of background levels, the use of MUG to quantify transient activity in white spruce embryos required that the assay be carried out for at least 5 hours.

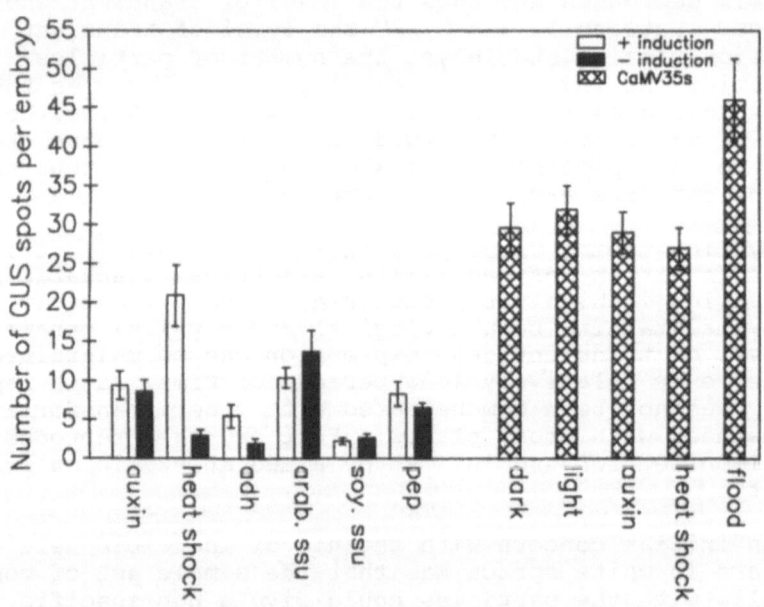

Fig. 6. Transient expression of heterologous promoters in white spruce embryos. Embryos were assayed for GUS enzyme activity six days after particle acceleration. Bars represent +/- SE.

276

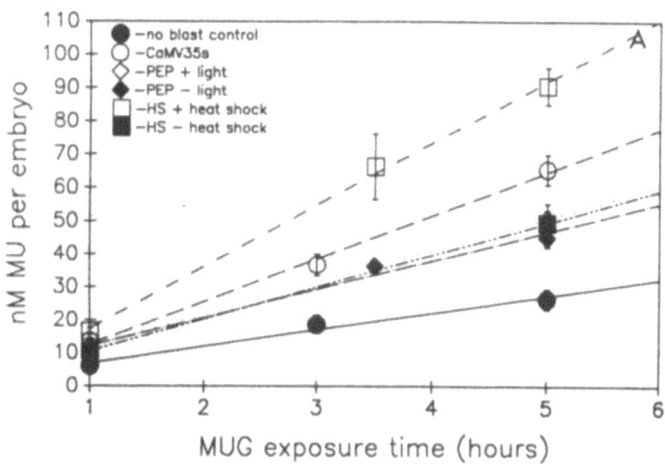

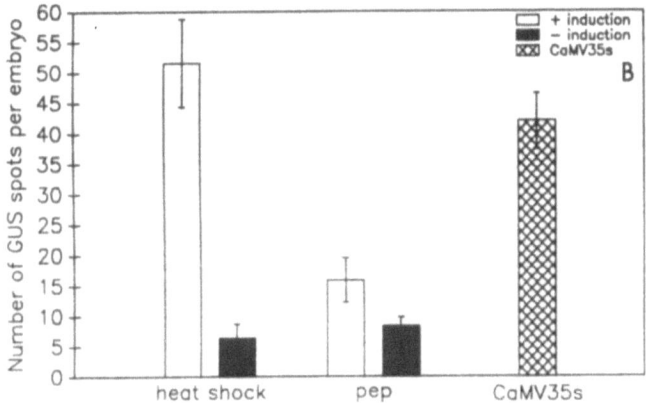

Fig. 7. Comparison of GUS enzyme activity in white spruce
embryos as quantified by A) MUG assay and B) counting
the GUS expressing spots after staining with x-gluc.
Note that the relative order of GUS enzyme activity
between the two methods is the same for the two
quantification methods. For the MUG assay, two embryos
were pooled for each extraction. Both MUG (A) and
x-gluc (B) assays were from embryos exposed to particle
acceleration at the same time. Embryos were assayed
for GUS enzyme activity six days following particle
acceleration. Bars represent +/- SE.

It was hoped that with seedlings, differences in gene expression relating to tissue specificity could be noticed. To facilitate this, the number of blue spots representing GUS expressing cells were counted separately in the roots, hypocotyls and cotyledons. The roots had the lowest level of expression of all tissues (Figure 8). This could be due to a general lack of transient gene expression by this tissue, a lack of expression by these promoters in roots or a lack of penetration of the roots by the particles. Since the parameters used for particle bombardment were optimized for embryos, the particles may not have penetrated the surface of the roots. Of the promoters tested, only the heat shock and CaMV 35s promoters were expressed in the roots demonstrating that at least some particles were penetrating to root cell layers.

GUS gene expression in the hypocotyls and cotyledons was evident with all promoters and, as with the embryos, the induction stimuli used did not affect the expression of CaMV 35s. The heat shock promoter was still inducible in seedlings but not at the level that it was in embryos. Interestingly, the auxin inducible promoter was negatively effected by induction. It could be that the seedlings already had a relatively high endogenous level of auxin and that additional auxin was detrimental to gene expression. As well, this auxin inducible promoter was isolated from soybean hypocotyls where it is strongly expressed relative to other seedling tissues (McClure and Guilfoyle 1989). In contrast, this promoter had a higher level of expression in the cotyledons than in the hypocotyls of white spruce seedlings.

Expression of the GUS gene in embryogenic callus was very low with all promoters tested except CaMV 35s (data not shown). As with the seedlings, this low level of expression could be related to a lack of fine tuning of the transformation system for embryogenic callus. The callus produces a mucilaginous coating which may inhibit particle penetration. Interestingly, the expression that was present was confined exclusively to the proembryonal head cells and the GUS gene was not expressed in other cell types such as the suspensors (Figure 9). This head cell specific expression is intriguing in that it offers a means to study transformation parameters which specifically effect gene expression in those cells which have the potential to be regenerated.

The lack of inducible gene expression by the rbcS, PEP and auxin inducible promoters does not imply that these promoters will not function in an inducible, tissue specific manner in vivo. Even when tested for transient expression in angiosperms, these promoters were not inducible in a transient assay, yet did function in the predicted manner when expressed in vivo in transformed angiosperms. It is clear that the expression and inducibility of these promoters depends on the tissue. For example, the heat shock promoter was inducible in both embryos and seedlings, yet was not inducible in embryogenic callus. This non-inducibility of heat shock in a relatively nondifferentiated tissue is not unique to white spruce as the heat shock promoter was also not inducible in poplar suspension cells (Sellmer, personal communication).

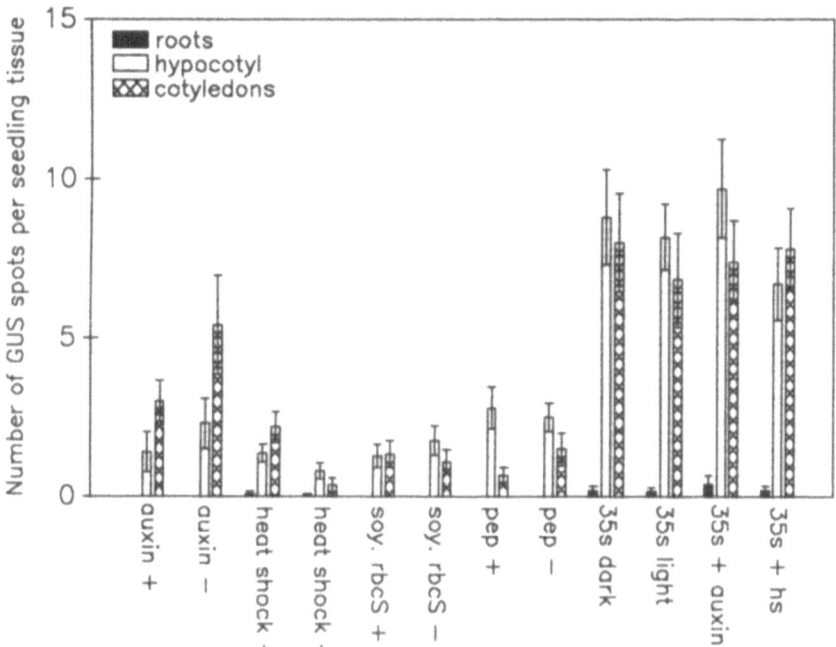

Fig. 8. Transient expression of heterologous promoters in white
spruce one-week-old seedlings. The number of GUS
expressing spots for each seedling is divided into the
tissues which were expressing GUS. Seedlings were
assayed for GUS enzyme activity six days after particle
acceleration. Bars represent +/- SE.

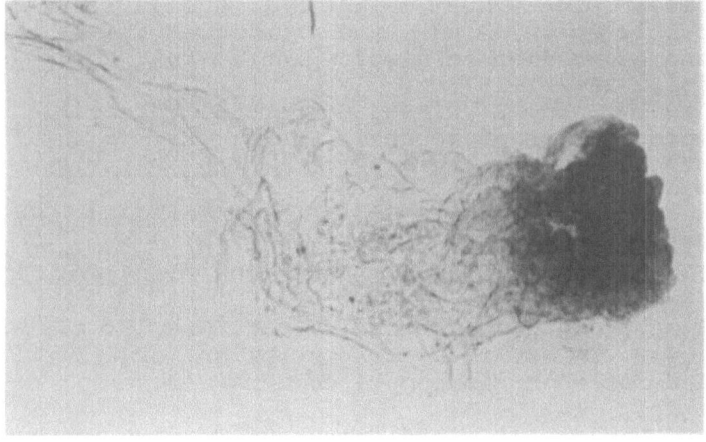

Fig. 9. A somatic proembryo expressing the CaMV 35s-GUS gene
construct only in the proembryonal head cells. Note
that there is no GUS activity in the suspensor cells.
Although all proembryonal head cells stain blue,
probably only one or a few cells are expressing the GUS
gene and diffusion is responsible for the other head
cells staining blue.

It is also interesting to note the conservation of mechanisms for gene regulation even in divergent groups like angiosperms and gymnosperms. In the case of the soybean heat shock promoter, for example, in order for the promoter to be properly induced in white spruce, the conifer would require the requisite *trans*-acting factors which would interact with the promoter for heat shock specific expression.

The use of electrical discharge particle acceleration enables the testing of numerous factors that effect the transient expression of introduced genes in white spruce. It is as yet unknown why biological factors such as the developmental state of the target tissues have such a large influence on transient gene expression. Although data such as these aid in answering specific questions regarding the expression of foreign genes, the precise regulatory mechanism controlling differential gene expression is lacking.

ACKNOWLEDGEMENTS

We thank Trent Marty of the Wisconsin Department of Natural Resources for supplying the white spruce seed used in this study, Jenny Hergenrother for assistance with preparation of the manuscript and Jim Sellmer for his critical review.

REFERENCES

Barton, K. A., Whiteley, H. R., and Yang, N.-S., 1987, *Bacillus thuringiensis* endotoxin expressed in transgenic *Nicotiana tabacum* provides resistance to lepidopteran insects, Plant Physiol., 85:1103-1109.

Berry-Lowe, S. L., McKnight, T. D., Shah, D. M., and Meagher, R. B., 1982, The nucleotide sequence, expression, and evolution of one member of a multigene family encoding the small subunit of ribulose-1,5-bisphosphate carboxylase in soybean, J. Mol. Appl. Genet., 1:483-498.

Jefferson, R. A., 1987, Assaying chimeric genes in plants: the GUS gene fusion system, Plant Mol. Biol., 5:387-405.

Kuhlemeier, C., Green, P. J., and Chua, N.-H., 1987, Regulation of gene expression in higher plants, Ann. Rev. Plant Physiol., 38:221-257.

Lloyd, G. and McCown, B., 1980, Commercially feasible micropropagation of mountain laurel (*Kalmia latifolia*) by use of shoot-tip culture, Proc. Inter. Plant Prop. Soc., 30:421-437.

McClure, B. A. and Guilfoyle, T. J., 1989, Tissue print hybridization. A simple technique for detecting organ- and tissue-specific gene expression, Plant Mol. Biol., 12:517-524.

McClure, B. A., Hagen, G., Brown, C. S., Gee, M. A., and Guilfoyle, T. J., 1989, Transcription, organization, and sequence of an auxin-regulated gene cluster in soybean, Plant Cell, 1:229-239.

Schoffl, F., Raschke, E., and Nagao, R. T., 1984, The DNA sequence analysis of soybean heat-shock genes and identification of possible regulatory promoter elements, EMBO J., 3:2491-2497.

Schoffl, F., Blaumann, G., and Raschke, E., 1988, The expression of heat shock genes--a model for environmental stress response, in: "Temporal and Spatial Regulation of Plant Genes," D. P. S. Verma and R. B. Goldberg, eds., Springer-Verlag, New York.

Simpson, J., Timko, M. P., Cashmore, A. R., Schell, J., Van Montagu, M. and Herrera-Estrella, L., 1985, Light-inducible and tissue-specific expression of a chimaeric gene under control of the 5'-flanking sequence of a pea chlorophyll a/b-binding protein gene, EMBO J., 4:2723-2729.

Timko, M. P., Herdies, L., de Almeida, E., Cashmore, A. R., Leemans, J., and Krebbers, E., 1988, Genetic engineering of nuclear-encoded components of the photosynthetic apparatus in *Arabidopsis*, in: "The Impact of Chemistry on Biotechnology," ACS Symp. Serv. 362:279-295.

Walker, J. C., Howard, E. A., Dennis, E. S., and Peacock, W. J., 1987, DNA sequences required for anaerobic expression of the maize alcohol dehydrogenase 1 gene, Proc. Natl. Acad. Sci., 84:6624-6628.

Webb, D. T., Webster, F., Flinn, B. S., Roberts, D. R., and Ellis, D. D., 1989, Factors influencing the induction of embryogenic and caulogenic callus from embryos of *Picea glauca* and *P. engelmanii*, Can. J. For. Res., 19:1303-1308.

INSERTION OF THE MAIZE TRANSPOSABLE ELEMENT *AC* INTO POPLAR

Glenn T. Howe, Steven H. Strauss, and Barry Goldfarb

Department of Forest Science
Peavy Hall 154
Oregon State University
Corvallis, Oregon, USA 97331-5705

ABSTRACT

The maize transposable element *Ac* (*Activator*) was inserted into hybrid poplar (*Populus alba* x *P. grandidentata*) using cocultivation of suspension cultures with *Agrobacterium tumefaciens*. Southern blot analysis indicates that six of eight hygromycin-resistant callus lines contain stably integrated foreign DNA. The number of inserted T-DNA copies ranges from approximately 1 to 10. Southern blot analysis suggests that *Ac* has excised from its original location in the T-DNA. We are currently studying transposition of *Ac* in transgenic shoots. We believe this to be the first report of transposon activity in a woody plant.

INTRODUCTION

Transposable elements can be used for isolating genes via "gene tagging" (reviewed by Weinand and Saedler[43]). If insertion of a transposon causes a gene to become non-functional, the mutant gene can be isolated using a cloned copy of the element as a gene probe. Subsequently, a functional gene can be isolated from a wild-type plant using the mutant gene as a probe. Gene tagging using transposable elements in their natural hosts has been used to clone genes from both maize and snapdragon (reviewed by Weinand and Saedler[43]).

By inserting transposable elements into new plants, the power of gene tagging can be extended to species lacking well-characterized transposable elements. Much of the current work on transposable elements is aimed at developing gene tagging systems for new species.[5, 14, 16, 18, 21, 23, 41, 44]

Abbreviations: BA, N[6]-benzylaminopurine; bp, base pairs; cfu's, colony forming units; CTAB, cetyltrimethylammonium bromide; MS, Murashige and Skoog medium;[29] PAR, photosynthetically active radiation; SDS, sodium dodecyl sulfate; T-DNA, transferred-DNA; WPM, Woody Plant Medium;[24] 2,4-D, 2,4-dichlorophenoxyacetic acid.

Insertion of transposable elements into new plants is also an important tool for studying the molecular biology of these elements. Because endogenous transposable elements are frequently present in many copies, characterization of individual elements is facilitated by studying their behavior in heterologous plants.[4, 7, 8, 12, 28, 32, 40]

Transposition of a plant transposable element in a foreign host was first demonstrated when the Ac (*Activator*) element from maize was shown to transpose in transgenic tobacco.[3] Transposition has been subsequently demonstrated by Ac in transgenic tomato,[44] potato,[21] *Arabidopsis* and carrot,[41] by the maize En/Spm element in tobacco,[27, 32] and potato,[14] and by the snapdragon Tam3 element in tobacco and petunia.[16]

Genes that have a visible phenotype can be used as indicators of transposition by tracking disruption of gene expression when transposons insert, or restoration of expression when they excise. Genes used for studying endogenous transposons include those coding for pigment biosynthetic enzymes in both maize (reviewed by Fedoroff[10]) and snapdragon,[6, 26] and starch and storage proteins that affect easily scored endosperm markers in maize (reviewed by Fedoroff[10]).

More recently, transposable elements have been inserted into reporter genes so that transposition can be easily scored in transgenic plants. Visual assays of excision have been developed using the ß-glucuronidase (GUS) reporter gene[13, 16, 27] and the streptomycin resistance gene.[19]

A relatively simple phenotypic assay for excision of Ac was developed by Baker et al[4] by inserting Ac into the untranslated leader of the neomycin phosphotransferase (NPT-II) gene. Excision of Ac restores NPT-II expression, and cells in which transposition has occurred can be identified by the acquisition of kanamycin resistance. Inclusion of the hygromycin phosphotransferase (HPT) gene[39] on the same Ti plasmid vector allows selection for transformed cells using hygromycin. We inserted Baker et al's[4] Ac::NPT-II/HPT construct into poplar using *Agrobacterium*-mediated transformation of suspension cultures.

MATERIALS AND METHODS

Plasmids

The Ac element, flanked by 60 bp of the waxy allele (wx-m7), was inserted into the untranslated leader of NPT-II to create plasmid pKU3 cloned in *Escherichia coli*[4] (Fig. 1). During this process, the nopaline synthase (nos) promoter which was upstream of the NPT-II coding region was replaced by the *Agrobacterium* 1' promoter (p1') from plasmid pOP4434.[42] Plasmid pKU2, from which pKU3 is derived, contains NPT-II without Ac. Ti plasmid pGV3850HPT was recombined with pKU3 to create the cointegrate transformation vector pGV3850HPT::pKU3 resident in *Agrobacterium tumefaciens* strain C58Cl rif.[4] pGV3850HPT is a disarmed cointegrate Ti plasmid vector[20, 45] that contains a hygromycin phosphotransferase gene (HPT)[39] within the T-DNA, allowing transformed plant cells to be selected using hygromycin. The presence of Ac in the untranslated leader of the NPT-II gene inhibits expression,[4] and because excision of Ac can restore expression, plant cells in which Ac has excised can be identified by their resistance to kanamycin.[4]

E. coli strains were grown at 37°C on LB agar medium (pH 7.5)[25] with 100 µg/ml ampicillin, whereas *Agrobacterium* was grown at 30°C on YEP agar

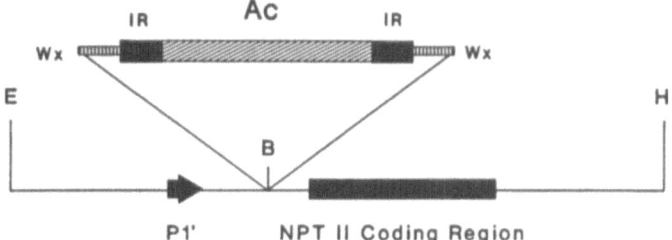

Figure 1. Partial restriction site map of the NPT-II region of plasmids pKU2, pKU3, and pGV3850HPT::pKU3. pKU3 and pGV3850HPT::pKU3 contain *Ac* inserted into the untranslated leader of NPT-II at a *Bam*HI site, whereas pKU2 contains NPT-II without the *Ac* insertion. *Ac*, which is delineated by the ends of the inverted repeats (IR), is flanked by 60 bp of the *waxy* (*wx*) allele from maize. The arrow represents the 1' promoter used to transcribe the NPT-II gene in transgenic plants. Restriction enzyme sites = E (*Eco*RI), B (*Bam*HI), and H (*Hind*III).

medium[2] (pH 7.5) with 100 μg/ml carbenicillin, 25 μg/ml kanamycin, 100 μg/ml rifampicin, and 100 μg/ml spectinomycin. Overnight cultures of *Agrobacterium* were grown at 30°C in MG/L liquid medium [2.5 g/l yeast extract, 5 g/l tryptone, 5 g/l NaCl, 5 g/l mannitol, 1.16 g/l monosodium glutamate, 0.25 g/l KH_2PO_4, 0.1 g/l $MgSO_4 \cdot 7H_2O$, and 0.001 g/l biotin].

Plant Cultures

Callus was initiated from leaves of *in vitro* shoot cultures of hybrid poplar clone NC-5339 (*Populus alba* L. x *P. grandidentata* Michx. cv.'Crandon')[11,36]. Leaves were cut along the midrib and placed adaxial side up on callus initiation medium (solidified MS medium containing 30 g/l sucrose, 1 mg/l 2,4-D, and 0.05 mg/l BA; pH 5.7-5.8). After 2-3 weeks of growth, the calli were transferred to liquid MS medium containing the same supplements. The suspension cultures were grown on a gyratory shaker (120 rpm) under cool white fluorescent light (<10 μE·m^{-2}·s^{-1} PAR) using a 16 hour photoperiod. Suspension cultures were maintained using weekly transfers of five to ten ml of inoculum to 50 ml of new media (either WPM- or MS-based) in a 250-ml flask. All callus cultures, including those derived from leaves or suspension cultures, were grown in the dark at 25°C.

Cocultivation

Log-phase suspension cultures were filtered through a 500 μm nylon screen (Small Parts Inc., Miami, FL) and 13.5 ml of the filtrate was transferred to each of six 50-ml Erlenmeyer flasks. After an additional 22 hours of growth, the flasks were inoculated with *Agrobacterium* to a final density of either 10^6, 10^7, or 10^8 cfu's/ml using an overnight culture that had been grown to a density of 10^9 cfu's/ml in MG/L medium (pH 5.6) containing 100 μg/ml carbenicillin and 39.2 μg/ml (200 μM) acetosyringone (Aldrich D13,440-6). After cocultivation for either 24 or 48 hours in low light with gentle shaking, the cells were pelleted at 250 rpm, washed with 25 ml of suspension culture medium, and resuspended to final densities of 1.25 x 10^4, 2.5 x 10^4, or 5 x 10^4 colonies/ml in suspension culture medium containing 500 μg/ml cefotaxime free-acid (Calbiochem 219380). Although filtered poplar suspension cultures contain colonies with as many as 40 cells, the average number of cells per colony is approximately five. Aliquots of 0.5 ml were plated to cellulose acetate filters (Millipore SMWP, 5.0 micron pore size, 47 mm dia.) on callus initiation medium containing 500 μg/ml cefotaxime and solidified with 0.5% agarose (BRL Ultrapure, Life Technologies Inc., Gaithersburg, MD). The resulting plating densities equaled 3.6, 7.2, and 14.4 colonies/mm^2. For each of the 18 treatments (i.e. 3 *Agrobacterium* cocultivation densities x 2 cocultivation times x 3 plating densities), cells were plated to four plates.

Selection

We used two phases of hygromycin selection. Four hygromycin concentrations (0, 7.5, 15, and 30 μg/ml) were tested during the initial phase of selection, whereas a single hygromycin concentration (15 μg/ml) was used for the second phase. During antibiotic selection all cultures were grown on WPM-based callus initiation medium containing 0.5% agarose.

The initial phase of selection was started five days after the suspension cultures were plated by transferring the filters to medium containing 500 μg/ml cefotaxime and either 0, 7.5, 15, or 30 μg/ml hygromycin. Filters were transferred to new antibiotic-containing medium every 4 to 16 days. Because the colonies on the filters were not growing as well as expected, the level of cefotaxime was reduced to 250 μg/ml for the second, and all subsequent transfers. Experiments conducted later indicated that cefotaxime inhibits callus growth at 500 μg/ml (unpublished data), and residual *Agrobacterium* has been completely controlled in more recent cocultivation experiments by plating to 250 μg/ml cefotaxime.

The second phase of selection began when individual colonies were large enough (>1 mm) to be transferred from the filters to medium containing 15 μg/ml hygromycin and 250 μg/ml cefotaxime. The first colonies were transferred 26 days after the cells were plated to the filters. After 1-2 months of selection on 15 μg/ml hygromycin, a few of the colonies were significantly larger than other colonies on the same plate. These antibiotic-resistant colonies were transferred to, and subsequently maintained on, callus initiation medium without antibiotics.

DNA Isolation

DNA was isolated from callus or suspension culture cells using a modified CTAB procedure.[30] Approximately 15 g of tissue was ground to a fine powder with a mortar and pestle in liquid nitrogen and dry ice. Twenty ml of ice-cold grinding buffer [0.35 M sorbitol, 50 mM Tris/HCl (pH 8.0), 5 mM EDTA, 10% PEG-8000, 0.5% 2-mercaptoethanol, 0.1% spermine tetrahydrochloride, 0.1% spermidine trihydrochloride] was added, and the homogenate was centrifuged at 14,000 rpm for 10 minutes at 3°C. The pellet was resuspended in 5 ml of ice-cold buffer [0.35 M sorbitol, 50 mM Tris/HCl (pH 8.0), 25 mM EDTA, 0.1% 2-mercaptoethanol] followed by the sequential addition of 1/10 volume 10% sarkosyl, 1/7 volume 5 M NaCl, and 1/10 volume 10% CTAB/0.7 M NaCl. The mixture was shaken vigorously and incubated at 65°C for 20 minutes. Following extraction with an equal volume of chloroform/isoamyl alcohol (24:1), an equal volume of isopropyl alcohol was added to the aqueous phase and the mixture was centrifuged at 15,000 rpm for 15 minutes. The pellet was resuspended in TE (10 mM Tris/HCl, 0.1 mM EDTA, pH 8.0), incubated with 20 μg/ml RNase A for 30 minutes at 37°C, and extracted once with an equal volume of phenol/chloroform/isoamyl alcohol (25:24:1), and once with an equal volume of chloroform/isoamyl alcohol.

The DNA was precipitated by adding 1/10 volume 3 M sodium acetate, 2 volumes of 95% ethanol, and centrifuging at 15,000 rpm for 30 minutes. The pellet was washed with 70% ethanol and resuspended in TE (10 mM Tris/HCl, 1 mM EDTA, pH 8.0). DNA concentration was measured fluorometrically using a DNA-specific dye[22] using a known concentration of CsCl-purified Douglas-fir DNA as a standard.

Southern Blot Analysis

DNA from eight hygromycin-resistant callus lines and control DNA from untreated suspension cultures were analyzed using Southern blot analysis. Five micrograms of poplar DNA and DNA from plasmids pKU2 and pKU3 equivalent to 0.1, 1, or 10 copies of inserted DNA per diploid genome of poplar (estimated to be 1.6 pg[31]) were digested with EcoRI and HindIII. Five micrograms of DNA from untransformed poplar suspension cultures was added to the plasmid reconstructions prior to digestion. The DNA samples were electrophoresed in a 1% agarose gel using Tris-borate (TBE) buffer[25] and transferred to a Zetabind nylon membrane (Cuno Inc., Meriden, CT) using alkaline transfer.[34] Probe DNA was prepared by gel-purifying the 506 bp EcoRI-HindIII fragment containing the 1' promoter from pOP4434.[42] Plasmid DNA was cut with restriction enzymes, separated by gel electrophoresis in 0.7% low gelling temperature agarose (SeaPlaque, FMC Bioproducts, Rockland, ME) and stained with ethidium bromide. Probe DNA (440 ng) was radioactively labeled "in gel" with [^{32}P]-dCTP by random-primed labeling following the manufacturer's instructions (Boehringer Mannheim Inc, Indianapolis, IN). The blot was hybridized[25] and washed at 65°C. Two 30-minute washes in 2X SSC (1X SSC is 150 mM NaCl, 15 mM sodium citrate) and 0.1% SDS were followed by two 30-minute washes in 0.1X SSC, 0.5% SDS. X-ray film was exposed to the filter for 11 days at -70°C using intensifying screens.

RESULTS

Cocultivation

Eight hygromycin-resistant colonies remained following the second phase of selection, 6 of which contain foreign DNA (Table 1). No hygromycin-resistant colonies were recovered from the 24-hour cocultivation treatment, and all but one of the transformants were recovered from a single Agrobacterium density treatment (10^7 cfu's/ml).

Selection

Because no hygromycin-resistant calli were recovered when the cells were cocultivated for 24 hours, Table 2 summarizes only the results from the 48-hour treatment. When averaged over all Agrobacterium density treatments, two transformants were recovered from a total of 861 calli grown in the absence of selection, resulting in a transformation frequency of 0.2%. In the best treatment combination (Agrobacterium density of 10^7 cfu's/ml and initial selection on 15 μg/ml hygromycin), 2 of 7 colonies (28.6%) were found to be transformed.

Initial selection on hygromycin reduced the number of surviving calli from 861 to less than 90. At the lower plating densities (3.6 and 7.2 colonies/mm^2), few calli survived initial selection on hygromycin, and all but one of the transformants were recovered from the highest plating density (14.4 colonies/mm^2, Table 1). Of a total of 171 calli surviving the initial phase of selection on hygromycin, only 33 came from either the 3.6 or 7.2 colonies/mm^2 plating densities.

Table 1. *Agrobacterium* cocultivation treatments and initial selection regimes from which hygromycin-resistant poplar calli were recovered.[a]

Callus line	Density of *Agrobacterium* (cfu's/ml)	Plating density (colonies/mm^2)	Initial hygromycin concentration (μg/ml)
--------------- Calli found to contain foreign DNA --------------			
1	10^7	14.4	15
2	10^7	14.4	15
4	10^7	14.4	0
5	10^7	14.4	30
6	10^6	7.2	0
7	10^7	14.4	30
--------------- Calli found to lack foreign DNA ---------------			
3	10^7	14.4	7.5
8	10^7	14.4	30

[a] hygromycin-resistant callus lines are from the 48-hour cocultivation treatment, no hygromycin-resistant calli were recovered from the 24-hour treatment.

Although a greater number of colonies were transferred from the highest initial hygromycin concentration (30 μg/ml) as compared to the two lower concentrations (7.5 and 15 μg/ml, Table 2), this is probably not indicative of enhanced growth. Instead, this probably results from the subjective nature of choosing resistant colonies for transfer. Of the 91 colonies transferred from the 30 μg/ml hygromycin treatment, 83 colonies were transferred on a single day by a single person.

In summary, the best treatment was 48-hour cocultivation using a starting *Agrobacterium* density of 10^7 cfu's/ml and plating of poplar cells onto media containing 15 μg/ml hygromycin at a density of 14.4 colonies/mm^2. Because the cocultivation and plating treatments were not replicated in a second experiment, however, it is difficult to say whether treatment differences are repeatable.

Southern Blot Analysis

Of the eight hygromycin-resistant callus lines, six (lanes 1,2,4,5, 6,7) contain DNA complementary to the pl' probe (Fig. 2, Panel C) and the NPT-II coding region (data not shown).

Figure 2 (Panel A) shows a portion of the T-DNA as it exists in the cointegrate Ti plasmid vector pGV3850HPT::pKU3, and as we expect it to appear in transformed poplar. If *Ac* is intact within the untranslated leader of NPT-II, a DNA fragment of 2.3 kb will be seen when transformed poplar DNA (or plasmid DNA from pKU3) is digested with *Eco*RI and *Hin*dIII

Table 2. Recovery of transformants following 48-hour cocultivation of poplar suspension cultures with *Agrobacterium* and initial selection on 0, 7.5, 15, or 30 µg/ml hygromycin.

Initial hygromycin concentration (µg/ml)	Number of colonies transferred to 15 µg/ml hygromycin	Number of colonies transformed	Percent transformed[a]
---------- Average of 3 *Agrobacterium* density treatments ----------			
(10^6, 10^7, and 10^8 cfu's/ml)			
0	861	2	0.2
7.5	49	0	0.0
15	31	2	6.5
30	91	2	2.2
------------- Best <u>Agrobacterium</u> density treatment --------------			
(10^7 cfu's/ml)			
0	416	1	0.2
7.5	25	0	0.0
15	7	2	28.6
30	91	2	2.2

[a]percent = [the number of calli found to contain foreign DNA (by Southern blot analysis) divided by the number of calli transferred to 15 ug/ml hygromycin] x 100.

and probed with the 1' promoter fragment from pOP4434. On the other hand, if *Ac* has excised, thus removing the *Eco*RI and *Hin*dIII sites internal to *Ac* (Fig. 2, Panel B), a fragment of 3.0 kb will be present when probed with the 1' promoter fragment. These diagnostic patterns can be seen on the autoradiograph in lanes which contain plasmid DNA reconstructions. The lanes labeled "I" (for intact) contain DNA from plasmid pKU3 which contains *Ac* in the untranslated leader of NPT-II. The lanes labeled "E" (for excised) contain the DNA from plasmid pKU2, which lacks *Ac*.

Five callus lines (lanes 1,2,4,5,6), have bands which comigrate with the 2.3 kb fragment from pKU3 ("I" lanes) indicative of *Ac* in its original location. Four callus lines (lanes 1,2,4,7) have bands that appear to comigrate with the 3.0 kb fragment from pKU2 ("E" lanes) indicative of T-DNA from which *Ac* has excised. Unfortunately, distortion of the gel in the 3.0 kb region makes it difficult to be certain that these bands resulted from excision of *Ac*. Nevertheless, by taking into account the pattern of distortion, which can be seen more clearly on the ethidium-stained gel (not shown), this conclusion is warranted.

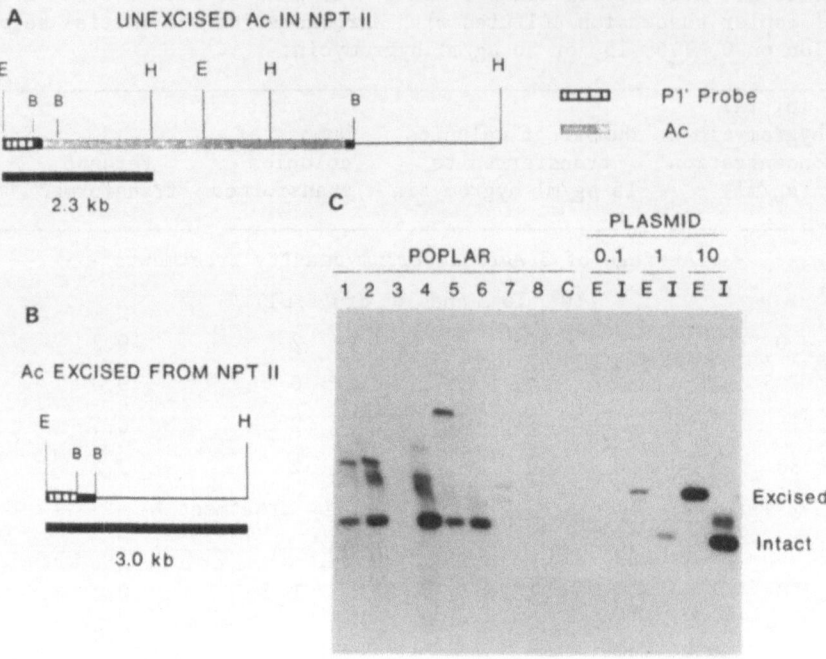

Figure 2. Southern blot analysis of DNA isolated from hygromycin-resistant poplar calli. A. Schematic of unexcised *Ac* in
NPT-II. When *Ac* is located at its original site in the untranslated leader of NPT-II, a 2.3 kb fragment is observed
when T-DNA (or plasmid DNA from pKU3) is digested with *Eco*RI
and *Hin*dIII and probed with the pl' fragment from pOP4434. B.
Schematic of *Ac* excised from NPT-II. If *Ac* has excised from
its original site, a 3.0 kb fragment is observed using the pl'
probe. C. Southern blot of DNA isolated from 8 hygromycin-
resistant callus lines (lanes 1-8), control DNA from untransformed poplar suspension cultures (lane C), and DNA from
plasmid pKU2 (lanes labeled "E" for excised *Ac*) or pKU3 (lanes
labeled "I" for intact *Ac*). Plasmid DNA samples equivalent to
0.1, 1, or 10 copies of inserted DNA per diploid genome of
poplar were used. Restriction enzyme sites are E (*Eco*RI), B
(*Bam*HI), and H (*Hin*dIII); Pl' probe is the 1' promoter fragment, and *Ac* is the *Ac* coding region.

Copies of T-DNA with *Ac* intact (2.3 kb band), and *Ac* excised (3.0 kb
band) may exist in the same callus line if there are multiple T-DNA inser-
tions and excision has occurred from only some of the T-DNA copies, or if
the callus is chimeric; that is *Ac* excision has occurred in some cells, but
not others. The presence of other, unexplained bands such as the top band
in lane 5 suggest that rearrangements may have occurred during T-DNA
transfer or *Ac* transposition.

Based on the relative intensity of the poplar DNA bands as compared to
the plasmid reconstructions, it appears that approximately 1 to 10 copies
of T-DNA have been inserted into the transformed lines.

DISCUSSION

Although *Agrobacterium*-mediated transformation of poplar has been successful using leaf disc cocultivation[11] (N. B. Klopfenstein pers. comm., D. E. Riemenschneider pers. comm.), this is the first report of transformation of poplar suspension cultures. By plating suspension cultures to antibiotic-containing medium, it is easy to kill residual *Agrobacterium* that is often a problem using leaf disc transformation.[38] The major disadvantage of suspension culture transformation is that it involves a two-step approach of selecting transformed calli, and regenerating plants from the transformed lines. Although this takes longer than leaf-disc transformation, use of embryogenic suspension cultures could speed the process considerably. Because an extremely large number of calli can be recovered from suspension cultures, this might be a good system for testing either strains of *Agrobacterium* or general cocultivation parameters. Although *Agrobacterium*-mediated transformation of suspension cultures has not been used extensively, it has been reported for other species.[1, 33, 37]

Because poplar has been transformed using armed[11] (N. B. Klopfenstein pers. comm., D. E. Riemenschneider pers. comm.), but not disarmed strains of *Agrobacterium*,[11] it has been suggested that Ti plasmid oncogenicity is required.[38] Transformation with the disarmed cointegrate vector pGV3850HPT::pKU3 demonstrates that oncogenicity is not necessary for transformation of poplar.

A chimeric gene consisting of the nopaline synthase (*nos*) promoter and the hygromycin B phosphotransferase (HPT) gene from *E. coli* has been used as a selectable marker in a number of plants, including tobacco,[16, 39] potato,[21] *Triticum monococcum*,[17] and petunia.[16] In some cases hygromycin works better than kanamycin for selecting transformants. For example, hygromycin may be a good alternative in species such as rice and other grasses, which have a high tolerance to kanamycin.[9]

We show that hygromycin can be used for selecting transformed poplar microcalli at concentrations ranging from 7.5 to 30 μg/ml. Two of eight calli which survived the second phase of selection on 15 μg/ml hygromycin did not contain foreign DNA. Therefore, a higher hygromycin concentration might be beneficial for the second phase of selection, although one of the "escapes" was grown on an initial hygromycin concentration of 30 μg/ml.

Because NC-5339 poplar callus is more sensitive to hygromycin than to kanamycin (data not shown), and because the response to kanamycin selection in leaf disc systems is variable,[38] the HPT gene may be a superior selectable marker than NPT-II in poplar.

We presented evidence suggesting that *Ac* excises in transgenic poplar. We have not yet determined, however, if *Ac* is reintegrating into the poplar genome. This work, as well as confirmation of *Ac* excision, is currently underway using shoots regenerated from transgenic callus.

ACKNOWLEDGEMENTS

We thank Dr. Barbara Baker of the USDA Plant Gene Expression Center in Albany, CA for providing the plasmids, and Dr. Brent McCown of the University of Wisconsin for providing the hybrid poplar shoot cultures. We also thank Nancy Murphy and Gail McGill for their excellent technical assistance. This research was supported by the U.S. Forest Service Pacific Northwest Research Station, under Cooperative Aid Agreement 87-414.

REFERENCES

1. An, G., 1985, High efficiency transformation of cultured tobacco cells. Plant Physiol. 79: 568-570.
2. An, G., Ebert, P.R., Mitra, A., and Ha, S.B., 1988, Binary vectors. In: "Plant Molecular Biology Manual," S.B. Gelvin and R.A. Schilperoort, eds., Kluwer Acad. Pub., Dordrecht, pp. A3/1-A3/19.
3. Baker, B., Schell, J., Lorz, H., and Fedoroff, N., 1986, Transposition of the maize controlling element *Activator* in tobacco. Proc. Natl. Acad. Sci. USA 83: 4844-4848.
4. Baker, B., Coupland, G., Fedoroff, N., Starlinger, P., and Schell, J., 1987, Phenotypic assay for excision of the maize controlling element *Ac* in tobacco. EMBO J. 6: 1547-1554.
5. Belzile, F., Lassner, M.W., Tong, Y., Khush, R., and Yoder, J.I., 1989, Sexual transmission of transposed *Activator* elements in transgenic tomatoes. Genetics 123: 181-189.
6. Bonas, U., Sommer, H., and Saedler, H., 1984, The 17-kb *Tam*1 element of *Antirrhinum majus* induces a 3-bp duplication upon integration into the chalcone synthase gene. EMBO J. 3: 1015-1019.
7. Coupland, G., Baker, B., Schell, J., and Starlinger, P., 1988, Characterization of the maize transposable element *Ac* by internal deletions. EMBO J. 7: 3653-3659.
8. Coupland, G., Plum, C., Chatterjee, S., Post, A., and Starlinger, P., 1989, Sequences near the termini are required for transposition of the maize transposon *Ac* in transgenic tobacco plants. Proc. Natl. Acad. Sci. USA 86: 9385-9388.
9. Dekeyser, R., Claes, B., Marichal, M., Van Montagu, M., and Caplan, A., 1989, Evaluation of selectable markers for rice transformation. Plant Physiol. 90: 217-223.
10. Fedoroff, N., 1983, Controlling elements in maize. In: "Mobile Genetic Elements," J.A. Shapiro, ed., Academic Press, New York, pp. 1-63.
11. Fillatti, J.J., Sellmer, J., McCown, B., Haissig, B., and Comai, L., 1987, *Agrobacterium* mediated transformation and regeneration of *Populus*. Mol. Gen. Genet. 206: 192-199.
12. Finnegan, E.J., Taylor, B.H., Dennis, E.S., and Peacock, W.J., 1988, Transcription of the maize transposable element *Ac* in maize seedlings and in transgenic tobacco. Mol. Gen. Genet. 212: 505-509.
13. Finnegan, E.J., Taylor, B.H., Craig, S., and Dennis, E.S., 1989, Transposable elements can be used to study cell lineages in transgenic plants. Plant Cell 1: 757-764.
14. Frey, M., Tavantzis, S.M., and Saedler, H., 1989, The maize *En-1/Spm* element transposes in potato. Mol. Gen. Genet., 217: 172-177.
15. Gritz, L., and Davies, J., 1983, Plasmid-encoded hygromycin B resistance: the sequence of hygromycin B phosphotransferase gene and its expression in *Escherichia coli* and *Saccharomyces cerevisiae*. Gene 25: 179-188.
16. Haring, M.A., Gao, J., Volbeda, T., Rommens, C.M.T., Nijkamp, H.J.J., and Hille, J., 1989, A comparative study of *Tam*3 and *Ac* transposition in transgenic tobacco and petunia plants. Plant Mol. Biol. 13: 189-201.
17. Hauptmann, R.M., Vasil, V., Ozias-Akins, P., Tabaeizadeh, Z., Rogers, S., Fraley, R.T., Horsch, R.B., and Vasil, I.K., 1988, Evaluation of selectable markers for obtaining stable transformants in the Gramineae. Plant Physiol. 86: 602-606.
18. Hehl, R. and Baker, B., 1989, Induced transposition of *Ds* by a stable *Ac* in crosses of transgenic tobacco plants. Mol. Gen. Genet. 217: 53-59.
19. Jones, J.D.G., Carland, F.M., Maliga, P. and Dooner, H.K., 1989, Visual detection of transposition of the maize element *Activator* (*Ac*) in tobacco seedlings. Science 244: 204-207.

20. Joos, H., Timmermann, B., Van Montagu, M., and Schell, J., 1983, Genetic analysis of transfer and stabilization of *Agrobacterium* DNA in plant cells. EMBO J. 2: 2151-2160.
21. Knapp, S., Coupland, G., Uhrig, H., Starlinger, P., and Salamini, F., 1988, Transposition of the maize transposable element *Ac* in *Solanum tuberosum*. Mol. Gen. Genet. 213: 285-290.
22. Labarca, C., and Paigen, K., 1980, A simple, rapid, and sensitive DNA assay procedure. Anal. Biochem. 102: 344-352.
23. Lassner, M.W., Palys, J.M., and Yoder, J.I., 1989, Genetic transactivation of *Dissociation* elements in transgenic tomato plants. Mol. Gen. Genet. 218: 25-32.
24. Lloyd, G.B., and McCown, B.H., 1980, Commercially-feasible micro-propagation of mountain laurel, *Kalmia latifolia*, by use of shoot-tip culture. Proc. Intl. Plant Prop. Soc. 30: 421-427.
25. Maniatis, T., Fritsch, E.F., and Sambrook, J., 1982, eds., "Molecular Cloning, a Laboratory Manual." Cold Spring Harbor, New York.
26. Martin, C., Carpenter, R., Sommer, H., Saedler, H., and Coen, E.S., 1985, Molecular analysis of instability in flower pigmentation of *Antirrinum majus*, following isolation of the *pallida* locus by transposon tagging. EMBO J. 4: 1625-1630.
27. Masson, P., and Fedoroff, N., 1989, Mobility of the maize Suppressor-mutator element in transgenic tobacco cells. Proc. Natl. Acad. Sci. USA 86: 2219-2223.
28. Masson, P., Rutherford, G., Banks, J.A., and Fedoroff, N., 1989, Essential large transcripts of the maize *Spm* transposable element are generated by alternative splicing. Cell 58: 755-765.
29. Murashige, T., and Skoog, F., 1962, A revised medium for rapid growth and bioassays with tobacco tissue cultures. Physiol. Plant. 15: 473-492.
30. Murray, M.G., and Thompson, W.F., 1980, Rapid isolation of high molecular weight plant DNA. Nucleic Acids Res. 8: 4321-4325.
31. Parsons, T.J., Sinkar, V.P., Stettler, R.F., Nester, E.W., and Gordon, M.P., 1986, Transformation of poplar by *Agrobacterium tumefaciens*. Bio/Technology 4: 533-536.
32. Pereira, A., and Saedler, H., 1989, Transpositional behavior of the maize *En/Spm* element in transgenic tobacco. EMBO J. 8: 1315-1321.
33. Pollock, K., Barfield, D.G., Robinson, S.J., and Shields, R., 1985, Transformation of protoplast-derived cell colonies and suspension cultures by *Agrobacterium tumefaciens*. Plant Cell Rep. 4: 202-205.
34. Reed, K.C., and Mann, D.A., 1985, Rapid transfer of DNA from agarose gels to nylon membranes. Nucleic Acids Res. 13: 7207-7221.
35. Reynaerts, A., De Block, M., Hernalsteens, J.-P., and Van Montagu, M., 1988, Selectable and screenable markers. In: "Plant Molecular Biology Manual," S.B. Gelvin and R.A. Schilperoort, eds., Kluwer Acad. Pub., Dordrecht, pp. A9/1-A9/16.
36. Russell, J.A., and McCown, B.H., 1986, Techniques for enhanced release of leaf protoplasts in *Populus*. Plant Cell Rep. 5: 284-287.
37. Scott, R.J., and Draper, J., 1987, Transformation of carrot tissues derived from proembryogenic suspension cells: A useful model system for gene expression studies in plants. Plant Mol. Biol. 8: 265-274.
38. Sellmer, J.C., and McCown, B.H., 1989, Transformation in *Populus* spp. In: "Biotechnology in Agriculture and Forestry vol. 9: Plant Protoplasts and Genetic Engineering II", Y.P.S. Bajaj, ed., Springer Verlag, Berlin, pp. 155-172.
39. van den Elzen, P.J.M., Townsend, J., Lee, K.Y., and Bedbrook, J.R., 1985, A chimaeric hygromycin resistance gene as a selectable marker in plant cells. Plant Mol. Biol. 5: 299-302.

40. Van Sluys, M.A., and Tempe, J., 1989, Behavior of the maize transposable element *Activator* in *Daucus carota*. Mol. Gen. Genet. 219: 313-319.

41. Van Sluys, M.A., Tempe, J., and Fedoroff, N., 1987, Studies on the introduction and mobility of the maize *Activator* element in *Arabidopsis thaliana* and *Daucus carota*. EMBO J. 6: 3881-3889.

42. Velten, J., Velten, L., Hain, R., and Schell, J., 1984, Isolation of a dual plant promoter fragment from the Ti plasmid of *Agrobacterium tumefaciens*. EMBO J. 3: 2723-2730.

43. Wienand, U., and Saedler, H., 1988, Plant transposable elements: unique structures for gene tagging and gene cloning. In: "Plant DNA Infectious Agents", T. Hohn and J. Schell, eds., Springer, New York, pp. 205-228.

44. Yoder, J.I., Palys, J., Alpert, K., and Lassner, M., 1988, *Ac* transposition in transgenic tomato plants. Mol. Gen. Genet. 213: 291-296.

45. Zambryski, P., Joos, H., Genetello, C., Leemans, J., Van Montagu, M., and Schell, J., 1983, Ti plasmid vector for the introduction of DNA into plant cells without alteration of their normal regeneration capacity. EMBO J. 2: 2143-2150.

APPROACHES TO STUDYING DNA ELEMENTS ASSOCIATED WITH LIGHT-REGULATED GENE

EXPRESSION IN CONIFERS

M.A. Campbell and D.B. Neale

Institute of Forest Genetics, Pacific Southwest
Research Station, U.S.D.A. Forest Service
Box 245, Berkeley, CA 94701 USA

ABSTRACT

In many angiosperms light regulates the expression of genes coding for
the small subunit of ribulose bis-phosphate carboxylase (rbcS) and
chlorophyll a/b-binding protein (cab). Upstream DNA sequences of the rbcS
and cab genes, associated with transcriptional regulation, are reviewed
and the possible role that these sequences have in conifer gene regulation
are discussed. A DNA sequence associated with light-regulated gene
expression in angiosperms, called box-II, was found in the upstream region
of a larch rbcS gene. Using a double reporter system, we present a method
that can assign a functional role to the box-II, as well as to other
sequences in the upstream region of the larch rbcS gene. This double
reporter system has potential in examining the upstream regions of both
rbcS and cab genes for sequences that are associated with transcriptional
regulation in conifers.

INTRODUCTION

Light plays a central role in plant growth and development. In
angiosperms, many genes have been identified whose expression is
controlled by quantitative and qualitative changes in light (25). In
conifers, however, little is known about how light regulates gene
expression. Although the fundamental mechanisms of light-regulated gene
expression can be inferred from annual plants (e.g. oats, Arabidopsis,
maize), certain processes, such as photoperiodic control of annual growth
cycles, are unique to perennial plants. Before the molecular basis of a
complicated process such as the control of the annual growth cycle can be
determined, some basic knowledge of light-regulated gene expression in
conifers is necessary.

Two families of light-regulated genes have been studied in detail in
angiosperms: the light-harvesting chlorophyll a/b-binding protein (cab)
and the small subunit of ribulose bis-phosphate carboxylase (rbcS). The
cab gene products form a complex with chlorophylls a and b in thylakoid
membranes for energy transfer to photosystem II. The rbcS gene codes for
the small subunit of ribulose bis-phosphate carboxylase, a multimeric

enzyme responsible for CO2 fixation in the Calvin cycle. The cab and rbcS genes are represented in the nuclear genome in more than one copy; i.e., they are represented as a gene family. The protein products from the genes are assembled on cytoplasmic ribosomes before transport into the chloroplast.

The DNA sequences that code for mRNA of the cab and rbcS genes are expressed in a tissue specific and light-regulated manner (25,19). The rbcS and cab gene products are produced at high levels in leaf tissues and are lacking in root tissues of many angiosperms. This tissue specific response is coupled to light-regulation in many plant species. In the absence of light, leaf tissues produce low or non-dectable levels of RNA from cab and rbcS genes. Upon illumination with wide-spectrum light the production (transcription) of mRNA from cab and rbcS genes increases dramatically in dark-grown leaves (6). This ability to respond to light, with an enhancment of cab and rbcS RNA production, has been linked to the presence of specific DNA sequences that are situated near the coding regions of the cab and rbcS genes. More specifically these DNA sequences, that are involved with light regulation of cab and rbcS, are situated 5', or upstream from the coding regions of these genes (Fig 1). It has been demonstrated that the level of light regulation for both cab and rbcS is highly variable within the angiosperms (25,19) and this variation is related to differences in the 5' DNA sequences of these genes. Recent reviews discuss the types and significance of upstream DNA variability to the light regulation of cab and rbcS genes in angiosperms (25,19,7) but there is little evidence of light regulated gene expression in woody perennials such as conifers. In this paper we will examine previous research into cab and rbcS gene expression in angiosperms in order to develop an approach to examine conifer DNA sequences that are responsible for the regulation of cab and rbcS genes.

Expression of cab and rbcS Genes in Conifers. A DNA copy of the coding region of a gene (cDNA) and a DNA copy of the coding region and flanking DNA sequences (genomic clone), have been isolated and sequenced for an rbcS gene from eastern larch (Larix laricina) (12). In addition a cDNA has been isolated and sequenced for a rbcS gene from Japanese black pine (Pinus tunbergii)(28).

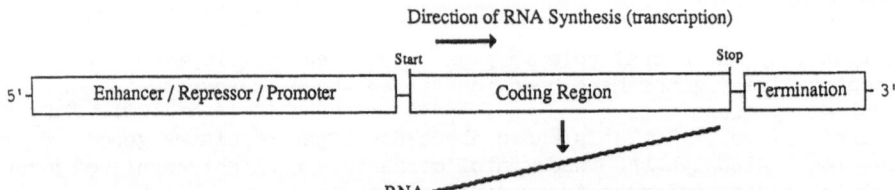

Fig 1. Basic structure of a functional eukaryotic gene. Rates of transcription of the coding region are controlled by the binding of enzymes and regulatory proteins to the 5' or upstream portion of the gene. Variability of the DNA sequence in this 5' enhancer/repressor/promoter (ERP) region may be responsible for different rates of transcription. Specific DNA sequences in the ERP region, by interactions with regulatory proteins, may limit gene transcription to specific tissue types or alter transcription rates in response to environmental stimuli (heat, light, ect.).

The cab gene has been studied in four conifers: Scots pine (Pinus sylvestris) (14), Japanese Black pine (29), Douglas fir (Pseudotsuga menziesii) (1) and eastern larch (11).

The ability to clone both the cab and rbcS genes from dark grown pine seedlings (28,29), plus recent evidence showing that steady-state RNA levels for the cab gene are not strongly light-regulated in dark grown seedlings of Douglas-fir (1), suggests that, in comparison to angiosperms, some conifers exhibit a lower level of light regulation of the cab and rbcS genes. However, cab and rbcS are increased 50-fold when dark grown larch tissues are exposed to light (11) which suggests that this conifer has a level of light-regulation of cab and rbcS genes which is similar to that of angiosperms. In addition, leaf specific expression of an rbcS gene in larch demonstrates a similarity between conifer and angiosperm with regards to tissue specific expression of rbcS. Futher studies into conifer cab and rbcS expression are therefore needed to elucidate the differences that exist in the mechanisms of light-regulated gene expression within the conifers and between conifers and angiosperms.

Upstream Region of rbcS

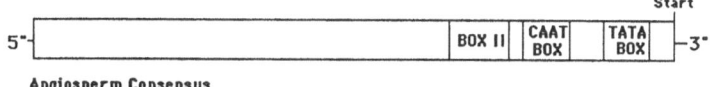

Angiosperm Consensus

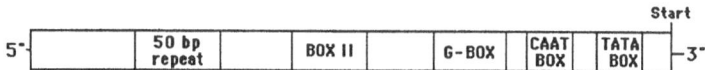

Larch

Fig 2. Comparison of the upstream portion of rbcS genes from angiosperms and larch. The consensus sequence for an angiosperm rbcS gene contains CAAT and TATA boxes, and a sequence identified as 'box II' (19). The upstream portion of a larch rbcS gene contains TATA and CAAT boxes, a sequence similar to the 'box II' element, a G-box element, and a novel 50 bp repeat (12).

Promoter and Upstream Elements of rbcS Genes. Specific DNA sequences, associated with transcriptional regulation of rbcS genes, have been characterized from angiosperms using the production and assay of gene fusions (Fig 2)(18). These fusions were produced by covalently linking the upstream portion of a rbcS gene, containing the promoter and DNA sequences that have a putative role in transcriptional regulation, to a DNA sequence that codes for an assayable protein product (a reporter). Through the assay of the reporter overall transcriptional activity of the rbcS promoter region has been determined (19). In addition, by sequential deletion of the rbcS upstream DNA, linked to the reporter, and introducing the fusion constructs into transgenic plants, it has been possible to assign light and tissue specific regulation to specific DNA sequences (19).

A number of rbcS genes have been cloned and sequences of rbcS genes from a range of angiosperms have been reported (4,19). A comparison of the upstream sequences from rbcS genes reveals that a large amount of variability exists interspecifically as well as within gene families (4). Although it is not possible to assign a regulatory role to the many unique or conserved upstream sequences described for rbcS, Dean et al. (4) have suggested that the large amount of sequence variability is associated with the many different expression patterns found among angiopserm rbcS genes.

```
               *      *      *
(A)        1  2  3  4  5  6  7  8  9  10 / 11 12 13 14 15
    5'- G  T  G  T  G  G  T  T  A  A  /  C  T  A  T  G  - 3'

                            *      *      *
(B)        11 12 13 14 15  1  2  3  4  5  6  7  8  9  10
    5'- C  T  A  T  G  T  T  T  T  T  G  T  T  A  A  - 3'
```

Fig 3. Comparison of the consensus sequence for the (A) angiosperm 'box-II' domain and the (B) 'box-II'-like element in the upstream sequence of larch rbcS. The angiosperm 'box-II' domain (A) is divided into two regions as reported by Dean et al. (4). The first part of the angiosperm 'box-II' element (bases 1-10) contains the GTTAA core, a region binding the GT-1 protein, and the second part is separate and located 3' from the GTTAA core. The larch upstream region (B)(12) contains a 'box-II'-like element which has a GTTAA core, the element is not divided into two separate domains, and the two domains, as compared to the angiosperm 'box-II' consensus sequence, are inverted. The presence of an asterisk denotes a base mismatch between the angiosperm and larch sequences.

Manzara and Gruissem (19) identified three essential regions for rbcS expression by comparative analysis of angiosperm rbcS genes. These essential regions include the TATA box, the CAAT box, and the box II region previously described by Fluhr et al. (5). The TATA and CAAT box regions are typical of many eukaryotic genes and are required for promoter function. The box II region appears to be unique to rbcS genes that have a reduced level of expression in the dark. However, it is not known whether the box II region is present in rbcS genes that are not light-regulated such as the constitutively expressed rbcS gene isolated from leaf tissue of mung-bean (24). Further examination of the box II region revealed that it is the binding site for a nuclear protein factor, GT-1, that acts as a silencer of transcription in dark-grown tissues (9,18).

The upstream region has been sequenced for a larch rbcS gene (12). Sequence analysis revealed similarities between the larch gene and angiosperm rbcS genes. The larch rbcS upstream region contains a consensus sequence, called the G-box, which is found in rbcS isolated from dicotyledonous plants having photosynthetically active cotyledons (12,7). We have examined the upstream sequence of the larch rbcS for sequence homology to the 17 regions described by Manzara and Gruissem for angiosperm rbcS genes (19). This analysis revealed larch sequences homologous to the box II region (Fig 3). Although the nucleotide sequences are conserved between angiosperms and gymnosperms, deletions will have to be analyzed to determine whether the regulatory role of these sequences is conserved as well.

Promoter and Upstream Elements of cab Genes. Upstream sequences associated with tissue-specific and phytochrome-mediated light-regulation of cab have been identified (20,23,2). In addition to these positively regulating elements, negatively regulating elements have also been described (2).

There are no reports of upstream sequences of conifer cab genes and it is not yet possible to determine the degree of base sequence conservation between angiosperms and conifers. With light-regulation of cab genes being different in Douglas-fir and larch, comparisons between the upstream region of this gene may yield common and unique sequences that are involved with transcriptional regulation. Upstream sequences of cab will soon be cloned and sequenced from larch and Douglas-fir, after which they can then be compared, and functional differences in upstream cab sequences can then be elucidated by deletion analysis. By placing deletions of the cab upstream region in front of a quantitatively assayable reporter gene, it will be possible to assign functional significance to a specific DNA sequence in transformed conifer tissues.

Approaches to the Study of Light-Regulated Gene Expression in Conifers. The use of a reporter system, such as E. coli beta-glucuronidase (GUS)(15) or firefly luciferase (LUC)(21), to quantify relative levels of gene expression regulated by a DNA sequence requires that these enzymes be easily assayable and have low background activity in conifer tissues. Both LUC and GUS have been used successfully as reporter genes in conifer tissues (10,27) and further studies into DNA sequences, associated with transcriptional control, can be accomplished using these genes.

Gene transfer into conifer cells has been accomplished by several direct methods, including electroporation, polyethylene glycol-mediated transfer, biolistics, and with the use of a biological vector such as Agrobacterium tumefaciens (22,3). However, examination of DNA sequences associated with light-regulated gene expression requires that assays for gene activity be accomplished with cell types and culture conditions that make some of the above gene-transfer methods impractical.

Electroporation and chemical-mediated transfer utilize protoplast systems as the target tissue. Isolation and culture of protoplasts has been reported from a number of conifers (16) but the ability to regenerate whole plants has yet to be demonstrated. In addition, conifer protoplast systems that exhibit relatively high rates of cell division are derived from dark-grown embryogenic cultures which have a translucent appearance and poor plastid development. Protoplasts can be derived from green, photosynthetically active tissues, but, once isolated they are typically sensitive to light and need to be cultured at low light intensities (18). Therefore, conifer protoplasts, isolated from embryogenic cultures or leaf tissue, are probably a poor system for the study of light-regulated gene expression.

The time required to inoculate plant cells or tissues with _Agrobacterium_ and the need to eliminate bacterial contamination, using antibiotic selection, make _Agrobacterium_ gene transfer undesirable for rapid reporter gene assays.

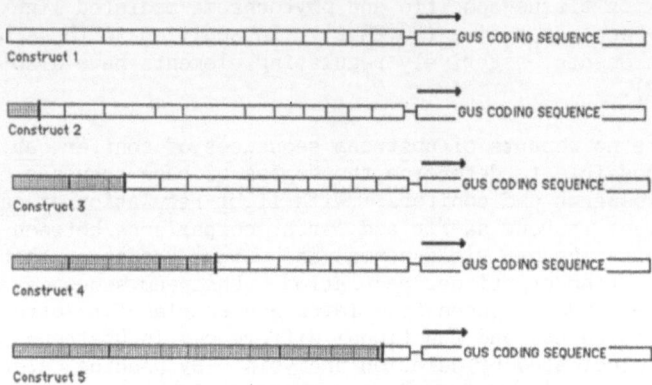

Fig 4. Different constructs demonstrating a series of deletions of upstream sequences isolated from conifer rbcS and cab genes. The top construct shows a full upstream sequence in front of the coding regions for both firefly luciferase (LUC) and beta-glucuronidase (GUS). The three lower constructs show deletions in the upstream sequences from rbcS and cab (shaded area) regulating GUS expression. As an internal control, the GUS gene is covalently linked to a full-length upstream region regulating LUC expression. Assays for LUC expression will determine the relative transformation efficiency between blasts. GUS activity is measured to evaluate the effects a specific deletion has on gene activity.

Biolistic gene transfer has been successfully used for transient and stable incorporation of foreign gene expression in plant tissues. The advantage of this approach is the ability to measure gene expression in a wide variety of cell and tissue types (17). However, gene transfer using particle bombardment results in a random pattern of transformed cells and cell types. Therefore, comparisons of gene activity must also take into consideration the variability inherent in each blast. In addition, histochemical localization of GUS activity (as reported in many laboratories), rather than a fluorometeric enzyme assay, is not a quantitative measure of gene activity.

We intend to use covalently linked GUS and LUC reporter genes to establish relative levels of gene activity, to overcome the problem of random blast patterns. The complete upstream region from a light-regulated gene will placed in front of the LUC gene, and deletions of the same upstream sequence will be inserted in front of the GUS sequence (Fig 4). With LUC activity serving as an internal standard to normalize for variation between blasts, we will use this dual reporter system to determine what effects specific DNA sequences have on the transcription of light-regulated genes in conifers. Deletions of upstream sequences, such as the box II-like and G-box elements in a larch rbcS gene, will be used to determine the effect that these regions have on gene activity.

Reciprocal transfers of genes between monocots and dicots has led to variable results that are dependent on the gene transferred and the plant material used in the transfer process (26). Now that foreign gene transfer and expression have been accomplished in a number of coniferous species, we believe that the strengths of gene transfer systems in conifers lie not in the transfer of heterologous controlling elements but in the ability to map homologous DNA sequences that control gene expression. Because of the variability reported in the many experiments using heterologous constructs, we believe that the underlying mechanisms for gene expression in conifers will be elucidated using detailed analysis of expression studies involving homologous gene constructs. Through the deletion and transfer of conifer rbcS and cab genes into conifer tissues we hope to elucidate the mechanism for light-regulated gene expression in conifers.

REFERENCES

1. Alosi, M.C., Neale, D.B. and Kinlaw, C.S., 1990, Steady State Expression of cab Genes in Douglas-fir is Not Strongly Regulated by Light. Plant Physiol 93:829-832.

2. Castresana, C., Garcia-Luque, I., Alonso, E., Malik, V.S. and Cashmore, A.R., 1988, Both Positive and Negative Regulatory Elements Mediate Expression of a Photoregulated CAB Gene From Nicotiana plumbaginifolia. EMBO J. 7:1929-1936.

3. Dandekar, A.M., Gupta, P.M., Durzan, D.J. and Knauf, V., 1987, Transformation and Foreign Gene Expression in Micropropagated Douglas-fir (Pseudotsuga menziesii). Bio/Tech. 5:587-590.

4. Dean, C., Pichersky, E. and Dunsmuir, P., 1989, Structure Evolution, and Regulation of RbcS Genes in Higher Plants. Ann. Rev. Plant Physiol. 40:415-439.

5. Fluhr, R., Kuhlemeier, C., Nagy, F. and Chua, N-H., 1986, Organ-Specific and Light-Induced Expression of Plant Genes. Science 232:1106-1112.

6. Gallager, T.F. and Ellis, R.J., 1982, Light-Stimulated Transcription of Genes for Two Chloroplast Polypeptides in Isolated Pea Leaf Nuclei. EMBO J. 1:1493-1498.

7.Gilmartin, P.M., Sarokin, L., Memelink, J., and Chua, N-H., 1990, Molecular Light Switches for Plant Genes. Plant Cell 2:369-378.

8. Giuliano, G., Pichersky, E., Malik, V.S., Timko, M.P., Scolnik, P.A. and Cashmore, A.R., 1988, An Evolutionarily Conserved Protein Binding

Sequence Upstream of a Plant Light-Regulated Gene. Proc. Natl. Acad. Sci.(USA) 85:7089-7093.

9. Green, P.J., Yong, M-H., Cuozzo, M., Kano-Murakami, Y. Silverstein, P., and Chua, N-H., 1988, Binding Site Requirements for Pea Nuclear Protein Factor GT-1 Correlate with Sequences Required for Light-Dependent Transcriptional Activation of the RbcS-3A Gene. EMBO J. 13:4035-4044.

10. Gupta, P.K., Dandekar, A.M. and Durzan, D.J., 1988, Somatic Proembryo Formation and Transient Expression of a Luciferase Gene in Douglas-fir and Loblolly Pine Protoplasts. Plant Science 58:85-92.

11. Hutchison, K.W. (personal communication).

12. Hutchison, K.W., Harvie, P.D., Singer, P.B., Brunner, A.F., and Greeenwood, M.S., 1990, Nucleotide Sequence of the Small Subunit of Ribulose-1,5-bisphosphate Carboxylase from the Conifer Larix laricina. Plant Molec. Biol. 14:281-284.

13. Hutchison, K.W., Harvie, P., Singer, P.B. and Greenwood, M.S., 1989, Structure and Expression of Light-regulated Genes During Growth and Maturation of Larix laricina. Abstract. IUFRO Workshop on Molecular Biology of Forest Trees. Riksgransen, Lappland, Sweden. June 11 to 14, 1989.

14. Jansson, S. and Gustafsson, P., 1990, Type I and Type II Genes for the Chlorophyll a/b-binding Protein in the Gymnosperm Pinus sylvestris (Scots pine): cDNA Cloning and Sequence Analysis. Plant Mol. Biol. 14:287-296.

15. Jefferson, R.A., 1987, Assaying Chimeric Genes in Plants: The Gus Gene Fusion System. Plant Molec. Biol. Rep. 5:387-405.

16. Kirby, E.G., Campbell, M.A., and Penchel, R.K., 1989, The Isolation and Culture of Protoplasts of Forest Tree Species. In: Biotechnology in Agriculture and Forestry: vol. 8. Plant Protoplasts and Engineering I. (Y.P.S. Baja ed.). Springer Verlag, Berlin pp. 262-274.

17. Klein, T.M., Gradziel, T., Fromm, M.E. and Sanford, J.C., 1988, Factors Influencing Gene Delivery into Zea mays Cells by High-Velocity Microprojectiles. Bio/Tech. 6:559-563.

18. Kuhlemeier, C., Green, P.J., and Chua, N-H., 1987, Regulation of Gene Expression in Higher Plants. Ann Rev. Plant Physiol. 38:221-257.

19. Manzara, T. and Gruissem, W., 1988, Organization and Expression of the Genes Encoding Ribulose-1,5-Bisphosphate Carboxylase in Higher Plants. Photosyn. Res. 16:117-139.

20. Nagy, F., Fluhr, R., Kuhlemeier, C., Kay, S., Boutry, M., Poulsen, C. and Chua, N-H., 1986, Cis-acting Elements for Selective Expression of Two Photosynthetic Genes in Transgenic Plants. Phil. Trans. R. Soc. Lond. 314:493-500.

21. Ow, D.W., Jacobs, J.D. and Howell, S.H., 1987, Functional Regions of the Cauliflower Mosaic Virus 35S RNA Promoter Determined by Use of the Firefly Luciferase Gene as a Reporter of Promoter Activity. Proc. Natl. Acad. Sci.(USA) 84:4870-4874.

22. Sederoff, R., Stomp, A.M., Chiton, W.C. and Moore, L.W., 1986, Gene Transfer into Loblolly Pine by Agrobacterium tumefaciens. Bio/Tech. 4:647-649.

23. Simpson, J. Van Montagu, M., Herrera-Estrella, L., 1986, Photosynthesis-Asociated Gene Families: Differences in Response to Tissue-Specific and Environmental Factors. Science 233:34-38.

24. Thompson, W.F., Everett, M.,Polans, N.O., Jorgensen, R.A., Palmer, J.D., 1983, Phytochrome Control of RNA Levels in Developing Pea and Mung-bean Leaves. Planta 158:487-500.

25. Tobin, E.M. and Silverthorne, J., 1985, Light Regulation of Gene Expression in Higher Plants. Annu. Rev. Plant Physiol. 36:569-593.

26. Weising, K., Schell, J. and Kahl, G., 1988, Foreign Genes in Plants: Transfer, Structure, Expression and Applications. Annu. Rev. Genet. 22:421-477.

27. Wilson, S.M., Thorpe, T.A. and Moloney, M.M., 1989, PEG-Mediated Expression of GUS and CAT Genes in Protplasts from Embryogenic Suspension Cultures of Picea glauca. Plant Cell Rep. 7:704-707.

28. Yamamoto, N. Kano-Murakami, Y., Matsuoka, M. Ohashi, Y., and Tanaka, Y., 1988, Nucleotide Sequence of a Full Length cDNA Clone of Ribulose Bisphosphate Carboxylase Small Subunit Gene form Green Dark-Grown Pine (Pinus tunbergii) Seedling (Abstract). Nucleic Acids Res. 16:11830.

29. Yamamoto, N. Matsuoka, M., Kano-Murakami, Y., Tanaka, Y. and Ohashi, Y., 1988, Nucleotide Sequence of a Full Length cDNA Clone of Light Harveting Chlorophyll a/b Binding Protein Gene From Green Dark-Grown Pine (Pinus tunbergii) Seedling (Abstract). Nucleic Acid Res. 16:11829.

GERMPLASM PRESERVATION

APPLICATION OF BIOTECHNOLOGY TO PRESERVATION OF FOREST TREE GERMPLASM

M. R. Ahuja

Federal Research Centre for Forestry and Forest Products
Institute of Forest Genetics and Forest Tree breeding
Sieker landstrasse 2, 2070 Grosshansdorf. Germany

ABSTRACT

In view of large-scale forest decline, world-wide, it is necessary to preserve forest tree germplasm for the present and future. Preservation may be carried out in situ (on site; by natural regeneration or by propagation via seed or vegetative parts) or ex situ (away from the original site; in seed or clonal orchards, or by storage of seed, pollen, vegetative parts, or DNA). In this paper approaches to cryopreservation (storage in liquid nitrogen) of seeds and dormant buds of forest tree species are presented.

INTRODUCTION

Forest decline is a global problem. In addition to deforestation, a number of other factors are involved in the forest decline. These include biotic and abiotic agents, and man-made stress factors, so-called anthropogenic factors. The last category includes gaseous, metalic and acid pollutants in the environment. These are believed to be involved in "Waldschaden" (forest damage) and "Waldsterben" (dying forests). Air pollution, in particular, acid rain may have initiated the process, but overall Waldsterben is most likely an end result of a multi-factor, multi-stage phenomenon. Some other factors that may be involved in forest decline are diverse and varied. These might include microbes, pests, drought, soil acidity, radiation, ozone, and perhaps viruses. In other words, it is a cumulative effect of several harmful factors acting synergistically to produce the present state of forest decline (5).

In order to preserve forest tree germplasm, various approaches have been proposed (6, 10, 11, 12, 13). Briefly, these include preservation by in situ and ex situ measures. In situ measures involve preservation on site of the original stand, by multiplication via natural regeneration by seed or vegetative propagation. Ex situ measures involve preservation away from the original site of the stand. I have outlined in Figure 1 some approaches to preservation of forest tree germplasm, emphasizing the ex situ measures under controlled conditions. This category includes storage of seed, pollen, plants, shoots, buds and meristems at low temperatures (3). Shoot cultures have been maintained at low temperatures

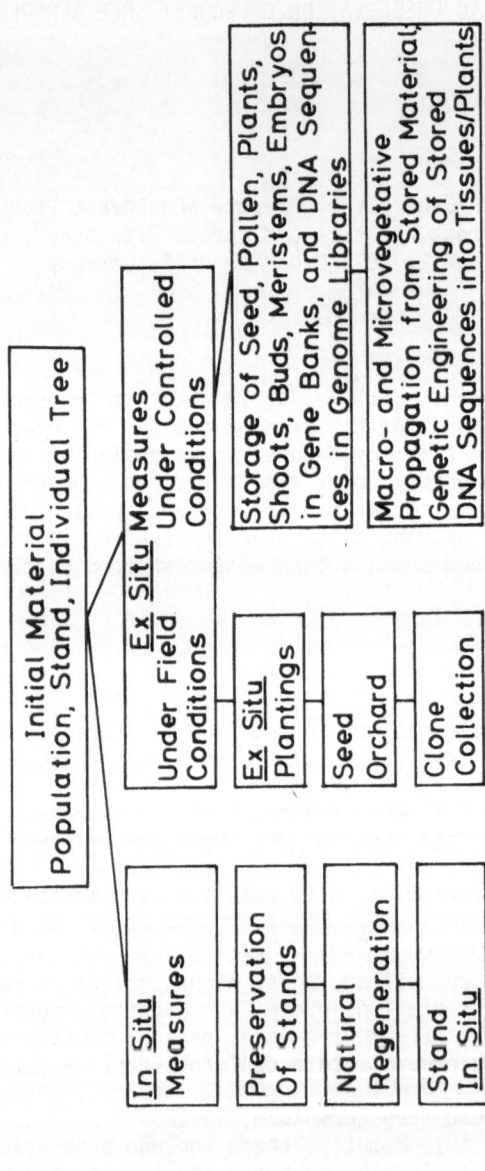

Figure 1. Strategies employed for preservation of forest tree germplasm. The initial material, for example, a population, stand, or an individual tree may be preserved by in situ (on site) or ex situ (away from the site) measures. The application of these measures depends on a tree species, the extent of damage (decline), and the aim and period of storage. The ex situ measures can be carried out under field or controlled conditions. Storage of seed, pollen, or vegetative parts (dormant buds, meristems, embryos, cells) at low temperatures, including cryopreservation (storage in liquid nitrogen) in gene banks, or DNA sequences in genome libraries, offer prospects for short-term or long-term preservation of forest tree germplasm.

(4-10 C) for several years (7, 14). In addition to storage of plant parts in gene banks, DNA sequences may be stored in genome libraries. The temperatures for storage may vary from 0 C to -196 C (liquid nitrogen). After the storage period, plants may be regenerated from the stored vegetative parts by macro- or micropropagation, and from the stored seed by germination. Plants may also be regenerated following genetic engineering of stored DNA sequences into cells/tissues; or the stored DNA may be directly introduced into whole plants (in vivo insertion).

We have explored the potential of seeds and dormant buds for storage in sub-zero temperatures, including 0 C, -5 C, -8 C, -18 C, -80 C and -196 C (1, 3, 4). In this paper, I shall discuss cryopreservation (storage in liquid nitrogen at -196 C) of seeds and dormant buds of forest tree species, and retrieval of plants from them after the storage period.

APPROACHES TO COLD STORAGE OF SEEDS AND BUDS

Seeds from tree species have been routinely stored at temperatures ranging from 4 C to -20 C. At these low temperatures, the seeds can be stored only for a limited period of time. In most cases, the seeds begin to lose viability after a few years. Seeds of beech (Fagus sylvatica), oak (Quercus robur, Q. petraea, and Q. rubra) and silver fir (Abies alba) can be stored at low temperatures only for one to 4 years.

We have stored seeds and dormant buds of forest tree species at low temperatures ranging from 0 C to -80 C for various lengths of time (3, 4, and unpublished data). Here only cryopreservation of seeds and dormant buds will be discussed.

Cryopreservation of material may be carried out by quick-freezing and storage in liquid nitrogen; or by a slow-freezing method. The latter involves slow freezing to approximately -40 C in a Cryostat at a controlled rate (for example, -1 C/minute), and then immersion and storage of material in liquid nitrogen. We have stored seeds and dormant buds of forest tree species for one to 6 days in liquid nitrogen by quick- and slow-freezing methods.

After storage in liquid nitrogen for a desired length of time, the material can be thawed and tested for viability and regeneration potential. The material may be thawed quickly by immersing the ampule in a water bath maintained at 40 C for a few minutes; or thawed slowly in a Cryostat at a controlled rate. Following thawing, seed viability may be tested by the tetrazolium method or germination. The tetrazolium test involves incubating seed in 1% triphenyltetrazolium chloride. Viable seed exhibit reddish couloration. Viability and regeneration potential of buds can be tested by culturing bud explants on a regeneration medium (generally a shoot-induction medium), and rooting of microshoots to yield viable plantlets.

CRYOPRESERVATION OF SEED

Before starting cryopreservation work with the seed (1), we asked the question if seeds from different forest tree species could be effectively stored in liquid nitrogen? A second question related to the size of the seed: whether larger seeds, for example from beech and oak, store equally well in liquid nitrogen as compared to smaller seeds, for example from Norway spruce or aspen? In order to answer these questions, we cryopreserved seeds from 9 forest tree species by a quick-freezing method for one and 6 days. These included European beech (Fagus sylvatica), European oak (Quercus robur, Q. petraea), American oak (Quercus rubra), silver fir (Abies alba), Norway spruce (Picea abies), scots pine (Pinus sylvestris), European larch (Larix decidua), and hybrid aspen (Populus tremula X P. tremuloides). The seeds were thawed after the desired storage period in liquid nitrogen, and their viability tested by germination. Seed germination data showed that following storage in liquid nitrogen for one or 6 days, there was practically no loss of viability in five (1) of 9 forest tree species , as compared to controls stored at 0 C.

Seeds from four large-seeded beech and oak species (Fagus sylvatica, Quercus robur, Q. petraea, Q. rubra) stored in liquid nitrogen for one or 6 days did not germinate following thawing. Tetrazolium test showed that cotyledons in beech seeds, although slightly red-coloured, were glossy and glistening in appearance, as compared to controls, and were probably damaged by the freezing-thawing procedures. However, the embryo showed dark red colouration, as in the controls, and were seemingly not damaged. On the other hand, both cotyledons and embryos showed no colouration in oak seeds stored by a quick-freeze in liquid nitrogen. Slow-freezing followed by storage in liquid nitrogen did not appreciably improve the storage quality oak seeds and beech cotyledons. Even slow-thawing followed by slow-freezing and storage in liquid nitrogen were not any more effective in cryopreservation of oak and beech seeds.

Since beech embryos exhibited red colouration with tetrazolium test, following freezing and thawing, we cultured beech embryos on a medium that would induce shoots on a normal non-cryopreserved embryo. Although cryopreserved beech embryos showed growth and differentiation, indicating that they were viable, there was high degree of bacterial contamination in these cultures, as compared to controls. This was a consistent observation with cryopreserved beech embryos. It would appear that freezing-thawing of beech seeds releases bacteria from presumably damaged/dead cotyledonary and/or embryonal cells under in vitro conditions. Bacterial contamination in beech embryo-cotyledon cultures is not uncommon, or for that matter in tissue cultures of other forest tree species. However, the endogenous bacteria are generally released in the cultures of tree species, when the cultures are stressed (for example, not regularly subculturing after 3-4 weeks, or changes in temperature, etc). In the circumstances, freezing-thawing of beech seeds may provide the kind of stress that would release a high titre of endogenous bacteria from the cultured cells.

We are continuing our work on beech and oak to improve on the existing cryopreservation technology so that the germplasm from large-seeded forest tree species may be effectively stored in liquid nitrogen.

CRYOPRESERVATION OF DORMANT BUDS

Production of dormant buds in the temperate hardwoods is a mechanism for survival and overcoming cold-freezing winters. We asked the question if cryopreservation of dormant buds would prolong the dormancy and provide a method for their long-term storage? We employed aspen as a model system to test this hypothesis. Thirteen different clones derived from European aspen, Populus tremula (5 clones), quaking aspen, P. tremuloides (1 clone), and hybrid aspen, P. tremula X P. tremuloides (7 clones) were employed in this study. Dormant buds from these genotypes were stored in liquid nitrogen for one day by a quick-freezing method. The buds were thawed and bud explants were cultured on a shoot induction medium (2). Bud explants from all 13 clones became brown within 48 hours in culture and exhibited no further growth and differentiation.They were essentially all killed by the quick-freezing cryopreservation method. Quick-thawing or slow-thawing following quick-freezing cryopreservation method did not make any difference: bud explants were dead following both retrieval methods.

After unsuccessful attempts with quick-freezing-thawing with dormant buds, we cryopreserved them by a slow-freezing method. In this protocol, dormant buds were first cooled to -40 C in a Cryostat (at -1 C/minute) and then they were immersed in liquid nitrogen abd stored for 24 hours. Following thawing, bud explants from 13 aspen clones were cultured on a shoot-induction medium. In contrast to buds cryopreserved by a quick-freezing method, bud explants derived from slow-freezing of dormant buds, exhibited growth and differentiation of microshoots. The number of micro-shoots per explant, as measured after 8-week growth period, were dependent on the genotype of the clone (data not shown). The microshoots were rooted on a rooting medium and plantlets were regenrated in each clone. Although there were some differences between growth potential of bud explants from cryopreserved material as compared to controls (Ahuja unpublished), the slow-freezing method clearly shows potential for a long-term cryopreservation of dormant buds of aspen.

PRESERVATION OF FOREST TREE GERMPLASM

In view of world-wide forest decline, caused by deforestation, air pollution and other harmful agents, there is an urgent need to preserve forest tree germplasm. Because of rapid population growth in the world, deforestation, particularly in the tropical and sub-tropical countries, has been carried out indiscriminately for making land available for industrial and agricultural use and for human settlements. On the other hand, Waldsterben (dying forests) in Europe and North America is probably caused by air pollution (acid rain) and other harmful agents in the environment. The loss of forests by deforestation and Waldsterben are ostensibly leading to loss of valuable gene resources. If the forest decline continues at the present pace, there is danger that some of the forest tree species may become extinct in a not too distant future. There-fore, we must act now to preserve the existing forests and plant more trees to maintain the biological and ecological balance essential for the continued existence of life on the Planet Earth.

In the Federal Republic of Germany, Waldsterben is a serious problem. Substantial data are available on the extent of forest damage in commer-cially important forest tree species since 1983. These studies carried out by the Federal Ministry of Food, Agriculture and Forestry (8, 9) have revealed that forest tree species show differential damage. For example, in 1986, 53.7 percent of the forest stands were damaged, and the magnitude of forest damage in different species was as follows: 82.9 percent in silver fir (Abies alba), 54.1 percent in Norway spruce (Picea abies), 54.0 percent in scots pine (Pinus sylvestris), 60.1 percent in oak (Quercus robur, Q. petraea), and 60.7 percent in beech (Fagus sylvatica). The pattern was somewhat different in 1987 with 52.3 percent (-1.4 percent from 1986) of the forest stands showing damage. The extent of damage in 1987 in the above species was as follows: 79.0 percent in silver fir (-3.9 percent from 1986), 48.9 percent in Norway spruce (-5.2 percent from 1986), 49.6 percent in scots pine (-4.4 percent from 1986), 65.7 in oak (+5.6 percent from 1986), and 64.5 percent in beech (+3.8 percent from 1986). These data show that while conifers are generally exhibiting decreased damage in 1987, the deciduous hardwoods, on the other hand, are showing increased damage. The age of a tree is also important in determining the extent of damage. As compared to trees that are below 60 years of age, those older than 60 years of age, in general, exhibit 2 to 2.5 times more damage. In silver fir, almost 94 percent of the stands with trees older than 60 years of age are damaged

or dying. Ostensibly this kind of damage in older trees would lead to
a loss of reproductive populations, and eventually to extinction of a
species. In the other important forest tree species, the damage in the
older tree populations ranges from 67 to 81 percent (8, 9), which is
also extensive.

As the present situation demands, we have started a programme on
gene conservation in forest tree species in Germany. As outlined in
Figure 1, both in situ (on site) and ex situ (away from the site) measures
are important for the preservation of forest tree germplasm. In the present
paper, we have highlighted the ex situ measures, under controlled conditions,
for the cryopreservation or storage in liquid nitrogen (-196 C) of seeds
and dormant buds of forest tree species. Conservation of seed is important
for maintenance of genetic variability of tree species. Storage of vegetative parts, for example dormant buds, meristems, or somatic embryos, on the
other hand, would be valuable for the preservation of elite genotypes
from a forest tree population. Cryopreservation of dormant buds is
relatively simple as compared to meristems or cells, which require in
vitro culture along with cryoprotectants, and further preparation before
storage in liquid nitrogen. Although we have carried out cryopreservation
of dormant buds successfully in aspen, these studies could be extended
to dormant buds of other forest tree species. In this regard, we have
tested the cryopreservation potential of dormant buds of beech and oak.
But so far we have not found the right conditions for storage of dormant
buds, or for that matter the seeds from these species in liquid nitrogen.
It appears that beech and oak germplasm may be sensitive to freezing-
thawing injury under the experimental conditions. Therefore, optimal
conditions for cold-storage of economically important forest tree species
need to be investigated.

REFERENCES

1. Ahuja, M. R., 1986, Storage of forest tree germplasm in liquid nitrogen
    (-196 C). Silvae Genet. 35: 249-251.
2. Ahuja, M. R., 1986, Aspen. In: Handbook of Plant Cell Culture, D. A.
    Evans W. R. Sharp and P. J. Ammirato (Eds.), Macmillan Publishing
    Company, New York, pp. 626-651.
3. Ahuja, M. R., 1988, Differential growth response of aspen clones
    stored at sub-zero temperatures. In: Somatic Cell Genetics of
    Woody Plants, M. R. Ahuja (Ed.), Kluwer Academic Publishers,
    Dordrecht, pp. 173-180.
4. Ahuja, M. R., 1989, Storage of forest tree germplasm at sub-zero
    temperatures. In: Application of Biotechnology in Forestry and
    Agriculture, V. Dhawan (Ed.), Plenum Press, New York, pp215-228.
5. Ahuja, M. R., 1990, Air pollution and mammal resistant genes. In:
    Genetics and Molecular Biology of Interactions between Harmful
    Agents and Trees, H. S. MaNabb (Ed.), Springer Verlag, Berlin,
    New York (in press).
6. Ahuja, M. R. and Muhs, H. J., 1989, Biotechnologische Verfahren zur
    besseren Evaluierung, Erhaltung und Nutzung von forstliche Gen-
    ressourcen. Berichte über Landwirtschaft, 201 Sonderheft.
    Verlag Paul Parey, Hamburg. 414-422.
7. Aitkin-Christie, J. and Singh, A. P., 1987, Cold storage of tissue
    cultures. In: Cell and Tissue Culture in Forestry, Volume 1,
    J. M. Bonga and D. J. Durzan (Eds.), Martinus Nijhoff Publishers,
    Dordrecht, pp. 285-304.
8. Berichte des bundesministers für Ernährung, Landwirtschaft und Forsten,
    1986, Waldzustandsbericht - Ergebnisse der Waldschadenserhebung.
    Landwirtschaftsverlag, Münster-Hiltrup.

9. Berichte des Bundesministers für Ernährung, Landwirtschaft und Forsten, 1987, Waldzustandsbericht - Ergebnisse der Waldschadenserhebung. Landwirtschaftsverlag, Münster-Hiltrup.

10. Krugman, S. L., 1985, Policies, strategies, and means of genetic conservation in forestry. In: Plant Genetic Resources, C. W. Yeatman, D. Kafton and G. Wilkes (Eds.), AAAAS Symposium, Westview Press, Boulder, Colorado, pp. 71-78.

11. Ledig, F. T., 1988, The conservation of diversity in forest trees. Bioscience 38: 471-479.

12. Melchior, G. H., Muhs, H. J. and Stephan, B. R., 1986, Tactics for the conservation of forest gene resources in the Federal Republic of Germany. For. Eco. Manag. 17: 73-81.

13. Millar, C. I., 1991, Conservation of germplasm in forest trees. In: Clonal Forestry: Genetics, Biotechnology and Application, M. R. Ahuja and W. J. Libby (Eds.), Springer Verlag (in press).

14. Preil, W. and Hoffmann, M., 1985, In vitro storage in Chrysanthemum breeding and propagation. In: In Vitro Techniques - Propagation and Long Term Storage, A. Schäfer-Menuhr (Ed.), Martinus Nijhoff Publishers, Dordrecht, pp. 161-165.

INFLUENCES OF SUBCULTURING PERIOD AND DIFFERENT CULTURE MEDIA ON COLD

STORAGE MAINTENANCE OF POPULUS ALBA X P. GRANDIDENTATA PLANTLETS

Young Woo Chun and Richard B. Hall

College of Forestry, Kookmin Univ., Seoul, Korea
Department of Forestry, Iowa State Univ., Ames, IA, USA

ABSTRACT

Continuous subculture is labor intensive and requires extensive culture space. Cold storage of in vitro cultured Populus plantlets could serve to alleviate the maintenance requirements of an established micropropagation system, as well as provide methods to facilitate germplasm conservation. In vitro cultured hybrid poplar, Populus alba X P. grandidentata could be stored at $4^{o}C$ air temperature in darkness for 24 months and still recover suitable multiplication potential. Subculturing period preceding cold storage, plantlet condition, and culturing medium all had an important influence on survival at $4^{o}C$ in darkness. A one-month subculturing period preceding cold storage was better than 0-month or 2-month subculturing period preceding cold storage. Shoot proliferation medium was better than rooting medium for long term cold storage. After 24 months storage, a 70% survival percentage was obtained with plantlets possessing 4-6 axillary branching shoots that were subcultured on shoot proliferation medium for one month preceding cold storage.

INTRODUCTION

Continuous subculturing is the initial requirement for establishing a micropropagation system. Maintenance of established cultures can require substantial space, chemicals, and labor. To minimize time and expenses for routine culture maintenance, preservation of plant material in a state of growth suspension has been attempted by various methods (16). Such growth suspension methods in plant tissue culture can be divided into three categories: non-frozen cold storage, alteration of the basic culture media and addition of growth retardants. Among these tissue culture storage methods, temperature reduction has been demonstrated to be very effective (17). There are many species that can be conserved by temperature reduction, especially with non-frozen, low temperature storage methods. These species include: Malus sp. (11), Musa sp. (2), Prunus sp. (12), Pinus radiata (1), Trifolium repens (3), Allium sativum (9) and Populus sp. (5).

The hybrid poplar, Populus alba X P. grandidentata, has attracted considerable attention from several workers because of its high productivity in naturally occurring stands and in a U.S. Forest Service trial plantation (10). Nevertheless, this hybrid poplar has been limited in its potential for planting because it is difficult to root by regular green-house or nursery techniques. For effective clonal propagation through in vitro culture of this hybrid poplar, we previously have established techniques for axillary bud culture and adventitious bud initiation from cultured explants (4, 6), and for survival and early growth of tissue cultured plantlets in soil (5). We have experineced some difficulties in the maintenance of this established micropropagation system, however, continuous subculture requires intensive labor and considerable expense. To alleviate these problesms, we attempted to determine the optimum medium and plantlet condition for non-frozen cold storage. In this paper, we report the effects of previous subculturing period and plantlet condition on maintenace of Populus alba X P. grandidentata plantlets in cold storage.

MATERIALS AND METHODS

The Crandon clone of Populus alba X P. grandidentata was used for this experiment. Procedures for micropropagation of this hybrid poplar are outlined by Chun and Hall (5) and Chun et al. (7). Axillary shoots were produced initially by serial subculture on MS (Murashige and Skoog, 14) basic salts supplemented with thiamine HCL 0.1 mg/l, nicotinic acid 0.5 mg/l, pyridoxine HCl 0.5 mg/l, glycine 2.0 mg/l, myo-inositol 100 mg/l sucrose 30 g/l, Bacto-agar 6 g/l, and benzyladenine (BA) 0.3 mg/l at pH 5.7.

Four sources of plant material were tested as follows:
1. Shootlets with 4-6 axillary branching shoots that were subcultured on MS media containing 0.3 mg/l BA for one month preceding cold storage (Figure 1).
2. Shootlets with 5-7 axillary branching shoots that were subcultured on MS media containing 0.3 mg/l BA for two months previous to cold storage (Figure 2).
3. Shoot segments, approximately 1 cm in length with one or two leaves, that were explanted on MS media containing 0.3 mg/l BA just prior to cold storage. These shoot segments were derived from plantlets grown on MS medium containing 0.2 mg/l indole-3-acetic acid (IBA) (Figure 3).
4. Plantlets that were subcultured on MS media containing 0.2 mg/l IBA for one month before cold storage (Figure 4).

For storing cultures at low temperature, each plantlet or shoot segment was transferred to a 15 X 200 mm glass culture tube which contained 15 ml medium. The culture tubes were covered with plastic caps and sealed with Parafilm to prevent excessive drying. Eighty tubes for each of the four culture treatments were placed in a cold room, at 4°C in the dark. Every 3 months, 10 culture tubes were removed for each different culture treatment. The percentage of culture tubes with viable shoots were determined 2 weeks following removal from cold storage. To determine proliferation potential after storage, these shoots were dissected and 10 of these dissected shoots were subcultured on the same MS medium with 0.2 mg/l BA under normal culture conditions which were controlled to maintain a daytime temperature of 25-28°C with a 16 h. photoperiod and a photosynthetically active radiation level of 50-60 μE/m^2/s from cool-white fluorescent tubes. After an additional one month culture, the number of developed shoots were counted.

RESULTS AND DISCUSSION

The effects of previous subculturing period and medium on survival and shoot proliferation of the Populus alba X P. grandidentata Crandon clone after cold storage are summarized in Table 1. During the first 12 months of cold storage, plantlets that were subcultured on rooting medium

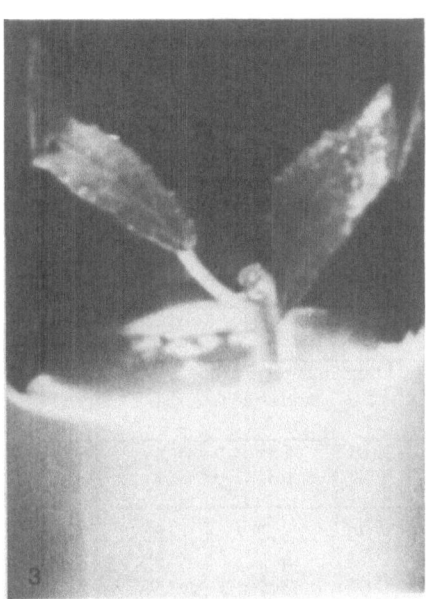 

Figure 1. Shootlets with 4-6 axillary shoots on shoot proliferation medium for one month subculturing period preceding cold storage.

Figure 2. Shootlets with 5-7 axillary shoots on shoot proliferation medium for two months subculturing period before cold storage.

Figure 3. Shoot segment on shoot proliferation medium just prior to cold storage.

Figure 4. Plantlets with 1-2 shoots on rooting medium for one month subculturing period before cold storage.

(MS+IBA 0.2 mg/l) for 1 month before cold storage and shoot segments with
one or two leaves that were placed on shoot proliferation medium (MS+BA
0.3 mg/l) immediately prior to cold storage exhibited similar survival
tendencies. The survival of both cultures decreased with the increased
duration of cold storage. They showed only 30% survival after 12 months
cold storage. With shootlets subcultured on shoot proliferation medium
for 2 months prior to cold storage, 50% of the cultures survived after 12
months cold storage; however, these cultures showed a rapid decrease in
their viability and vigor as cold storage periods increased beyond 12
months. The shootlets that were subcultured on shoot proliferation medium
for 1 month before cold storage obtained the highest survival percentage
and also did not change in their viability and vigor until after 12 months
of cold storage. The survival percentage and multiplication potential of
these cultures began to decrease after 15 months cold storage with the
survival percentage of these shootlets dropping to 70%. Among the
cultures that were subcultured on shoot proliferation medium for varying
periods before cold storage, a 1 month subculture period prior to cold
storage generated the highest survival percentage and multiplication
potential. The previous subculturing period is, therefore, a determining
factor in culture survival during cold storage, and in subsequent culture
proliferation potential. Our results with _Populus_ are in accordance with
those with _Prunus_ where Marino et al. (12) reported that of three
different previous subculturing period, 14 days of previous subculturing
period resulted in significantly better survival and regenerative capacity
than did 0 or 7 days of previous subculturing period. Lundergran and
Janick (11) also demonstrated that with a previous subculturing period,
shoot tips of apple (_Malus domestica_) could be stored for one year.

The shootlets with axillary branching shoots that were subcultured on
shoot proliferation medium for one month preceding cold storage were not
affected in their potential for rapid shoot multiplication with storage
periods up to 15 months. After 15 months of cold storage, however, the
average shoot proliferation rate decreased from 6.7 to 4.4 shoots. In

Table 1. Effects of previous subculturing period and culture medium on
survival and shoot proliferation of _Populus_ _alba_ X _P._
grandidentata plantlets

Storage period (month)	Shoot proliferation med.[a]						Rooting med.[b]	
	Previous subculturing period							
	0 month		1 month		2 months		1 month	
	Surv.[c] %	Shoot no.	Surv. %	Shoot no.	Surv. %	Shoot no.	Surv. %	Shoot no.
3	80	1.4	100	7.3	90	5.3	80	3.6
6	80	1.6	100	6.5	70	4.2	50	3.2
9	60	1.2	100	6.8	60	3.8	50	3.2
12	30	1.3	100	6.4	50	3.8	10	3.1
15	20	1.5	80	6.6	20	4.2	10	3.1
18	0		90	4.6	0		0	
21	0		80	4.4	0		0	
24	0		70	4.3	0		0	
S.E[d]		0.13		0.26		0.20		0.15
Mean		1.40		5.86		4.26		3.34

[a]Shoot proliferation med. = MS+BA 0.3 mg/l.
[b]Rooting med. = MS+IBA 0.2 mg/l.
[c]Surv. % = Survival percentage.
[d]S.E. = Standard error of means.

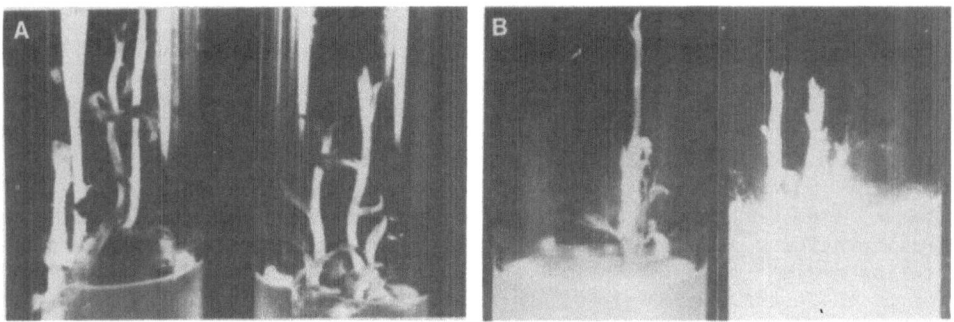

Figure 5.   After 12 months (A), and 24 months (B) cold storage of treatment 1.

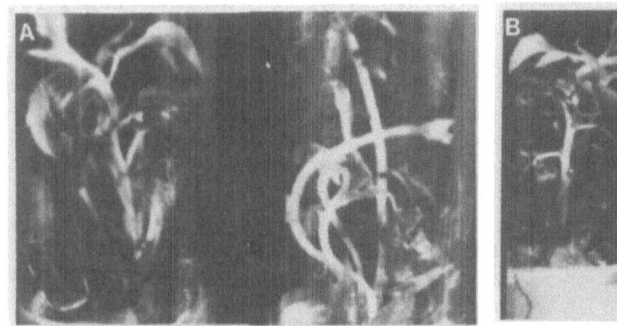 

Figure 6.   After nine months (A), and 12 months (B) cold storage of treatment 2.

 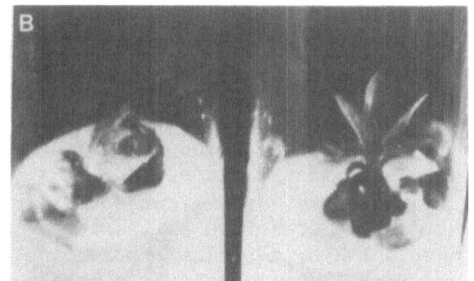

Figure 7.   After six months (A), and 12 months (B) cold storage of treatment 3.

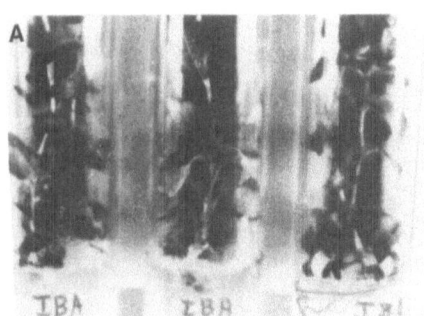 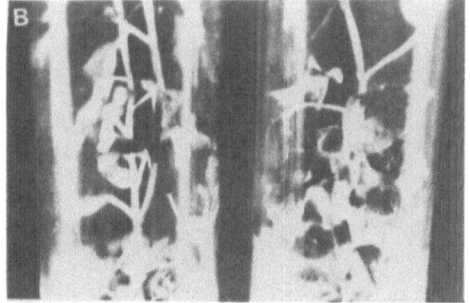

Figure 8.   After three months (A), and six months (B) cold storage of treatment 4.

contrast, the rate of shoot proliferation shootlets subcultured for 2 months on shoot proliferation medium and plantlets subcultured for 1 month on rooting medium appeared to remain consistent through all storage periods. Cultures with one or two months previous subculture on shoot proliferation medium produced more multiple shoots than did those with previous subculture on rooting medium. Similar responses have been reported in an earlier study in which the number of axillary shoots induced from axillary branching shoots that were subcultured on shoot proliferation medium was more than that from axillary buds subcultured on rooting medium (5). This result is perhaps attributable to preconditioning of the plantlets that were previously subcultured on a cytokinin medium before cold storage. The shoot segments that were explanted on shoot proliferation medium just before cold storage produced only a few axillary branching shoots after cold storage because these shoot segments were also derived from plantlets grown on rooting medium.

In relation to other aspects of cold-stored plantlets, the shootlets which were subcultured for 1 month on shoot proliferation medium before cold storage had small, green leaves and healthy stems until 9 months of cold storage. After 12 months of cold storage, the shootlets had developed elongated, thin stems and small, yellowish leaves (Figure 5). The lower leaves on stems in these cultures became necrotic after 12 months of cold storage. From this time, etiolated stem with only a few yellowish leaves were observed. After 24 months of cold storage, these etiolated and elongated stems turned green when returned to normal culture conditions. Stem and leaf necrosis started to occur after 3 to 6 months of cold storage on plantlets in three other treatments (Figure 6-8). This damage may be due in part to an accumulation of oxidative substances in the medium due to defoliation and topping of stored plantlets and also to the limited availability of nutrients for the plantlets which were subcultured for two months before cold storage (8).

Our cold storage test was conducted in complete darkness because there is evidence that complete darkness is more suitable for successful cold storage of in vitro cultures of Prunus rootstocks than 16 h. photoperiod with high light intensity (12). However, Preil and Hoffmann (15) have demonstrated that low level illumination (10-50 lux) is much more effective than high level illumination (500 lux) for successful longterm storage of Chrysanthemum at 2-3°C. In contrast, Mix (13) reported that the potato varieties could be stored with high illumination during a 16 h. photoperiod for 3 years.

This study demonstrates that in vitro cultured hybrid poplar plantlets can be preserved at low temperatures. Because clones of this important hybrid poplar are routinely grown in greenhouses as a source of germplasm, this relatively simple and inexpensive cold storage method could reduce the costly greenhouse maintenance of these hybrid poplar stocks. The technique could also facilitate germplasm shipments, as well as alleviate the maintenance problems of established micropropagation systems.

ACKNOWLEDGEMENTS

This work was supported by Project 2210, a contributing project to North Central Regional Research Project NC-99, Regional Tree Improvement. The authors wish to thank Dr. Ned B. Klopfenstein at the Forestry Sciences Laboratory, USDA Forest Service, East Campus, University of Nebraska for his critical reading of the manuscript.

REFERENCES

1. Aitken-Christie, J., and Gleed, J. A., 1984, Uses for micropropagat-

ions of juvenile radiata pine in New Zealand. Pages 47-57 in Proc. International Symp. Recent Advances in Forest Biotechnology, Traverse City, MI.

2. Banerjee, N., and de Langhe, E., 1985, A tissue culture technique for rapid clonal propagation and storage under minimal growth conditions of Musa (banana and plantain). Plant Cell Reports 4:351-354.

3. Bhojwani, S. S., 1981, A tissue culture method for propagation and low temperature storage of Trifolium repens genotypes. Physiol. Plant. 52:187-190.

4. Chun, Y. W., 1987, Biotechnology applications of Populus micropropagation. Ph.D. Dissertation. Iowa State Univ., Ames, Iowa 149 pp.

5. Chun, Y. W., and Hall, R. B., 1984, Survival and early growth of Populus alba X P. grandidentata in vitro cultured plantlets in soil. J. Korean For. Soc. 66:1-7.

6. Chun, Y. W., and Hall, R. B., 1986, Low temperature storage of in vitro cultured hybrid poplar, Populus alba X P. grandidentata plantlets. Page 13 in D. A. Sommers, B. G. Gengenbach, D. D. Biesboer, W. P. Hackett, and G. E. Green, eds. Proc. Sixth International Congress of Plant Tissue and Cell Culture. Univ. of Minn., Minneapolis, MN.

7. Chun, Y. W., Hall, R. B., and Stephens, L. C., 1986, Influences of medium consistency and shoot density on in vitro proliferation of Populus alba X P. grandidentata. Plant Cell Tissue Organ Culture 5: 179-185.

8. Druart, P., 1985, In vitro germplasm preservation technique for fruit trees. Pages 149-154 in A. Schafer-Menuhr, ed. In vitro techniques: Propagation and long term storage. Martinus Nijohoff/Dr W. Junk Publishers, Boston.

9. El-Gizawy, A. M., and Ford-Lloyd, B. V., 1987, An in vitro method for the conservation and storage of garlic (Allium sativum) germplasm. Plant Cell Tissue Organ Culture 9:147-150.

10. Hall, R. B., Hilton, G. D., and Maynard, C. A., 1982, Construction lumber from hybrid aspen plantations in the Central States. J. of Forestry 80:291-294.

11. Lundergan, C., and Janick, J., 1979, Low temperature storage of in vitro apple shoots. HortScience 14:514.

12. Marino, G., Posati, P., and Sagrati, F., 1985, Storage of in vitro culture of Prunus rootstocks. Plant Cell Tissue Organ Culture 5: 73-78.

13. Mix, G., 1985, Preservation of old potato varieties. Pages 149-154 in A. Schafer-Menuhr, ed. In vitro techniques: Propagation and long term storage. Martinus Nijohoff/Dr W. Junk Publishers, Boston.

14. Murashige, T., and Skoog, F., 1962, A revised medium for rapid growth and bio assays with tobacco tissue cultures. Physiol. Plant. 15:473-497.

15. Preil, W., and Hoffmann, M., 1985, In vitro storage in chrysanthemum breeding and propagaion. Pages 161-166 in A. Schafer-Menuhr, ed. In vitro techniques: Propagation and long term storage. Martinus Nijohoff/Dr W. Junk Publishers, Boston.

16. Wilkins, C. P., and Dodds, J. H., 1983, The application of tissue culture techniques to plant genetic conservation. Sci. Prog. Oxf. 68:259-284.

17. Withers, L. A., 1985, Long-term storage of in vitro culture. Pages 137-148 in A. Schafer-Menuhr, ed. In vitro techniques: Propagation and long term storage. Martinus Nijohoff/Dr W. Junk Publishers, Boston.

THE PROBLEM OF BIOTECHNOLOGICAL CONSTIPATION

W. J. Libby

Dept. of Forestry & Resource Management
Univ. California,
Berkeley, 94720 USA

ABSTRACT

This is not a technical paper. Rather, it is the
Banquet Address, covering two problems. These are:
1. During the missionary phase of biotechnology
development, the promises made, or implied, or perhaps
perceived or assumed by the listeners, have been fulfilled
in only a few cases.
2. The real or assumed diversion of funds and
facilities to biotechnology are perceived as having slowed
the flow of information, techniques, activities, and
advanced plant material from more conventional disciplines.
Some details of these problems are offered, and some
solutions offered. Finally, a summary of the hour-long
discussion following the address is included.

BIOTECHNOLOGICAL CONSTIPATION

I first heard the term "biotechnological constipation"
during a forest genetics meeting this summer. It was
contained in a strong complaint about the delivery of
research results, techniques and improved plant stock to the
field. Being curious about this, and a little worried, I
have discussed it with many colleagues and have been alert
for it at other gatherings. Given this modest sample, it
now seems to me likely that these opinions are common and
widespread.

In short, foresters and managers charged with the
everyday practice of forestry are wondering and even
complaining about biotechnology on two main counts:
1. Most can thus far identify no positive impact from
biotechnology on their operations. Some are aware of
specific promises that were either made or perceived, but
many are simply curious about whether and when something
useful is going to come out.

2. Many others, however, cite real or imagined negative impacts that they associate with the attention recently paid to forest biotechnology. These impacts mostly take the form of perceived reductions in information production and in information transfer on such topics as the genetic architecture of tree species and of associated species; in particular, on the nature and magnitude of genotype-x-environment interactions, which are not well served by isozymes, RFLPs, etc. This concern also extends to a perceived reduction in attention given to areas such as pest management, etc. They suspect, often with some evidence, that attention to and support of these fields, activities and experiments has been reduced, in both relative and absolute terms, and furthermore, that at least some of those reductions were executed by managers in search of funds and facilities for highly promising ventures in biotechnology. Some suspect that a stampede to the biotechnology bandwagon has occurred.

There are possibilites of anything from retrenchment to reprisal with respect to funding for much of forest science, and particularly for biotechnology. For biotechnology to flourish, it needs if anything greater funding. It also needs other healthy forest-science disciplines to put its products in perspective and to use.

With respect to point (1), above, not all failures of delivery are due to the producers. You may have noticed the poster on a possible clonal response to a short supply of appropriate seeds following California's big fires of 1987 and 1988. Expanding from the inadequate numbers of appropriate seedlings available to the numbers of propagules needed for emergency reforestation following these devastating fires isn't very high tech. It is well-researched technology. But responsible decision-makers either weren't aware of the availability of this relatively simple technology, or they were too timid to commit to its large-scale application to solve a short-term emergency need. The constipation here was not in the availability of appropriate technology, but in its aggressive application. Interestingly, much of this technology was developed and paid for by the research arm of the U. S. Forest Service, the very agency whose management arm needed it but failed to use it.

In a plenary talk at the 1989 National Meeting of the Society of American Foresters in Spokane, Dan Keathley made the following points, among others. He indicated that forest biotechnology was built on a funding base established by forest genetics and tree improvement, which had histories of being content with low levels of funding. Such funding levels, even if wholly coopted for biotechnical purposes, are seriously inadequate for such purposes. Unlike more classical forest genetics and tree improvement, biotechnology has a need for expensive equipment, which often needs to be updated or replaced as new techniques are developed, and much of biotechnology involves labor-intensive activities. Recognizing the developing impatience among clients and supporters of forest biotechnology, Dan ended his address with a nice analogy. He likened this new field of endeavor to a garden, where most of the plants are

324

now well rooted, but few have begun to produce fruit. They
still need water, fertilizer, and perhaps some weeding.

There is another factor causing problems for forest
biotechnology, both with respect to overall program
effectiveness and to funding efficiency. That is, as the
promise of biotechnology was recognized, most forest
research organizations wanted a piece of the action, and
many bought in at a token level. By that I mean that many
universities, research labs, etc. hired only three, or two,
or even just one scientist well trained in the concepts and
techniques of molecular biology and related biotechnical
disciplines. This produced localized problems with respect
to scientific and intellectual ferment in these small semi-
isolated groups, and problems of availability of expensive
and sometimes exotic equipment and supplies. A logical
solution is to have a modest number of forest biotechnology
centers, each with 12-15 or so scientists, adequate support
staff, and a full range of appropriate equipment and
supplies. There's probably not much that we can do about
that problem at this meeting.

But there are a couple of strategies that I will
suggest, and that I hope we can discuss tonight.

One, perhaps to a degree cosmetic, is to get some trees
out where people can see them. Not all of you can do that,
but some of you can. For example, I'm aware of some forest
biotechnology experiments that generate shoots from culture,
and even go on to get roots on these shoots, and then these
plantlets are converted to data and discarded. It might be
wise to take the extra intensive-care-ward steps necessary
to allow them to survive in a normal plantation environment,
and then to deliver them to a colleague who can put them in
a seedling-plantling test or demonstration plantation. Just
as Carl Syrach-Larsen used "tree shows" during the
missionary phases of forest genetics to demonstrate examples
of genetic variation, such plantings could not only
demonstrate your developing capabilities to deliver real
plants, but perhaps more important, they may serve to dispel
some apprehension that you are about to loose some
genetically engineered dendrological monsters on our
innocent forests.

It is clear that, in the realm of forest biology in
many countries, the late 1980s were the years of commitment
to biotechnology. This may or may not continue into the
1990s. If one listens at meetings, or reads pending
legislation, or watches the activities of various public-
interest environmental groups, or tracks the frequency of
various legal injunctions and initiatives, the term
"biodiversity" seems to be eclipsing "biotechnology".
Perhaps biotechnology can keep a larger part of center stage
if it responds to the needs and concerns associated with the
as-yet ill-defined but nevertheless generally supported
concepts of biodiversity. Some suggestions: (a) Genetic
conservation, using techniques ranging from low-temperature
tissue-culture to cryopreservation to DNA libraries. (b)
Population characterization, using such techniques as mtDNA
polymorphisms, microcalorimetry, and other high-tech methods
of describing both the structural and functional genetic

architecture of populations of interest. (c) Surely, we can think of several others.

Finally, there is a danger that, as more scientists compete for constant or even reduced funds, the battle lines will deepen divisions among us. Molecular biologists are, I think deservedly, viewed as being highly competitive and even secretive. It is thought that, sometimes, you even fail to share information and discoveries in ways that would allow the public and private funds that you are clamoring for to be used more effectively. We are not used to such behavior in most of the more traditional fields of forest biology. To maintain and increase your levels of support, let me suggest that you need to make common cause, not only with each other, but also with other fields of forest science, particularly with those that are important to you, and perhaps also with those that are suspected of having been injured by a diversion of funds from them to you.

DISCUSSION OF BIOTECHNOLOGICAL CONSTIPATION

Following the main address, a discussion lasting about an hour was organized under five topics. R. Ahuja led and recorded the discussion, while W. Libby took notes. Summaries of these are presented below, along with a listing of those participating in each topic's discussion.

Does This Restiveness Occur Outside Of North America?

E. Birnbaum, A. Diner, R. Dinus, M. Greenwood, H. McNabb, D. Smith, K. von Weissenberg

There are signs of impatience with biotechnology evident in other regions as well. The demands for quick results are unreasonable, and indicate a failure to understand the time that it takes to accomplish many of these tasks. Part of this is probably due to some of us making promises that were too optimistic. Many of our grant proposals contrive objectives within "easy reach" that are in fact unlikely to be reached easily or at all. But, the way the game is currently played is to oversell...is there a better way? (None was offered.) It is important to explain to our larger public the nature of our work, and why much of it takes many years. Taking some of our time to present talks and demonstrations (at, for example, high schools) is in the long run a good use of what in the short run is precious time. The research with poplar is a good example of a team approach, where scientists from other disciplines are working with the biotechnologists, to their mutual benefit and support. Too many of us have a slave mentality, and when things get tough, our response is "Oh Massuh, please don't sell me!" There is, in today's scientific world, a greater demand for accounting by the marketplace. There are three possible responses to this new wave: (1) hold your breath and dive under; (2) attempt to walk on water; and (3) ride the wave.

Could We, Or Should We, Give More Attention To Demonstration & Test Plantings?

R. Dinus, A. Diner, M. Greenwood, W. Libby, D. Smith, S. Strauss

Some of our material is more difficult to get to the field than were Syrach-Larsen's "tree shows", because of legal restrictions, regulations, and public sensitivity with respect to genetically-engineered plants. There are also dangers of prematurely putting in demonstrations of plants that then behave badly, thus losing rather than creating confidence. Still, most contributors indicated that putting out tests and demonstrations is not only politically useful, but ethically correct. We must then carefully deal with those that fail to perform up to expectations. Besides deploying demonstration plantings of biotechnically-created fast-growing healthy trees, we need to emphasize the impacts of these new trees on product quality.

How Can Biotechnology Serve The Purposes And Research Questions Of Biodiversity?

S. Friedman, M. Greenwood, D. Neale, S. Strauss

Biodiversity seems to be the latest sacred cow. While it may be tempting to chase biodiversity funds, many of us are well advised to do what we do well, stick to "pure biotechnology", and don't be diluted by attempting to coopt the latest bandwagon. If we do this well, an important indirect contribution that "pure biotechnologists" may make to the concerns of the new "biodiversity movement" will be to increase wood productivity on some forest areas. This increase in wood productivity on some forests should free other forested areas (that would have been heavily logged) for dedication to biodiversity goals. In addition, there are some real and important issues in "biodiversity" that we can directly contribute to, and it would be a waste of our talents and abilities not to do so. Some suggested contributions include screening for genetic diversity, and finding markers correlated with traits of interest and importance. Finally, it was noted that those who are paying the bills are not thrilled to pay to solve problems that the investigator perceives, but that we don't really have.

What Are Some Ways To More Effectively Make Common Cause, With Each Other And With Other Forest-science Disciplines, With Respect To the Need For Better And Perhaps More-rational Funding?

R. Dinus, W. Hackett, D. Neale, D. Paton, D. Smith

There was general agreement on the importance of long-term research, and that we would all love secure long-term funding, but that that's not the way things are going. Organizations responding to yearly or quarterly financial analyses are not going to fund much research that doesn't have a payoff in less than 5 years. When arguing for long-term research, we should support appropriate review, but within a framework of clear long-term expectations.

Governments seem the likely sponsors of long-term research, and one response of government (in New Zealand) was to privatise operations with short-term goals, and then focus the much-reduced government support on operations with long-term goals. But stating such a rational-appearing strategy and achieving it are two different things, and it has been difficult (in the U.S.) to get balanced funding. The U.S. Office of Management and Budget likes "quality", and a funding strategy relying heavily on short-term competitive grants is a response to this preference. But needed "maintenance" research typically fails in the competitive-grants arena. Somebody needs to address such thorny issues as there being too many forestry schools (in the U.S. and elsewhere). Finally, there is more money around than many of us realise, both at the national and international levels, and we have not been effectively tapping into it.

Anything Else?

R. Ahuja, E. Birnbaum, W. Libby, D. Paton, D. Smith

Funding strategies were further discussed. Each of us acting alone perhaps has a feeling of loneliness and powerlessness. IUFRO seems a logical source of joint effort, but IUFRO hasn't been and probably shouldn't be a fund-raising or fund-disbursing organization. When it comes to lobbying, the Australian Minister of Science called scientists "wimps". (For non-Australians: a "wimp" is a timid unassertive person. For Americans: Recall the last presidential contest, characterized as "the shrimp vs the wimp". Thus, it is implied that scientists should be trees, not bushes.) We should join others in ethical issues, such as "biodiversity". (Where are trees on the "furry index", on which koalas score 10 and snakes 0?) We should join the debates on national priorities, and try to make forestry more visible and important.

POSTER PRESENTATIONS

PLANT REGENERATION OF HYBRID POPLARS USING NODULE CULTURE SYSTEM

Kyoung Ho Chung and Young Woo Chun

College of Forestry, Kookmin Univ., Seoul, Korea

Development of micropropagation method for hybrid poplars, Populus euramericana, P. nigra X P. maximowiczii, was established using nodule culture system. Callus was initiated from in vitro cultured leaf explants of hybrid poplars on Murashige and Skoog's medium supplemented sucrose 20 mg/l, agar 7 g/l, and 2,4-D 0.5 mg/l. To raise cell suspension, the induced calli from leaf explant cultures were transferred to liquid medium of the same composition. Among 4 tested BA concentrations, BA 0.2 mg/l gave the best formation of fine nodules (about 1-2mm diameter) after 2 weeks of culture. Using the developed liquid nodule culture system, the fine nodules grow via nodule enlargement (about 5-7mm diameter) and nodule multiplication. Plant regeneration could be obtained from the nodule (3-5mm diameter) in liquid medium or on agar solidified medium. In particular, the regeneration medium with BA 0.5 mg/l and adenine sulfate 40 mg/l stimulated shoot differentiation on hybrid poplar nodules.

POSITIONAL AND REJUVENATION EFFECTS OF MICROPROPAGATION OF MATUE FAGUS

SYLVATICA L.

Karin Meier and Gerhard Reuther

Forschungsanstalt Geisenheim, Division of Botany
6222 Geisenheim, Fed. Rep. of Germany

Large scale regeneration of superior genotypes of Fagus sylvatica L. (European beech) by conventional propagation methods is not yet feasible. On the other hand, tissue culture technology has been successfully exploited for clonal mass propagation of many other hardwood species. It was the aim of this study to determine the influence of explant position and age of stock plant on multiplication rate during the initial phase of shoot tip culture of beach.

In spring, just before budbreak, buds from one year old long-shoots were taken and surface sterilized. After removal of the bud scales, the shoot tips were cultured on Woody Plant Medium supplemented with 2 % glucose and 1 ppm BA. Shoot elongation and multiplication rate of the explants depended on the positional origin of the bud. Comparing apical and axillary buds, the former showed highest response with respect to propagation.

Additionally, the age of the stock plant played an important role as found also in other deciduous trees. In order to facilitate the establishment of mature tissues, scions of 6 mature beech genotypes were rejuvenated by grafting them on juvenile seedling rootstocks. In all cases the promoting effect of this measure could be demonstrated as buds taken from these graftings showed a much greater regenerative capacity than buds taken directly from the mother plant.

While growth and development during the initial culture period was strongly influenced by the factors investigated, this was no longer evident in the following subcultures.

Woody Plant Biotechnology, Edited by M.R. Ahuja
Plenum Press, New York, 1991

ESTABLISHMENT OF A SHOOT CULTURE SYSTEM FOR LARIX DECIDUA

David Ellis, Brent McCown, Darroll Skilling[1], Melanie Barker[1], Rodney Serres and Mike Ostry[1]

University of Wisconsin-Madison, Department of Horticulture, Madison, WI and [1]USDA-Forest Service, North Central Forest Experiment Station, St. Paul, MN

One limiting aspect in the micropropagation of many conifers is the inability to establish a sustained shoot culture based on axillary bud proliferation. In contrast, all successful commercial micropropagation systems of dicotyledonous woody plants, are based on the establishment of such a shoot culture system. Most conifer microculture systems rely on the one time proliferation of adventitious buds from embryonic tissue, with very few species being capable of reliable and sustained axillary bud proliferation from these adventitious buds.

The identification of Larix decidua (European larch) somaclonal variants resistant to Gremmeniella abietina (the casual agent of scleroderris canker) and Mycosphaerella laricina (the casual agent of larch needle cast) would greatly aid in the acceptance of this tree as a potential timber species for the Lake States. One drawback to this has been the inability to clonally mass propagate European larch so that large scale disease screening can be done. The objectives of our research were 1) to initiate sustained culture systems for European larch, 2) to identify factors limiting the establishment of sustained shoot cultures and 3) to develop methods to overcome these limitations.

Adventitious buds were induced on 5 day-old seedling cotyledons by placing excised cotyledons on BLG medium [1] (supplemented with 10mM glutamine) containing 10 mg/l (44.4 μM) benzyladenine (BA) for 2 weeks. These cotyledons were then placed on the basal medium without hormones for bud elongation. Buds were excised from the cotyledons and placed on WPM medium supplemented with a wide range of cytokinin concentrations. The objectives of this stage of the research was 1) to determine at what cytokinin level the shoots continued to elongate and 2) at what level axillary bud production was enhanced. Further adventitious bud production was avoided so that lines of clonal material could be obtained from each of the original adventitious cotyledonary buds. WPM supplemented with 0.1 uM BA was chosen as this allowed shoot elongation similar to hormone free medium. With this system, axillary bud production occurred only at 6-10 needle axils, rather than at every axil. Unfortunately, no method to date has been developed which induces the production of axillary buds at each axil. It is interesting to note that seedlings have this spaced axillary bud scheme, suggesting that the spacing of axillary buds is strongly pre-determined.

Woody Plant Biotechnology, Edited by M.R. Ahuja
Plenum Press, New York, 1991

As shoots elongated they could be divided into discrete groups of explants (Figure 1). An elongated shoot was divided into apical and basal portions (*tips* and *bases* respectively, of approx. 2-3 cm each), stems containing 1-3 axillary buds, and new flushed axillary buds (*new fl ax*). The *tip* of an elongated shoot was the most productive explant because every 6 weeks a 220% increase in the number of explants could be realized. Both the *stems* and the *bases* would yield roughly 60% increases in explant number over a 6 week period, predominantly by the flushing of pre-existing axillary buds to form *new fl ax*. Probably due to the developmental fate in larch of the axillary buds becoming short shoots, it is not surprising that *new fl ax* had a relatively low rate (13%) of elongation. After 6 weeks as *new fl ax*, those that did not elongate, became old flushed axillary buds (*old fl ax*) and were maintained as such until elongation occurred. Once elongation occurred, the buds could then be termed elongated axillary buds (*elong ax*). The *elong ax* would slowly elongate, going through a developmental change over a 1-6 month period until they were morphologically similar to elongated shoots produced by *tips*.

Unfortunately, once buds became *old fl ax*, most (95%) reached a developmental dead-end and could remain in this non-elongated stage of over a year. Hormonal modifications were unsuccessful in inducing elongation of this class of axillary bud. Efforts have currently centered on the use of flooding (or overlaying with liquid) the *old fl ax* to induce elongation. Overlay of the buds with liquid induced up to 50% to elongate compared to no elongation of the control. In addition, many axillary buds which flush on the stem will by-pass the *new fl ax* stage and form directly into an *elong ax* (Figure 2). Interestingly, this effect can be seen over 6 months after the treatment. The physiological and developmental basis for this phenomenon is unknown, yet it is probably related to some stress response rather than nutritional because a distilled H_2O overlay yielded similar results compared to the basal medium overlay. In addition, non-flushed axillary buds (*stems*) will respond to exogenous cytokinins in the medium by elongating, yet at a lower percentage than with flooding. Recent work at the USDA-Forest Service in St. Paul, MN has shown that rooting of the flushed axillary buds will also induce elongation.

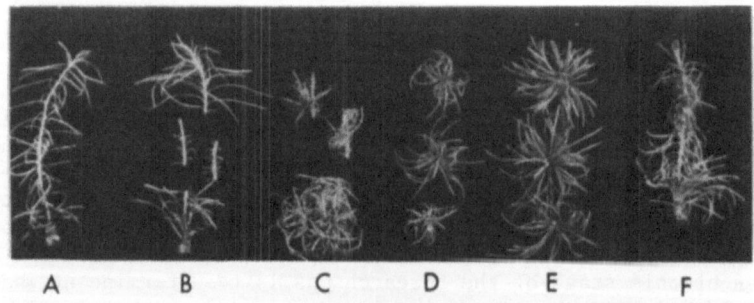

Figure 1. Developmental sequence of explants for European larch. A) Elongating tip B) Elongating tip divided into: a tip which recyles every 6 weeks, 2 stems containing 1-3 axillary buds and a base C) New flushed axillary buds (*new fl ax*) from stems after 6 weeks (above) and the base with numerous new flushed axillary buds D) Isolated new flushed axillary buds are subcultured for 6 weeks E) Old flushed axillary buds (*old fl ax*) are any flushed axillary bud over 6 weeks old F) Elongating axillary bud (*elong ax*) will develop in 1-6 months into a tip.

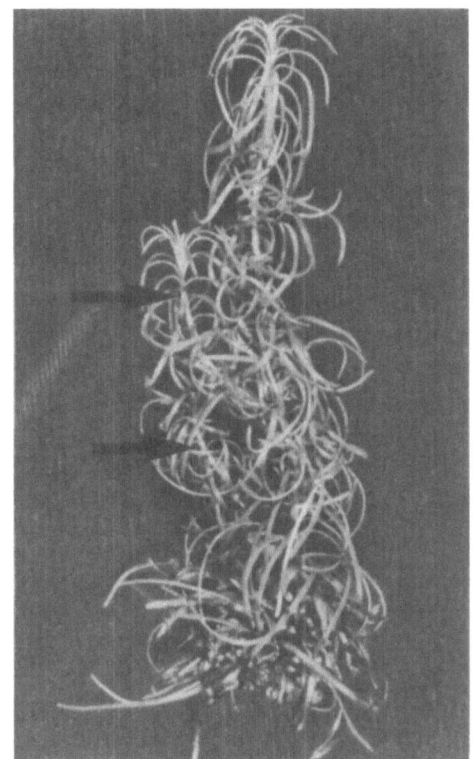

Figure 2. An elongated tip from a flooded old flushed axillary bud demonstrating excellent elongation of the tip. Note the good elongation of the axillary buds (arrows) still attached to the stem. The vitrified basal bud mass is not unusual for flooded buds.

LIMITATIONS FOR THE ESTABLISHMENT OF A SHOOT CULTURE SYSTEM FOR LARCH

1. Axillary bud formation only at every 6th-10th node.
2. Lack of elongation of flushed axillary buds.
3. Reliance on juvenile material.

SUMMARY OF RESULTS

1. Elongating tips are essential for a large increase in the number of explants.
2. Conversion of flushed axillary buds into elongating tips can be enhanced by flooding.
3. Elongating tips originating from axillary buds may need to undergo an establishment period prior to normal elongation.
4. WPM + 0.1 µM BA results in a 30% increase in the number of explants every 6-8 weeks.

This project was funded by a Cooperative Research agreement with the North Central Forest Experiment Station, USDA-Forest Service, St. Paul, MN.

REFERENCE

1. C.L. Brown and R.H. Lawrence. 1968. Culture of pine callus on a defined medium. For. Sci., 14:62-64.

PROTOPLAST CULTURE FROM *FAGUS, ULMUS* AND *ABIES*

Hermann Lang and Hans Willy Kohlenbach

Botanisches Institut der J.W. Goethe-Universität,
Siesmayerstr. 70, D-6000 Frankfurt am Main, F.R.G.

In times of progressing forest decline efforts on the preservation of genetic resources and on new approaches to the vegetative propagation from proven mature trees gain in significance. Since most forest trees are difficult to propagate from stem cuttings past their juvenile stage in-vitro-culture methods are currently being applied. Protoplast culture would allow the establishment of in-vitro-techniques for genetic preservation, vegetative propagation and somatic hybridization, if in addition methods for plant regeneration from protoplast-derived clones could be developed.

Leaf tissue or proliferating callus cultures can serve as source material for protoplast isolation. Mesophyll protoplasts were isolated from young leaves of juvenile and adult beech trees (*Fagus sylvatica*) during a short period following bud break. The isolation of callus protoplasts is demonstrated with cultures derived from *Ulmus* leaf explants. Protoplasts of *Abies alba* were isolated from embryogenic callus which originated from immature seed explants (6).

The protoplast isolation is accomplished in an enzyme solution containing 0.5% (w/v) Pectinol and 2% (w/v) Cellulase R-10. By this method for example about 3×10^7 protoplasts can be obtained from 1 g leaf material of *Fagus sylvatica*(Fig.1). A culture medium modified from Kao and Michayluk (2, KM8p) proved suitable for mesophyll protoplasts of *Fagus* and callus protoplasts of *Ulmus* and *Abies* as well.

For successful protoplast isolation leaves can be used only for a short time after bud break. Two weeks after leaf expansion the cell walls were resistant to the usual enzymic digestion. Isolated protoplasts in culture first regenerate a cell wall. Cell divisions were observed after 8-10 days for *Fagus* (Fig.2) and after 6 days for callus protoplasts of *Ulmus* (3,4). Sustained divisions resulted in the formation of cell clusters and colonies within a period comparable to protoplast cultures of herbaceous species. Subsequent development of calli has been achieved.

In protoplast cultures derived from embryogenic callus of *Abies alba* (Fig.3), clusters and colonies develop by an increase in cytoplasmic material and subsequent cell division (Fig.4) predominantly from the small protoplasts within the suspension. An early differentiation of small cytoplasmic cells and large vacuolated "suspensor-type" cells has been observed in developing colonies, after 2-3 weeks in culture (5). Similar structures in morphogenic protoplast cultures of *Pseudotsuga menziesii* have been interpreted as "somatic proembryos" (1).

Woody Plant Biotechnology, Edited by M.R. Ahuja
Plenum Press, New York, 1991

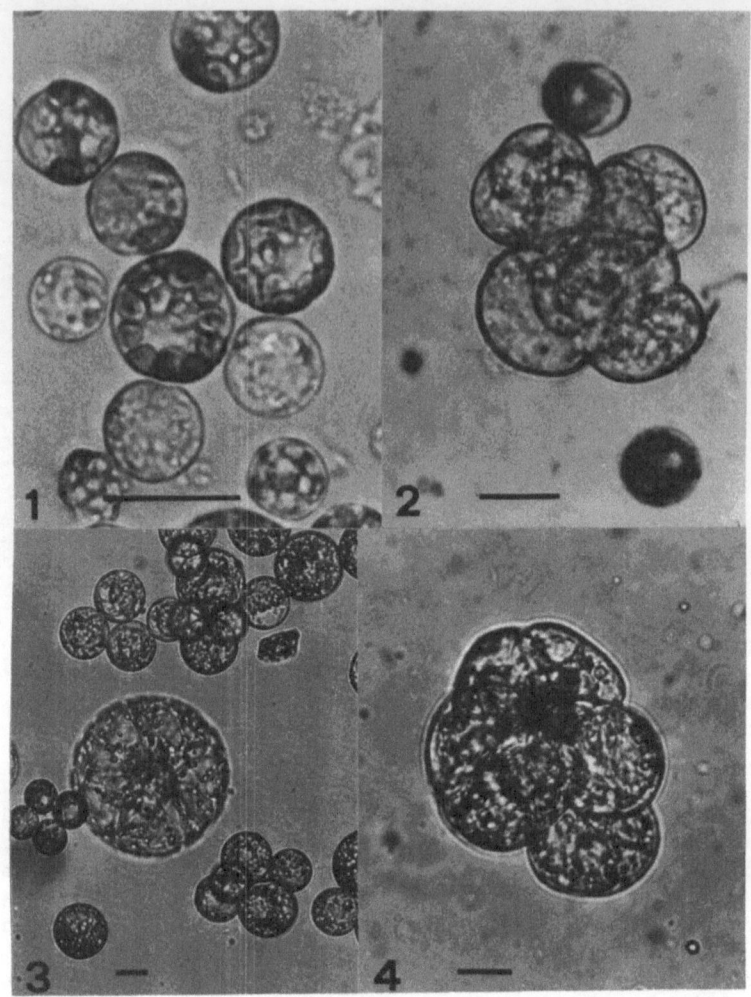

Fig. 1.   Protoplasts freshly isolated from leaves of *Fagus sylvatica*
Fig. 2.   Microcolony formation from leaf protoplast of *Fagus sylvatica*
Fig. 3.   Protoplast release from embryogenic cultures of *Abies alba*
Fig. 4.   Cell division from a protoplast of *Abies alba*. Bars represent 20 µm

REFERENCES

1. Gupta, P.K., Dandekar, A.M., and Durzan, D.J., 1988, Somatic embryo for-
    mation and transient expression of a luciferase gene in Douglas fir and
    loblolly pine protoplasts. Plant Science 58: 85-92.
2. Kao, K.N., and Michayluk, M.R., 1975, Nutritional requirements for growth
    of *Vicia hajastana* cells and protoplasts at a very low population den-
    sity in liquid media. Planta 126: 105-110.
3. Kohlenbach, H.W., and Lang, H., 1988, Protoplasten-Technik für Forst-
    pflanzen? Allg. Forst Zeitschr. Heft 49, 1355-1357.
4. Lang, H., and Kohlenbach, H.W., 1988, Callus formation from mesophyll
    protoplasts of *Fagus sylvatica* L. Plant Cell Reports 7: 485-488.
5. Lang, H., and Kohlenbach, H.W., 1989, Cell differentiation in protoplast
    cultures from embryogenic callus of *Abies alba* L. Plant Cell Reports
    8: 120-123.
6. Schuller, A., Reuther, G., and Geier,T., 1989, Somatic embryogenesis from
    seed explants of *Abies alba.* Plant Cell Tiss. Org. Cult. 17: 53-58.

STUMP SPROUTS: A VALUABLE SOURCE FOR CLONAL PROPAGATION AND

GERMPLASM PRESERVATION OF ADULT SESSILE OAK

Karl Gebhardt, Birgit Heineker and Horst Weisgerber

Forschungsinstitut für schnellwachsende Baumarten
Prof. Oelkersstr. 6
D-3510 Hann. Münden, F.R. Germany

ABSTRACT

From a natural stand of sessile oak ten continuously proliferating shoot tip cultures were established from stump sprouts
of originally 24 selected genotypes. The rate of multiplication
per subculture period (4 weeks) varied between 1.08 to 1.70.
After six subculture intervals a total number of 440 elongated
shoots proliferated from 79 established shoot tips. Shoot tip
cultures were stored aseptically at growth retarding conditions.
From one genotype rooted plantlets were produced from elongated
shoots by a one minute dip treatment using 500 ppm indolebutyric
acid.

INTRODUCTION

Over the past few decades there has been a gradual but
steady change in public and professional attidues towards our
natural resources. At many natural stands the oak trees are
severely damaged even before they come to flower. Also because of
the sylvicultural practice of thinning individual trees which are
well adapted to the biotic environment cannot contribute more
offspring to the next generation. This disadvantage is overcome
by the _in vitro_ propagation of individual trees which allows
both, the long term storage of germplasm as well as the clonal
multiplication (3).

MATERIALS AND METHODS

In a forest district nearby Kassel a stand of sessile oak
(Quercus petraea (Matt.) Liebl.) was thinned in 1987. Stump
sprouts developed within the following vegetation period and were
collected in Jan. and Oct. 1988. Shoot tips (0.5 mm) were prepared from dormant buds after surface sterilization using 3 %
sodium hypochlorite for 10 Min..After washing (tap water) buds
were immersed for 5 Min. in 70 % ethylalcohol with an addition of
0.1 % ascorbic acid. After 5 Min. immersed buds were passed
through the flame of a petroleum-burner for 5 Sec.. The bud scales were removed and shoot tips of a size of 0.5 mm were placed
on a modified Gresshoff-Doy-medium (4) supplemented with 0.2 ppm

benzyladenine, 2 % sucrose, 100 ppm myo-inositol. Prior to auto-claving the pH was adjusted to 5.7. The media were solidified with 0.3 % Gelrite (Kelco). Shoot tip cultures were kept in a growth chamber at 26 ^{o}C in a 16 hour photoperiod supplied by cool white fluorescent lamps at an light intensity of 1500 Lux. From developing shoot clusters elongated shoots were dissected and subcultured monthly.

RESULTS

Using chemical as well as heat sterilization techniques it was possible to establish aseptically growing shoot tip cultures from closed winter buds of stump sprouts. A size of 0.3 to 0.6 mm in diameter was favourable. By the addition of ascorbic acid to the disinfectant the browning of shoot tips was prevented. Ten continuously proliferating shoot tip cultures were established from originally 24 genotypes. After the first subculture on modified GD-medium (4) the number of established shoot tips was reduced within a range of 25 - 94 %. The leaf condition of sur-viving shoots was scored from 1 (best) to 6 (brown, necrotic) and varied between genotypes. An average of 3.07 with a standard deviation of 1.31 was counted. Independent of an initial reduc-tion in the number of growing shoots new shoots proliferated during the course of the following six subcultures. The rate of multiplication per subculture period (1 month) varied between 1.08 to 1.70. After six subculture intervals a total number of 440 elongated shoots (5 -12 mm) proliferated from 79 established shoot tips (0.5 mm). The insertion of new shoots at stem base or at the apical region of the stems varied within subculture peri-ods and between genotypes. Callus tissue proliferated at stem base and remained partly green. If shoots were dissected in nodal segments the rate of multiplication was enhanced. Shoot elonga-tion was favoured by subculturing shoots for 2 weeks in the dark. Shoots can be stored aseptically if the growth is retarded by low temperature (10°C) and reduced light (500 Lux) in a growth cham-ber. Depending on the health and nutritional status of genotypes the subculture period might be prolonged 2 to 12 times. From one genotype rooted plantlets were produced by a dip treatment (1 Min.) of elongated shoots with 500 ppm IBA and subsequent subcul-ture on hormone-free medium.

DISCUSSION

The genetic makeup of a population will change by natural selection in a long period which is measured in generations. In contrast an increasing selection pressure caused by environmental stress conditions may account for losses of even the fittest genotypes within one generation. This would lead to a drastic reduction of silviculture-stability of the forest stands. By the method of shoot tip culture of oaks it is possible to preserve oak germplasm. Successful establishment of shoot tip cultures depends on the method of sterilization, specific requirements of nutrients, hormones and cultural conditions (1, 5). Also the degree of juvenility in the starting material determines the regeneration capacity (2, 5, 6). Using shoot tips from stump sprouts a large number of genotypes was established and multi-plied in vitro. The origin, performance and genetic variation of genotypes is well known. By a large number of genotypes a rather high degree of variation is guaranteed. After the establishment of aseptic shoot cultures it is possible to screen for resistances in vitro. Low temperature stored material would allow clonal masspropagation after reactivation and rooting. Rooting was successfully done with one genotype but still more research is necessary in order to obtain rooted plantlets from shoot tip cultures more easily. Rooted plantlets could be used to compensate the depletion of genetic resources at the original forest stands.

Acknowledgement: This work was supported by the German Federal Ministry for Research and Technology (BMFT), Project 0318920.

LITERATURE

1.   Chalupa, V., 1984, In vitro propagation of oak (Quercus robur.
        L.) and linden (Tilia cordata Mill.). Biol. Plant. 26: 374-377.
2.   Civinova, B. and Sladky, Z., 1987, A study of the regeneration capacity of oak (Quercus robur L.). Scripta FAC.SCI.NAT. Univ. PURK. BRUN (Biologia) 17: 103-110.
3.   Gebhardt, K., 1989, Fortschritte und Perspektiven der Mikrovermehrung von Forstpflanzen. Die Holzzucht 43: 6-12.
4.   Gresshoff, P.M. and Doy, C.H., 1972, Development and differentation of haploid Lycopersicon esculentum. Planta 107: 161-170.
5.   Meier-Dinkel, A., 1987, In-vitro-Kultur und Weiterkultur von Stieleiche (Quercus robur L.) und Traubeneiche (Quercus petraea (Matt.) Liebl.). Allg. Forst u. Jagdzeitung 158: 199-204.
6.   Vieitez, A.M., San-Jose, M.C. and Vieitez, E., 1985, In vitro plantlet regeneration from juvenile and mature Quercus robur L. J. Hort. Science 60: 99-106.

RECOVERY OF JUVENILE CHARACTERISTICS THROUGH IN VITRO

PROPAGATION OF MATURE FAST-GROWING BIRCH HYBRIDS

Andreas Meier-Dinkel

Niedersächsische Forstliche Versuchsanstalt
Abteilung Forstpflanzenzüchtung
W-3513 Staufenberg-Escherode, Federal Republic of Germany

ABSTRACT

Mature hybrids of *Betula platyphylla var. japonica x B. pendula* were micro-propagated using dormant winter buds as explant source. The in vitro propagated plants showed characteristics of juvenile seedlings such as vigorous orthotropic growth, absence of flowering, and a high rooting capacity of cuttings taken from 2-year-old in vitro propagated plants. When reestablished in vitro, explants of the micropropagated plants gave better results compared to the original mature material.

INTRODUCTION

Hybrids between *Betula platyphylla var. japonica* and *Betula pendula* are fast-growing trees with an outstanding performance. In an experimental plot the volume increment of these hybrids at the age of 28 was double that of the indigenous *Betula pendula*. The best trees were chosen for micropropagation in order to use this genetic gain directly and to avoid the genetic variation typical for seedling progeny.

MATERIAL AND METHODS

Mature birch hybrids were micropropagated as described by Meier-Dinkel (3). In order to ascertain the degree of juvenility of the micropropagated plants two experiments were carried out. Green cuttings of the mature ortets and of 2-year-old micropropagated ramets were inserted for rooting in a plastic greenhouse. Winter bud explants were isolated from the mature trees as well as from the micropropagated plants growing in the nursery in order to compare the response in vitro.

RESULTS

The in vitro establishment was a critical phase. In different experiments with explants of 9 genotypes finally only 8 % of the explants of 6 genotypes gave rise to proliferating shoot cultures. The surviving explants developed a short shoot with a few leaves. After 8 to 10 weeks the shoots did not grow on and gradually turned yellow. However, a callus-like tissue with adventitious shoot buds was formed at the base of the explants after 8 to 12 weeks. Adventitious shoots were separated, cut into nodal segments and shoot tips, and further multiplied via axillary bud breaking. The best multiplication rates were obtained on MS medium or WPM containing 2.0 mg/l BAP.

Calculated as the number of new segments per subcultured segment they averaged 3.0 on MS medium and 4.7 on WPM. The in vitro rooting of more than 5000 micro-cuttings of different clones averagerd 87.5 %. After transfer to a greenhouse 72.4 % of these plantlets grew on vigorously. The micropropagated plants derived from mature material showed characteristics of juvenile seedlings, such as vigorous orthotropic growth and absence of flowering. Another typical juvenile characteristic, a high potential to form adventitious roots, was regained through in vitro propagation. Green cuttings taken from two-year-old micropropagated plants in the nursery rooted with high percentages, between 73.0 and 87.5 %, whereas the cuttings taken from the mother trees did not root at all. The cuttings from micropropagated plants grew on vigorously. We compared the capacity for establishment in vitro of buds from in vitro derived trees with those from the original mother trees. It was found that 65 % of axillary bud explants from micropropagated plants could be reestablished in vitro (12 % dead, 23 % contaminated), whereas only 20 % of the explants from the mature mother trees survived (69 % dead, 11 % contaminated). The growth of the explants from the mature trees was slower than the growth of the explants from the rejuvenated plants.

DISCUSSION

A rejuvenating effect through in vitro propagation techniques has been described for several woody plant species. Mullins et al. (5) reported the formation of juvenile plants after several multiplication cycles via axillary bud breaking on BAP containing media in a very old cultivar of *Vitis vinifera*. Struve and Lineberger (6) found that high rooting potential in mature birch (*Betula papyrifera*) could be restored via tissue culture. The microcuttings resulted from adventitious shoot initiation on leaf and stem tissue. The results obtained with *Betula papyrifera* and our results with birch hybrids support the suggestion that adventitious shoots are juvenile (1) or undergo at least a partial reversion to a more juvenile state (2). For practical application in tree improvement programs based on vegetative propagation, rejuvenation methods have to reverse several phase-related characteristics: a fast orthotropic juvenile growth has to be restored completely, a high adventitious rooting potential has to be restored, and no precocious flowering should occur (4). Our results show that large scale propagation of selected outstanding mature birch hybrids is possible by in vitro techniques. Juvenile characteristics have been recovered. The plants are now 4 years old and their field performance so far is very good. The commercial feasibility of the method may be increased through propagation of the in vitro derived plants by cuttings.

REFERENCES

1. Bonga, J.M., 1982, Vegetative propagation in relation to juvenility, maturity, and rejuvenation. In: Tissue Culture in Forestry, J.M. Bonga and D.J. Durzan, (Eds.), Nijhoff/Junk, The Hague, pp. 387 - 412.
2. Hackett, W.P., 1985, Juvenility, maturation, and rejuvenation in woody plants. Horticultural Reviews, 7: 109 - 155.
3. Meier-Dinkel, A., (in press), Micropropagation of Birches (Betula spp.). In: Biotechnology in Agriculture and Forestry, Vol. 18, "High-Tech and Micropropagation II", Y.P.S. Bajaj (Ed.), Springer, Berlin, New York.
4. Meier-Dinkel, A. and Kleinschmit J., (1990), Aging in Tree Species: Present Knowledge, In: Plant Aging: Basic and Applied Approaches, NATO ASI Series, Series A: Life Sciences Vol. 186, R. Rodriguez, R.S. Tamés and D.J. Durzan (Eds.), Plenum Press, New York, pp. 51 - 63.
5. Mullins, M.G., Nair, Y., and Sampet, P., 1979, Rejuvenation in vitro: Induction of juvenile characters in an adult clone of *Vitis vinifera* L., Ann. Bot., 44: 623 - 627.
6. Struve, D.K., and Lineberger, R.D., 1988, Restoration of high adventitious root regeneration potential in mature *Betula papyrifera* Marsh. softwood stem cuttings, Can. J. For. Res., 18: 265 - 269.

SOMATIC EMBRYOGENESIS IN LARIX DECIDUA

Daniel Cornu and Christophe Geoffrion

Station d'Amelrioration des Arbres Forestiers

INRA-CRO, Ardon 45160 OLIVET France

Success in somatic embryogenesis has been reported for conifer species since few years. Picea abies, P. glauca, P. mariana, Pinus taeda and P. lambertiana as well as Pseudo-tsuga menziesii have given embryogenic callus from immature and, for some species, mature zygotic embryos. Recently, it has been reported (3,4) that haploid somatic embryos develop on the female gametophytes of larch (Larix decidua). Embryogenic callus lines were also derived (2) from immature zygotic embryos of hybrid larch (Larix decidua x L. kaempferi).

At the Forest Tree Improvement Station, we have established embryogenic callus from immature seeds of Larix decidua. Numerous embryogenic lines were obtained in 1989 on a BM medium (1) supplemented with 0.2 mg/l 2,4-D. These lines characteristically showed white, translucent and mucilaginous callus. Embryogenic calli show active growth centres consisting of agregates of small cytoplasmically dense cells, and long, ' irregular and vacuolated cells. Lines are differentiated by the proportion of the two types of cells. Most of callus (75%) are derived from the zygotic embryos or the suspensors, and were found to be diploid (2n=2).

Embryogenic lines could be subcultured over several months on the same medium without growth regulators in dark. In the liquid medium the callus double in size in 7 days. The first stage of multiplication of somatic embryo is characterized by an assymmetrical division of isolated long cells.

Proembryos appear at the surface of the callus at the tip of a bundle of elongated cells which resemble the suspensor cells of the zygotic embryo. Following the globular stage, somatic embryos elongate progressively and hypocotyl and cotyledon turn green. Contnuous light increases the number of embryos, as does high concentration of Gelrite (4g/l) instead of agar (6g/l). In this way, the embryos are able to complete their developmental process for germination.

REFERENCES

1. Gupta P.K. and Durzan, D.J., 1986, Plantlet regeneration via somatic embryogenesis from subcultured callus of mature embryos of Picea abies. In Vitro 22:685-688

2. Klimaszewska, K., 1989, Plantlet development from immature zygotic embryos of hybrid larch through somatic embryogenesis. Plant Sci 63:95-103
3. Nagmani, R. and Bonga, J.M., 1985, Embryogenesis in subcultured callus of Larix decidua. Can.J.For.Res.15:1088-91
4. Von Aderkas, P., and Bonga, J.M., 1988, Formation of haploid embroids of Larix decidua: early embryogenesis. Amer. J. Bot. 75:690-700

RESPONSE OF ABIES ALBA EMBRYONAL-SUSPENSOR MASS TO VARIOUS

CARBOHYDRATE TREATMENTS

Astrid Schuller and Gerhard Reuther

Forshungsanstalt Geisenheim, Division of Botany
6222 Geisenheim, Fed. Rep. of Germany

Cultures of Abies alba embryonal-suspensor mass (EMS) were
screened for their response to various sucrose concentrations
ranging from 1%-6% (w/v) and various carbohydrate sources in
equimolar concentrations, i.e. glucose, fructose, galactose,
sucrose, and soluble starch.

Nutritional factors and sequential relationships of growth
regulators control the course of embryogenesis (1). In order
to stimulate maturation processes in various stages of somatic
embryogenesis a sequence of culture steps was applied. The
proliferation medium (2) was modified by gradual omission of
BA and subsequent addition of ABA in combination with a higher
concentration of the respective carbohydrates. These medium
variants influenced the structure of the EMS markedly. The
microscopic examination of acetocarmine stained EMS revealed
great differences concerning size and development stages of
the somatic embryos (SE) in the different carbohydrate treatments. While fructose was unable to promote any SE maturation,
glucose, sucrose and galactose so far promoted early stages.
SE development from early globular stage to precotyledonary
(torpedo) stage was observed only on media with soluble starch
or lactose. In contrast to the results with soluble starch,
the lactose treatment led to more synchronized maturation,
compable to the zygotic embryogenesis. There is evidence
that lactose has a morphogenic effect on the somatic embryogenesis of Abies alba.

REFERENCES

1. Gupta, P.K., and Durzan, D.J., 1986, Somatic polyembryo-
    genesis from callus of mature sugar pine embryos. Bio/
    Technology 4:643-645
2. Schuller, A., Reuther, G, and Geier, T., 1989, Somatic
    embryogenesis from seed explants of Abies alba. Plant
    Cell, Tissue and Organ Culture 17:53-58

SOMATIC EMBRYOGENESIS IN AESCULUS HIPPOCASTANUM AND

QUERCUS PETRAEA FROM OLD TREES (10 to 140 YEARS)

Jörg Jörgensen

Niedersächsische Forstliche Versuchsanstalt
3513 Staufenberg-Escherode, FRG

ABSTRACT

Tissues of the flower of oak (anthers) and horse-chestnut (filaments) were placed on modified WPM. On these media calli developed, which were transferred to embryos on modified WPM with 2.5μM BAP. Zygotic embryos of oak placed on modified WPM regenerated embryos on their surface. Some of the horse-chestnut embryos were potted in soil and are growing in the greenhouse.

INTRODUCTION

Somatic embryogenesis in woody plants is very rare and only possible when using young material like embryos or embryo segments (1,5). The technique described here is only useful for the mass-propagation of elite trees which have reached reproductive maturity. Thus in the Lower Saxony Forest Research Institute, experiments have been carried out concerning induction of somatic embryos from 10 to 140-year-old trees (3). On the other hand experiments have been realised with immature seeds in order to induce somatic embryogenesis (JÖRGENSEN in preparation).

MATERIAL AND METHODS

Aesculus filaments and Quercus anthers were excised from sterilized flower buds and were placed on WPM (4) with 1000 mg/l glutamine, 50 mg/l glycine, 50 mg/l serine, 0.0 to 5.0 μM BAP, and 0.0, 1.0, 2.5, or 5.0 μM 2,4 D. After two to four weeks, the calli and embryos were transferred to the same medium, containing 0.0, 0.25, 0.5, 1.0, or 2.5 μM BAP. The anthers were placed on the same medium as Aesculus, but with 0.5 μM BAP and 2.5, 5.0 or 10 μM 2,4 D.

RESULTS

A great number of calli were formed on the media with 0 to 5 μM BAP and 2.5 to 5.0 μM 2,4 D. When transferred on media without 2,4 D the calli produced embryos. In oak and in some cases of Aesculus all of the calli developed into embryos. These were cloned like androgenic embryos (2) by adding 2.5 to 5.0 μM BAP to the medium. The embryos resemble androgenic embryos, both showing the same reactions in different culture conditions. In the case of oak, pollen was found the second day in the one-nucleus stage when the anthers were excised from sterilized flower buds. Isoenzyme patterns, made by Dr. Müller-Starck, showed conformity of the genotype of mother tree and embryos.

DISCUSSION

It is common knowledge in in vitro culture that definite tissues have a greater organogenic potential than other tissues. Zygotic embryos (in this experiment shown with Quercus petraea) seem to be most useful for somatic embryogenesis of trees. Nevertheless, the embryos or plants from these tissues are too expensive for masspropagation or as a basic material to produce artificial seed. For in vitro techniques tissues from old trees are more suitable, because the genotype is relatively well-known. In the experiments described here, somatic embryogenesis was induced by culture of filaments and anthers. In this case the genotypes of the embryos are identical to those of the mother trees. Dr. Müller-Stark proved this by testing the isoenzyme pattern.

ACKNOWLEDGEMENTS

I wish to thank the technical assistants Andrea Madeheim, Kerstin Küchemann and Ellen Sahling for their help in the practical part of the experiments.
The study was financed by the Ministerium für Ernährung, Landwirtschaft und Forsten von Nordrhein-Westfalen.

REFERENCES

1. DAMERI, R. M., CAFFARO, L., GASTALDO, P. and PROFUMO, P. (1986) Callus fomation and embryogenesis with leaf explants of Aesculus hippocastanum L. J. Plant Physiol., 126, 93-96.
2. JÖRGENSEN, J. (1988) Embryogenesis in Quercus petraea and Fagus sylvatica, J. Plant Physiol. 132, 638-640.
3. JÖRGENSEN, J. (1989) Somatic embryogenesis in Aesculus hippocastanum L. by culture of filament callus, J. Plant Physiol, 135, 240-241.
4. LLOYD, G. and McCOWN, B. H. (1981) Commercially feasible micropropation of mountain laurel (Kalmia latifolia) by use of shoot tip-culture, Proc. Int. Plant Prop. Soc. 30, 421-427.
5. RADOJEVIC, L. (1987) Plant regeneration of Aesculus hippocastanum L. (Horse Chestnut) through somatic embryogenesis. J. Plant Physiol. 132, 322-326.

ANDROGENESIS IN QUERCUS PETRAEA, FAGUS SYLVATICA
AND AESCULUS HIPPOCASTANUM

Jörg Jörgensen

Niedersächsische Forstliche Versuchsanstalt
3513 Staufenberg-Escherode, FRG

ABSTRACT

Anthers of oak, beech, and horse-chestnut were cultured on modified MS-media or WPM with different BAP and 2,4D combinations. On these media embryos (horse-chestnut) or calli (oak and beech) developed. On media with 2.5 μM BAP the calli were transformed into embryos. On media without hormones oak and horse chestnut plants developed, which were potted in soil in the greenhouse.

INTRODUCTION

Embryogenesis of material obtained from pollen is especially important for tree-breeders with regard to the production of haploids and homozygous clones which are useful for cell fusion, gene transfer etc. The techniques used to produce embryos from pollen and anthers enable the storage of many genotypes in liquid nitrogen in the case of oak and beech. In the Lower Saxony Forest Research Institute experiments were carried out to develop haploid plants.

MATERIAL AND METHODS

Anthers were excised from 30-year-old oak trees, from 137-year-old beech trees, as well as from 10 to 100-year-old horse-chestnut trees and then cultured on MS-(3) or WPM (2). Both media contained 1000 mg/l glutamine, 50 mg/l serine, 50 mg/l glycine for oak and beech, whereas for horse-chestnut the vitamins were replaced by those described by Radojevic (4). BAP and 2,4-D were added in the following concentrations:
for oak: 0.0 - 10.0 μM BAP and 2.5 - 5.0 μM 2,4-D
for beech: 2.5 μM BAP and 5.0 μM 2,4-D
for horse-chestnut: 0.0 - 5.0 μM BAP and 0.0 - 5.0 μM 2,4-D.
The developed calli of oak and beech were placed on the same medium with 2.5 μM BAP but without 2,4D. The embryos were multiplied on the same medium. For induction of shoots and roots the embryos were put on WPM described above without hormones. Developed plants of oak and horse-chestnut were transferred to soil in foil funnel with a moisture content of 90 %. For hardening-off the plants the moisture content was reduced in steps of 10 % per week down to 60 %. Chromosome numbers were counted after staining root tips with carmin acetic acid. In addition Dr. Müller-Starck examined the isoenzymes to identify homozygote and haploid plants.

RESULTS AND DISCUSSION

In oak and beech callus was formed two to three months after preparation, but only from pollen which was in the one nucleus stage at the first day. Aesculus anthers formed embryos within the same period when the pollen was in the tetraed or the one nucleus stage. Oak and beech calli developed completely into embryos on the same media with 2.5 μM BAP and without 2,4-D. The embryos of all three species could be cloned on media described for oak and beech containing 2.5 μM BAP (1). Aesculus and Quercus formed roots and shoots on the same media without hormones. In Fagus, which was cultivated one year later, the formation of one shoot could be observed, so that the developments of Quercus and Fagus seem to be comparable. The plants and rooted embryos (Aesculus and Quercus) were transferred to soil. Now plants of Aesculus and Quercus are growing outdoors.

Aesculus chromosome counting and isoenzyme examination showed that at this time both, haploid plants (developed from pollen) and heterozygote plants with the same isoenzyme pattern as the mother tree (developed from the anther wall) were obtained. In Quercus all haploid plants developed from pollen of the first day in the one-nucleus stage, whereas all embryos derived from anthers whose pollen had been the second day in the one-nucleus stage are somatic in their origin.

Acknowledgements

I wish to thank the technical assistants Andrea Madeheim and Kerstin Küchemann for their help in the practical part of the experiments.
The study was financed by the Ministerium für Ernährung, Landwirtschaft und Forsten von Nordrhein-Westfalen.

REFERENCES

1. JÖRGENSEN, J. (1988) Embryogenesis in Quercus petraea and Fagus sylvatica, J. Plant Physiol,132, 638-640.
2. LLOYD, G. and McCOWN, B. H. (1981) Commercially feasible micropropation of mountain laurel (Kalmia latifolia) by use of shoot tip culture, Proc. Int. Plant. Prop. Soc., 30, 421-427.
3. MURASHIGE, T. and SKOOG, F. (1962) A revised medium for rapid growth and bioassay with tobacco tissue culture, Physiol. Plant. 15, 473-497.
4. RADOJEVIC, L. (1978) In vitro induction of androgenic plantlets in Aesculus hippocastanum, Protoplasma, 96, 369-374.

CRYOPRESERVATION OF HAPLOID EMBRYOS OF
QUERCUS PETRAEA, FAGUS SYLVATICA AND SOMATIC
EMBRYOS OF AESCULUS HIPPOCASTANUM

Jörg Jörgensen

Niedersächsische Forstliche Versuchsanstalt
3513 Staufenberg-Escherode, FRG

ABSTRACT

Globular or heart-shaped haploid embryos of oak and beech and somatic horse-chestnut embryos in the same stage were deep-frozen with a modified method of SAKAI. After a few days the cryopreserved embryos were rewarmed and placed on modified MS-media or WPM. After one month the embryos grew in the same manner as unfrozen embryos.

INTRODUCTION

Seeds of oak and beech cannot be stored for more than five years. In the Lower Saxony Forest Research Institute, cryopreservation experiments were carried out on Quercus petraea, Fagus sylvatica and Aesculus hippocastanum with the aim to conserve the present genetic variability in these species. Due to different temperatures in big seeds and the resulting dangers of destruction mature seeds cannot be stored in liquid nitrogen.

MATERIAL AND METHODS

A technique described by SAKAI (4) was modified and very young, globular or heart-shaped embryos were cryopreserved. Cryoprotectant solutions were mixed of equal volumes of 4, 10, or 20 % DMSO or 5 or 10 % glycerol with 5 % sucrose or 10 % glucose, mannitol, sorbitol, or inositol. The young embryos were incubated in 2 ml tubes of these solutions where they remained for one hour at 0°C. Afterwards, the embryos were quickly cooled down to -10°C and then in steps of 0.5 to 1.0°C/minute down to -40 to -50°C. Then the tubes were fastened on racks and dipped in liquid nitrogen. After rewarming the samples in 40°C warm water, the embryos were washed twice with a solution of 0.3 to 0.5 mol of an untoxic carbohydrate and then placed on WPM (3) with 1000 mg/l glutamine, 50 mg/l glycine, 50 mg/l serine, and 2.5 μM BAP.

RESULTS

After one month without further development, the deep-frozen embryos grew and developed very differently. Those who were incubated in 2 % DMSO and 2.5 % sucrose had a survival rate of more than 90 % in oak and beech. Other DMSO concentrations and other types of carbohydrates led to lower survival rates. In some cases there was no further development and the embryos died. Two months later no difference could be observed between cryopreserved and non-cryopreserved embryos

Woody Plant Biotechnology, Edited by M.R. Ahuja
Plenum Press, New York, 1991

when a good cryoprotectant had been used. In Aesculus 90 % of the embryos remained alive, indicated by lack of turning brown, but did not show any further development within five months.

DISCUSSION

When storing at -196°C little or no physiological processes occur and we suppose that tissues can be stored for a long time. This is especially important for species such as beech and oak which have seeds that cannot be stored for a long time, even with special preparation methods (STEINHOFF, pers. comm.). Acorns and beechnuts can only be stored for four to five years. In this case in vitro production of either androgenic (oak and beech) (1) or somatic embryos (horse-chestnut (2), oak, JÖRGENSEN in preparation; beech, when a method is found) combined with the cryopreservation technique described above, will allow long- term storage of material in a genebank and circumvent the loss of genes induced by the injuries of air pollution.

ACKNOWLEDGEMENTS

I wish to thank the technical assistants Judith Gabriel, Andrea Madeheim, Ellen Sahling, and Kerstin Küchemann for their help in the practical part of the experiments. The study was financed by the Ministerium für Ernährung, Landwirtschaft und Forsten von Nordrhein-Westfalen.

REFERENCES

1. JÖRGENSEN, J. (1988) Embryogenesis in Quercus petraea and Fagus sylvatica, J. Plant. Physiol. 132, 638-640.
2. JÖRGENSEN, J. (1989) Somatic embryogenesis in Aesculus hippocastanum L. by culture of filament callus, J. Plant. Physiol. in press.
3 .LLOYD, G. and McCOWN, B. H. (1981) Commercially feasible micropropagation of mountain laurel (Kalmia latifolia) by use of shoot-tip culture, Proc. Int. Plant Prop. Soc., 30, 421-427.
4. SAKAI, A. (1985) Cryoprservation of shoot-tips of fruit trees and herbaceous plants, In: Cryopreservation of Plant Cells and Organs (KARTHA, K.K. ed.) CRC Press, Inc., 135-158.

TRANSFORMATION OF POPULUS - FROM SYSTEM DEVELOPMENT TO FIELD PLANTINGS

N. B. Klopfenstein, R. W. Thornburg, H. S. McNabb, Jr.,
R. B. Hall, E. R. Hart, Y. W. Chun, A. Kernan and N.-Q. Shi

Depts. Forestry, Plant Pathology, Entomology, and Biochem-
istry and Biophysics
Iowa State University
Ames, IA 50011 USA

Populus hybrids are being used for woody plant transformation in
attempts to increase insect and disease resistance. Refined in vitro
micropropagation systems were the initial basis for Populus alba L. X P.
grandidentata Michx. transformation system development (Chun et al., 1988b).

An optimum concentration of N^6-benzyladenine (1 mg/l) in combination
with alpha-naphthaleneacetic acid (0.1 mg/l) was determined for regenerating
adventitious shoots from leaves. An Agrobacterium tumefaciens host range
study established that strain A281 (containing pTiBo542) was capable of
forming galls on the Populus hybrids (Chun et al., 1988a).

Kanamycin sensitivity studies determined that 10 mg/l kanamycin sulfate
inhibited adventitious shoot regeneration from leaves and inhibited root formation from explanted shoot segments (Chun et al., 1988a). This allowed the
use of Neomycin Phosphotransferase II (NPT II) as a selectable marker gene.
A binary vector plasmid, pRT45, was constructed containing 2 chimeric gene
constucts: 1)Nopaline synthase (nos) promoter-NPT II-nos terminator and 2)
potato Proteinase Inhibitor II (pin2) promoter-Chloramphenicol Acetyltransferase (CAT)-pin2 terminator for studying wound-inducible gene expression.
The binary vector was transferred to A. tumefaciens strain A281 by triparental
mating (Thornburg et al., 1987).

Excised leaves of in vitro cultured poplar were co-cultured for 2 days
with the Agrobacterium binary vector system. Following 10 days of preselec-
tive culture, co-cultured leaves were transferred to selective (40 mg/l kanamycin) shoot regeneration medium. After at least 24 days, kanamycin-resistant shoots developed to a stage that allowed transfer and secondary selection on root-inducing medium containing kanamycin (20 mg/l). Root initiation
was observed after 10 days. Kanamycin-resistant, rooted plantlets were subsequently adapted to soil conditions and greenhouse grown.

NPT II expression in transgenic poplars was confirmed with NPT assays
of leaf protein extracts. Southern hybridization was used to confirm CAT
gene insertion into the hybrid poplar genome. Assays of wound-inducible
CAT expression are hampered by a compound interfering with CAT assays that
is wound inducible in the hybrid poplar leaves. Northern hybridization does
demonstrate, however, that the wound-inducible promoter of pin2 functions in

Woody Plant Biotechnology, Edited by M.R. Ahuja
Plenum Press, New York, 1991

poplar by increasing levels of CAT mRNA. Attempts to further characterize wound-inducible CAT expression by Western blotting are in progress.

A 4-year field test planting of transgenic poplar, Tr15, was established 28 July 1989 in accordance with guidelines set by USDA/APHIS Biotechnology Permit Unit (permit number 89-109-03). This field test is designed to determine the effects of the transgenes on poplar growth and development, as well as characterizing transgene expression in tissue developed under field conditions.

REFERENCES

Chun, Y. W., Klopfenstein, N. B., McNabb, H. S., Jr., and Hall, R. B., 1988a, Transformation of Populus species by an Agrobacterium binary vector system, J. Korean For. Soc., 77:199.
Chun, Y. W., Klopfenstein, N. B., McNabb, H. S., Jr., and Hall, R. B., 1988b, Biotechnological applications in Populus species, J. Korean For. Soc., 77:467.
Thornburg, R. W., An, G., Cleveland, T. E., Johnson, R., and Ryan, C. A., 1987, Wound-inducible expression of a potato inhibitor II-chloramphenicol acetyltransferase gene fusion in transgenic tobacco plants, Proc. Natl. Acad. Sci. USA, 84:744.

TRANSIENT EXPRESSION OF B-GLUCURONIDASE AND LUCIFERASE IN

ELECTROPORATED CONIFER PROTOPLASTS

Michael Campbell, Claire Kinlaw and David Neale

Institute of Forest Genetics
Pacific Southwest Research Station
USDA Forest Service
Berkeley, CA 94701

Suspension-derived protoplasts of Monterey pine (Pinus radiata
Don.) and Douglas-fir (Pseudotsuga menziesii (Mirb.) Franco) were
electroporated with the B-glucuronidase (GUS) and firefly
luciferase reporter genes driven by the 35S-cauliflower mosaic
virus promoter. Low levels of transient expression were obtained
using the GUS reporter gene and maximal activities were found
within 48 hrs of electroporation using a variety of conditions.
Transient expression of luciferase was not detectable using the
electroporation, culture, and extraction procedures established
for maximal GUS expression. Extracts of both protoplasts and
suspension cells of Monterey pine and Douglas-fir were shown to
inhibit GUS activity in extracts of Escherichia coli by up to
2-fold and completely inhibit luciferase activity in firefly tail
extracts.

Woody Plant Biotechnology, Edited by M.R. Ahuja
Plenum Press, New York, 1991

CAB EXPRESSION IN DOUGLAS-FIR DEMONSTRATES WEAK LIGHT REGULATION

M. Carol Alosi and David B. Neale

Institute of Forest Genetics
Pacific Southwest Research Station
USDA Forest Service
Berkeley, CA 94701

The cab gene family encodes for the major light-harvesting
chlorophyll a/b binding protein of photosystem II. Cab genes are
strongly up-regulated by light in all angiosperms studied. It
has also been shown that cab transcription rates and steady state
levels of cab mRNA fluctuates dramatically during dark and light
cycles in a manner that shows a pronounced, circadian rhythm. We
find that Douglas-fir (Pseudotsuga menziesii) responds
differently to light in comparison to what is generally true for
angiosperms. Douglas-fir seedlings appear to maintain relatively
high levels of cab mRNA regardless of light and dark treatments.
For example, cab mRNA in dark grown seedling is found at about
40% of amounts produced under natural light conditions. We also
determined the steady-state mRNA levels in Douglas-fir seedlings
at 4-hour intervals throughout a diurnal cycle of 12 hours light
and 12 hours dark, and during extended dark and light periods.
Our results show that relatively minor fluctuations occur in
steady state cab mRNA during light and dark phases and no
circadian rhythm is apparent. We hypothesize that there are
basic differences between Douglas-fir and angiosperms with
respect to factors effecting the photoregulation of certain light
responsive genes.

PATERNAL INHERITANCE OF ORGANELLE DNA IN CONIFERS

Kimberly A. Marshall and David B. Neale

Institute of Forest Genetics
Pacific Southwest Research Station
USDA Forest Service
Berkeley, CA 94701

We have been investigating the modes of inheritance of chloroplast DNA (cpDNA) and mitochondrial DNA (mtDNA) in conifers using RFLP markers in controlled crosses. Four species were chosen: Pinus taeda (Pinaceae), Psuedotsuga menziesii (Pinacease), Sequoia sempervirens (Taxodiacease) and Calocedrus decurrens (Cupressaceae). Controlled crosses of the four conifers tested showed strict paternal inheritance of cpDNA. MtDNA is maternally inherited in Pinus taeda, just as it is in angiosperms. In Sequoia sempervirens and Calocedrus decurrens, however, mtDNA is paternally inherited.

The inheritance of organelle DNA is clearly very different in conifers than in angiosperms. Ultrastructural observations of conifer fertilization show transmission of paternal plastids and mitochondria and exclusion of maternal organelles. We can not explain why paternal inheritance of organelle DNA has evolved in this group of plants. Paternal inheritance of cpDNA is common to all conifers studied, whereas only the closely related Taxodiaceae and Cupressaceae show paternal inheritance of mtDNA. Data from other families is needed to trace the evolution of mtDNA inheritance in conifers.

RFLP MAPPING OF THE LOBLOLLY PINE (Pinus taeda L.) GENOME

David M. Gorzo and David B. Neale

Institute of Forest Genetics
Pacific Southwest Research Station
USDA Forest Service
Placerville, CA 95667

A genetic linkage map of loblolly pine (Pinus taeda L.), a
commercially important forest species, is being developed using
restriction fragment length polymorphisms (RFLPs) as molecular
markers. The genome of loblolly pine (2x=2n=24) is very large
(10E10bp), which presents some difficulty in detecting
single-copy sequences on Southern blots. Recombination size has
been conservatively assessed at 2,000 cM; therefore, 20
cM-saturation can be achieved with 200-300 RFLPs. A single
3-generation pedigree, obtained from a commercial tree breeding
program, was selected to construct the map. This pedigree was
chosen on the basis of its high level of heterozygosity, as
determined from isozyme and RFLP analyses. Several restriction
enzymes were chosen for their ability to digest loblolly pine
genomic DNA and reveal polymorphisms. Random cDNA clones are
currently being screened to identify RFLPs in the parent lines.
Thus far, approximately 30% of the clones have revealed
polymorphisms. These clones will be used for segregation
analysis of RFLPs in progeny DNAs. The completed map will be
valuable in understanding the structure and organization of the
loblolly pine genome, as well as assisting in the genetic
improvement of this species.

PARTICIPANTS

Ahuja, M. Raj (S): Federal Research Centre for Forestry
    and Forest Products, Institute of Forest Genetics and
    Forest Tree Breeding, Sieker Landstrasse 2, 2070
    Grosshansdorf, Federal Republic of Germany

Becwar, Michael (O): Westvaco Corp. Forest Research,
    Box 1950, Summerville, SC 29484, USA

Birnbaum, Elliot (O): CELBI, Quinta do Furadouro, Amdreira,
    2510 Obidos, Portugal

Bonga, J.M. (S): Canadian Forestry Service Maritimes,
    Box 4000, Fredericton, N.B. E3B 5P7, Canada

Bradshaw, Harvey. D. (S): Department of Biochemistry SJ-70,
    University of Washington, Seattle, WA 98195, USA

Campbell, Michael (P): Institute of Forest Genetics,
    USDA Forest Service, 1960 Addison Street, Berkeley,
    CA 94701, USA

Chun, Young Woo (S): College of Forestry, Kookmin University,
    861-1, Chongnung-dong, Songbuk-gu, Seoul, 136-702,
    Korea

Cornu, Daniel (P): Station d'Amelioration des Arbres
    Forestieres, INRA, CRO Ardon, 45160 Olivet, France

Diner, Alex M. (P): USFS Southern Forest Experiment Station,
    Department of Plant & Soil Sciences, Alabama A&M
    University, Normal, AL 35762, USA

Dinus, Ronald J. (S): Forest Biology Division, Institute of
    Paper Science and Technology, 575 14th Street, N.W.,
    Atlanta, GA 30318, USA

Durzan, Don J. (S): Department of Environmental Horticulture,
    University of California, Davis, CA 95616, USA

Ellis, David D. (S): Department of Horticulture, 1575
    Linden Drive, University of Wisconsin-Madison, Madison,
    WI 53706, USA

Ernst, Stephen G. (S): Department of Forestry, Fisheries, and Wildlife, 101 Plant Industry, East Campus, University of Nebraska, Lincoln, NE 68583-0814, USA

Eshita, Steve (O): USDA Forest Service, 359 Main Road, Delaware, OH 43015, USA

Goldfarb, Barry (S): Department of Forest Science, Peavy Hall 154, Oregon State University, Corvallis, OR 97331-5705, USA

Greenwood, Michael S. (S): Department of Forest Biology, 122 Nutting Hall, University of Maine, Orono, ME 04469, USA

Gupta, Pramod K. (O): WTC-G-30, Weyerhaeuser Technical Centre, Tacoma, WA 98477, USA

Hackett, Wesley P. (S): Department of Horticultural Science, University of Minnesota, St. Paul, MN 55108, USA

Haggman, Hely M. (O): Department of Forestry, North Carolina State University, Raleigh, NC 27695, USA

Haggman, Juhani H.O. (O): Department of Forestry, North Carolina State University, Raleigh, NC 27695, USA

Handro, Walter (S): Department of Botany, Institute of Biosciences, University of Sao Paulo, C.P. 11461, 05499 Sao Paulo, Brazil

Heth, Dan M. (O): Department of Natural Resources, Agricultural Research Organization, The Volcani Centre, P.O. Box 6, Bet-Dagan 50250, Israel

Hiramath, Shivanand (O): USDA Forest Service, 359 Main Road, Delaware, OH 43105, USA

Howe, Glenn (S): Department of Forest Science, Peavy Hall 154. Oregon State University, Corvallis, OR 97331-5705, USA

Hutchison, Keith W. (S): Department of Biochemistry, Hitchner Hall, University of Maine, Orono, ME 04469, USA

James, David D. (S): Institute of Horticultural Research, East Malling, Maidstone, Kent ME19 6BJ, United Kingdom

Jelaska, Sibila (O): Department of Molecular Biology, Faculty of Science, University of Zagreb, P.O. Box 933, Yu-41001 Zagreb, Yugoslavia

Jokinene, Kari (S): Kemira Oy, Espoo Research Cntre, P.O. Box 44, 02271 Espoo, Finland

Jörgensen, Jörg (P): Lower Saxony Forest Research Institute, Dept. of Forest Tree Breeding, Forstamtstrasse 6, D-3513 Staufenberg-Escherode, Federal Republic of Germany

Karnosky, David F. (S): School of Forestry, Michigan
    Technological University, Houghton, MI 49931, USA

Kinlaw, Clair S. (P): Institute of Forest Genetics, USDA
    Forest Service, 1960 Addison Street, Berkeley, CA
    94701, USA

Klopfenstein, N.B. (P): Department of Forestry, 221 Bessey
    Hall, Iowa State University, Ames, IA 50011-1021, USA

Kreitinger, Mary (O): WTC-G-30, Weyerhaeuser Technical
    Centre, Tacoma, WA 98477, USA

Lang, Hermann (P): Botanisches Institut der J.W. Goethe-
    Universität, Siesmayerstrasse 70, D-6000 Frankfurt/M,
    Federal Republic of Germany

Libby, William J. (S): Department of Forestry and Resource
    Management, 145 Hulford Hall, University of California,
    Berkeley, CA 94720, USA

Liu, Ming Chin, 13-Lane 430, 2nd Section, Ta-Tung Road,
    Taiwan City, Taiwan

Lee-Stadelmann, Ok Young (P): Department of Horticultural
    Science, University of Minnesota, St. Paul, MN 55108,
    USA

Marshall, Kimberly, A. (P): Institute of Forest Genetics,
    USDA Forest Service, 1960 Addison Street, Berkeley,
    CA 94701, USA

McCown, Brent H. (S): Department of Horticulture, 1575
    Linden Drive, University of Wisconsin-Madison, Madison,
    WI 53706, USA

McNabb, Harold S. (S): Department of Forestry and Plant
    Pathology, 221 Bessey Hall, Iowa State University,
    Ames, IA 50011-1021, USA

Meier-Dinkel, Andreas (P): Lower Saxony Forest Research
    Institute, Dept. of Forest Tree Breeding, Forstamt-
    strasse 6, D-3513 Staufenberg-Escherode, Federal
    Republic of Germany

Merkle, Scott A. (S): School of Forest Resources,
    University of Georgia, Athens, GA 30602, USA

Mo, Hakan (P): Department of Forest Genetics, Swedish
    University of Agricultural Science, Box 7027,
    S-750 07 Uppsala, Sweden

Neale, David B. (P): Institute of Forest Genetics, USDA
    Forest Service, 1960 Addison Street, Berkeley, CA
    94701, USA

Paton, Dugald M. (S): Forestry Department, Australian
    National University, GPRO Box 4, Canberra ACT 2601,
    Australia

Preece, John E. (P): Department of Pomology, Uinversity
of California, Davis, CA 95616, USA

Pullman, Jerry (O): WTC-G-30, Weyerhaeuser Technical
Centre, Tacoma, WA 98477, USA

Rawat, J.S. (O): School of For-stry, Auburn University,
Auburn, Alabama, 36849-5418, USA

Redenbaugh, Keith (O): Calgene, 1920 5th Street, Davis,
CA 95616, USA

Riemenschneider, Don E. (S): North Central Forest
Experiment Station, USDA Forest Service, Forestry
Science Laboratory, Box 898, Rhinelander, WI 54501,
USA

Schmidt, Josef (P): Austrian Research Centre Ltd,
A2444 Seibersdorf, Austria

Schuller, Astrid (P): Forschungsanstalt Giesenheim,
Institut für Biologie, Fachgebiet Botanik, Von
Ladestrasse 1, 6222 Geisenheim, Federal Republic
of Germany

Smith, Dale (P): Forest Research Institute, Private
Bag, Rotorua, New Zealand

Strauss, Steve (O): Department of Forest Science,
Peavy Hall 154, Oregon State University, Corvallis,
OR 97331-5705, USA

Thimma Raju, K.R. (S): Division of Horticulture,
University of Agricultural Sciences, Bangalore-
560065, India

Von Aderkas, Patrick (S): Department of Biology,
University of Victoria, Victoria, B.C. V8W 2Y2,
Canada

Von Weissenberg, Kim (O): University of Joensuu, School
of Forestry, P.O. Box 111, SF-80101 Joensuu, Finland

Webster, Fiona (S): BC Research, 3650 Wesbrook Mall,
Vancouver, B.C. V6S 2L2, Canada

Welander, Margareta (S): Department of Horticultural
Science, Swedish University of Agricultural Sciences,
Box 55, S-230 53 Alnarp, Sweden

Wilde, Dayton (S): School of Forest Resources, University
of Georgia, Athens, GA 30602, USA

S = Speaker;   P = Poster;   O = Observer

INDEX